# Measurable and Composable Security, Privacy, and Dependability for Cyberphysical Systems

## The SHIELD Methodology

# Measurable and Composable Security, Privacy, and Dependability for Cyberphysical Systems

## The SHIELD Methodology

Edited by
Andrea Fiaschetti, Josef Noll,
Paolo Azzoni, and
Roberto Uribeetxeberria

**CRC Press**
Taylor & Francis Group
Boca Raton  London  New York

CRC Press is an imprint of the
Taylor & Francis Group, an **informa** business

MATLAB® and Simulink® are trademarks of the MathWorks, Inc. and are used with permission. The MathWorks does not warrant the accuracy of the text or exercises in this book. This book's use or discussion of MATLAB® and Simulink® software or related products does not constitute endorsement or sponsorship by the MathWorks of a particular pedagogical approach or particular use of the MATLAB® and Simulink® software.

CRC Press
Taylor & Francis Group
6000 Broken Sound Parkway NW, Suite 300
Boca Raton, FL 33487-2742

© 2018 by Taylor & Francis Group, LLC

CRC Press is an imprint of Taylor & Francis Group, an Informa business

No claim to original U.S. Government works

Printed on acid-free paper

International Standard Book Number-13: 978-1-138-04275-9 (Hardback)

**Visit the Taylor & Francis Web site at**
**http://www.taylorandfrancis.com**

**and the CRC Press Web site at**
**http://www.crcpress.com**

# Contents

## Section I  SHIELD Technologies and Methodology for Security, Privacy, and Dependability

## Section II  SHIELD Application Scenarios, New Domains, and Perspectives

# *Foreword*

The Internet of Things (IoT), cyber-physical systems (CPSs), Industry 4.0, Swarm Systems, and the Human Intranet are all based on similar principles: interaction between the physical world and cyberspace, and the massive use of distributed, wireless devices, machine to machine, and machine to humans. They are all safety critical, heterogeneous, and of a scale never seen before. The technical hurdles to overcome are many, but possibly the most serious challenge to the vision of the instrumented world is security. We need to guarantee protection from intrusion or unauthorized access so that we can help prevent an adversary from affecting safety and bringing a level of chaos to human society of unimaginable consequences as the following examples show.

On October 21, 2016, the domain name service provider Dyn suffered a distributed denial of service (DDoS) attack leading to a significant collapse of fundamental infrastructures comprising the Internet. In a DDoS attack, a number of compromised or zombie computers, forming a botnet, send a flood of traffic to the target server, causing a denial of service by exhausting computation and/or communication resources. The relevance of this attack for our arguments is the fact that many of the compromised computers launching the attack were relatively small devices including printers, webcams, residential gateways, and baby monitors, that is, the IoT.

On February 23, 2017, many of Google's smart gateway (router) devices called OnHub failed for about 45 minutes mainly because of the failure of the connection of the devices to the cloud, making it impossible to authenticate them on Google's servers. This led to Internet connection problems to major websites including Twitter, Netflix, Spotify, and the *Financial Times*. The lesson learned in the Google OnHub case is that depending on remote cloud servers for authentication and authorization may be a liability.

The first known example of an attack on a physical plant was a uranium enrichment plant in Iran whose centrifuges were monitored and controlled with Siemens process control systems (PCS). This attack was perpetrated via a worm named Stuxnet. Stuxnet consisted of two parts: one part of the program was designed to lie dormant for long periods, then speed up the centrifuges so that the spinning rotors in the centrifuges wobbled and then destroyed themselves. The other part sent out false sensor signals to make the system believe everything was running smoothly. This part prevented a safety system from kicking in, which would have shut down the plant before it could self-destruct. This attack did not use the Internet to spread the worm inside the plant since the plant was physically isolated. The point of entry was memory sticks containing apparently innocuous data that were read inside the plant perimeter by unaware workers.

In 2010, an important study by the University of California, San Diego, and the University of Washington outlined the many points of attack in the electronic control system of automobiles. Examples of hijacked cars abound on YouTube. Intruders could create havoc in transportation systems (trains and ships are as amenable to intrusion), yielding catastrophic events that may jeopardize the lives of many people.

On December 23, 2015, a cyberattack on the Ukrainian power grid demonstrated another instance of the gravity of security issues in networks of things. The attackers gained control over the supervisory control and data acquisition (SCADA) system of the Ukrainian power grid and caused a blackout for several hours across a large area of Ukraine's Ivano-Frankivsk region populated by 1.4 million residents.

These last three cases show that, unlike cyberattacks in the past, the consequences from attacks on the IoT could be much more devastating than information theft or financial loss.

The safety and security characteristics of a CPS and the IoT are quite different from those of "standard" Internet applications. Indeed, in the CPS community, it is widely agreed that the main challenges to the security of a CPS/IoT include *heterogeneity, scarce computing and power resources, scalability* and *operation in open environments.*

It is impractical to adopt security measures that are successfully used in the Internet such as transport layer security/secure sockets layer (SSL/TLS), and in the wireless sensor network (WSN) and the mobile ad hoc network (MANET) due to their heavy computational burden. For sensor networks, for example, performance is not at a premium as in mobile payment applications, but being able to connect thousands, if not millions, of devices and to keep them "alive" for a long time is (this is related to the scarce computing and power resources issue).

Many expect that there will be tens of billions of connected devices by 2020, far exceeding the world population, and billions of terabytes (1012 bytes) of Internet Protocol (IP) traffic, a significant portion of which will be generated by the instrumented world. Hence, the security solution must *scale* accordingly; in particular, the overhead of adding and removing devices to/from the security solution should be minimal.

In CPSs and the IoT, the lifetime and availability of the devices and of their interconnections are as important as data security, if not more. For some sensors such as environment data, privacy is of no interest, in others such as medical data it is of paramount importance. This *heterogeneity* requires different privacy and security mechanisms for the same network while traditional security mechanisms are valid for homogeneous networks.

CPSs and the IoT must support safety-critical components in *open, untrusted, and even hostile environments.* Due to their open nature, CPS and IoT networks are susceptible to entirely new classes of attacks, which may include illegitimate access through mediums other than traditional networks (e.g., physical access, Bluetooth, and radios). For instance, an attack on a traffic controller on the streets of Ann Arbor, Michigan, including the manipulation of actual traffic lights was demonstrated; this attack was made possible via direct radio communication with the traffic controller. Jamming attacks on wireless communication channels can also be a threat to the availability of the IoT operating over wireless networks.

To the best of my knowledge, integrated security solutions for CPS/IoT have not been widely accepted; only a few have carried out research on the topic. This book is an important step to fill this gap in research: it presents a methodology, called SHIELD, and an innovative, modular, composable, expandable, and highly dependable architectural framework conceived and designed with this methodology. This framework allows to achieve the desired security, privacy, and dependability (SPD) level in the context of integrated and interoperating heterogeneous services, applications, systems, and devices in a CPS framework.

**Alberto Sangiovanni Vincentelli**
**The Edgar L. and Harold H. Buttner Chair**
Department of EECS, University of California, Berkeley

# *Preface*

We are at the beginning of a new age of business, where dynamic interaction is the driving force for whatever kind of business. To draw from a known analogy, "bring your own device" (BYOD) exemplifies the trends of devices accessing processes and information on enterprises. In the upcoming years, not only phones, tablets, and computers will demand access, but also *sensors* and *embedded systems* will deliver and request information. In the traditional way of handling dynamic interaction, the attempt was to secure the whole infrastructure of a company. To follow the analogy, BYOD is often seen as a threat, and answered in the classical way by preventing employees from using their devices, as security cannot be ensured. A second variant of counteracting classic threats such as insufficient authentication and loss of devices is addressed through an approach of integrating, managing, and securing mobile devices. But these strategies *cannot be* applied to *sensors* and other kinds of *cyber-physical systems*. Companies cannot stop integrating embedded systems into their infrastructures, as their businesses and processes need them to remain competitive. So, they need to be able to assess the dynamic interaction impact of integrating a new system into their infrastructure in a manageable way, which conventionally suffers from two aspects:

i. Secure interaction issues in current systems are described through an integrated approach, and do not open for scalability.
ii. Measurable security in terms of quantifiable results is not industry.

A paradigm shift in handling dynamic interaction is required, addressing the need for securing information instead of securing infrastructure. The paradigm shift includes *the need for a security methodology definition*, first, and for the consequent *measurability*.

SHIELD addresses both these shortcomings, providing the methodology and the means of integrating new infrastructures, new ways of communication, and new devices. It thereby answers the upcoming trends of wireless sensors, sensor networks, and automated processes. Though the focus of SHIELD is on introducing security for cyber-physical systems, we see that these security measures need to be the basis for running automated processes. Consequently, the solution proposed in this book addresses a metrics-based approach for a quantitative assessment of both the potential attack scenario and the security measures of the information, and outlines the methodology of measurable security for systems of cyber-physical systems.

Measurable security is often misinterpreted as a good risk analysis. The SHIELD approach works toward measuring security in terms of cardinal numbers, representing the application of specific security methods as compared to the specific threat scenario. The approach is based on the semantic description of a potential attack scenario, the security-related aspects of sensors/systems, and security policies that should be applied irrespective of the scenario.

Through SHIELD, we address measurable security and introduce countable numbers for the security components of systems. We also address the scalability aspect by using composition techniques that are able to build a security representation of the composed system (system of systems) based on the individual security representations of each individual element. This simplifies the process of measuring the security of the composed system, and opens up the opportunity to build the system in an incremental way.

This approach is particularly indicated to manage all the security aspects of *cyber-physical systems*, embedded systems that are interconnected, interdependent, collaborative, and smart. They provide computing and communication, monitoring, and control of physical components and processes in various applications. Many of the products and services that we use in our daily lives are increasingly determined by cyber-physical systems and, the software that is built into them is the connection between the real physical world and the built-in intelligence. The SHIELD approach also represents an answer to dependability aspects.

Dependability is a key aspect of cyber-physical systems, in particular in safety-critical environments that may often require 24/7 reliability, 100% availability, and 100% connectivity, in addition to real-time response. Moreover, security and privacy are both important criteria that affect the dependability of a system; therefore, this book focuses on *security, privacy, and dependability* issues within the context of embedded cyber-physical systems, considering security, privacy, and dependability both as distinct properties of a cyber-physical system and as a single property by *composition*.

Increasing security, privacy, and dependability requirements introduce new challenges in emerging *Internet of Things* and *Machine to Machine* scenarios, where heterogeneous cyber-physical systems are massively deployed to pervasively collect, store, process, and transmit data of a sensitive nature. Industry demands solutions to these challenges—solutions that will provide measurable security, privacy, and dependability, risk assessment of security critical products, and configurable/composable security. Security is frequently misconstrued as the hardware or software implementation of cryptographic algorithms and security protocols. On the contrary, security, privacy, and dependability represent a new and challenging set of requirements that should be considered in the design process, along with cost, performance, power, and so on.

The SHIELD methodology addresses security, privacy, and dependability in the context of cyber-physical systems as "built in" rather than as "add-on" functionalities, proposing and perceiving with this strategy the first step toward security, privacy, and dependability certification for future cyber-physical systems.

The SHIELD general framework consists of a four-layered system architecture and an application layer in which four scenarios are considered: (1) airborne domain, (2) railways, (3) biometric-based surveillance, and (4) smart environments.

Starting from the current security, privacy, and dependability solutions in cyber-physical systems, new technologies have been developed and the existing ones have been consolidated in a solid basement that is expected to become the reference milestone for a new generation of "security, privacy, and dependability-ready" cyber-physical systems. SHIELD approaches security, privacy, and dependability at four different levels: *node, network, middleware,* and *overlay*. For each level, the state of the art in security, privacy, and dependability of individual technologies and solutions has been improved and integrated (hardware and communication technologies, cryptography, middleware, smart security, privacy, and dependability applications).

The leading concept has been the demonstration of the *composability* of security, privacy, and dependability technologies and the composition of security, depending on the application need or the attack surrounding.

To achieve these challenging goals, we developed and evaluated an innovative, modular, composable, expandable, and highly dependable architectural framework, concrete tools, and common security, privacy, and dependability metrics capable of improving the overall security, privacy, and dependability level in any specific application domain, with minimum engineering effort.

Through SHIELD, we have (i) achieved a *de facto* standard for measurable security, privacy, and dependability; (ii) developed, implemented, and tested roughly 40 security-enhancing prototypes in response to specific industrial requests; and (iii) applied the methodology in four different domains, proving how generic the approach is.

The book's main objective is to provide an *innovative, modular, composable, expandable and high-dependable architectural framework* conceived and designed with the SHIELD methodology, which allows to achieve the desired security, privacy, and dependability level in the context of integrated and interoperating heterogeneous services, applications, systems, and devices; and to develop concrete solutions capable of achieving this objective in specific application scenarios with minimum engineering effort.

The book is organized in two parts:

**Section I: SHIELD Technologies and Methodology for Security, Privacy, and Dependability** is dedicated to the SHIELD methodology, to technical aspects of new and innovative security, privacy, and dependability technologies and solutions, and to the SHIELD framework.

**Section II: SHIELD Application Scenarios, New Domains, and Perspectives** covers four different application scenarios for SHIELD in the airborne domain, railway domain, biometric security, and smart environments security (smart grid, smart vehicles, smart cities, etc.). This section also describes some domain-independent technology demonstrators and provides an overview of the industrial perspectives of security, privacy, and dependability and of the results obtained by adopting the SHIELD methodology in other European research projects.

This book is foreseen for system integrators, software engineers, security engineers, electronics engineers, and many other engineering disciplines involved in the extremely rapidly digitalizing world. But also, managers and policy makers in industry and public administration can make use of it to get awareness on the security challenges of this massive digitalization. The book is intended to be written in a language as plain as possible to reach a wide audience. The goal is to raise awareness on security aspects of the cyber-physical systems that are increasingly being connected to the rest of the world. Systems are often responsible for critical infrastructures that provide the foundations of our modern society. It provides the shortcomings of current approaches, indicates the advances coming from the distributed approach as suggested by SHIELD, and addresses the state of the art in security in various market segments.

Finally, it must be acknowledged that *Measurable and Composable Security, Privacy, and Dependability for Cyberphysical Systems: The SHIELD Methodology* is the result of the two SHIELD projects co-funded by the ARTEMIS Joint Undertaking (https://www.artemis-ju.eu/). Several institutions of different European countries have participated in SHIELD and this book would not have been possible without all the work carried out during all those years by this team of highly professional researchers. The participation by major European industry players in embedded systems security, privacy, and dependability, also made possible the commercial exploitation of the results developed in the SHIELD projects.

MATLAB® is a registered trademark of The MathWorks, Inc. For product information, please contact:

The MathWorks, Inc.
3 Apple Hill Drive
Natick, MA 01760-2098 USA
Tel: 508 647 7000
Fax: 508-647-7001
E-mail: info@mathworks.com
Web: www.mathworks.com

# Editors

**Andrea Fiaschetti** is honorary fellow (Cultore della Materia) at the University of Rome "La Sapienza" in the Department of Computer, Control, and Management Engineering "A. Ruberti," promoting research and teaching activities in the field of automatic control. He is teaching assistant in several courses within the control engineering, system engineering, and computer science degrees, as well as supervisor of dozens of BSc/MSc thesis on innovative topics.

Since 2007, he has been actively involved in several European projects, mainly in the security domain, including but not limited to SatSix, MICIE, MONET, and TASS, as well as pSHIELD and nSHIELD (on which this book is based). His main research interests are in the field of applied automatic control, pursuing a cross-fertilization between control theory and computer science, with a particular focus on innovative solutions for security and manufacturing domains; in this perspective, his major achievement is the formalization of the so-called composable security theory, an innovative methodology born from a collaboration with a restricted pool of academic and industrial experts, which represents the foundation of the SHIELD roadmap.

He is author of several papers on these topics. On an industrial perspective, Andrea Fiaschetti is a certified Project Management Professional (PMP®) and works as R&D project manager at Thales Alenia Space Italia S.p.A. (a Thales/Leonardo company), within the observation and navigation business domain. Last, but not least, he is actively involved in the Engineers Association of Rome, where he has recently been appointed as president of the Smart Cities and Internet of Things Committee.

**Josef Noll** is visionary at the Basic Internet Foundation and professor at the University of Oslo (UiO). Through the foundation, he addresses "information for all" as the basis for sustainable development and digital inclusion. Regarding sustainable infrastructures, where communication and security are key topics for the transfer to a digital society, he leads the national initiative "Security in IoT for Smart Grids" (IoTSec.no), Norway's largest research project within IoT security. In 2017, the 20 partner opened the Smart Grid Security Centre to contribute to trusted and more secure power grids and smart home/city services.

He is also head of research in Movation, Norway's open innovation company for mobile services. The company supported more than 200 start-ups in the last 10 years. He is co-founder of the Center for Wireless Innovation and Mobile Monday in Norway. He is IARIA fellow, reviewer of EU FP7/H2020 projects, and evaluator of national and EU research programs.

Previously, he was senior advisor at Telenor R&I in the products and markets group, and project leader of the JU ARTEMIS pSHIELD project on "Measurable Security for Embedded Systems," Eurescom's "Broadband Services in the Intelligent Home," and use-case leader in the EU FP6 "Adaptive Services Grid (ASG)" projects, and has initiated a.o. the EU's 6th FP ePerSpace and several Eurescom projects.

He joined UiO in 2005 and Telenor R&D in 1997, coming from the European Space Agency, where he was staff member (1993–1997) in the Electromagnetics Division of ESA ESTEC. He received his Diplom-Ingenieur and PhD degree in electrical engineering from the University of Bochum in 1985 and 1993. He worked as an integrated circuit designer in 1985 with Siemens in Munich, Germany, and returned to the Institute for Radio Frequency at the University of Bochum as a research assistant from 1986 to 1990.

**Paolo Azzoni** is the research program manager at Eurotech Group. He is responsible for planning and directing industrial research projects, investigating technologies beyond the state of the art in computer science, developing a wide network of academic research groups, and providing the financial support to company research activities. His main working areas include cyber-physical systems (CPSs), intelligent systems, machine-to-machine distributed systems, device to cloud solutions, and Internet of Things. He has participated in several European research projects in the contexts of FP7, ARTEMIS, Aeneas, ECSEL, and H2020, and he is a European Community independent expert.

He is one of the founders and promoters of the SHIELD initiative (pSHIELD and nSHIELD ARTEMIS projects), from the early stage of concepts definition to the development of the entire roadmap. He has represented Eurotech in the ARTEMIS Industrial Association (ARTEMIS-IA) since 2007. He is currently a member of the ARTEMIS-IA steering board and was recently appointed to the ARTEMIS-IA presidium as guest member.

Previously, he was involved in academic lecturing and research in the areas of hardware formal verification, hardware/software co-design and co-simulation, and advanced hardware architectures and operating systems. In 2006, he joined ETHLab (Eurotech Research Center) as research project manager and he has been responsible for the research projects in the area of embedded systems.

He is an accomplished researcher and author of publications focusing on the latest trends in IoT, intelligent systems, and CPSs, with a wide experience matured over more than 20 years of direct involvement in European research, technology transfer, and ICT innovation. He holds a master's degree in computer science and a second master's degree in intelligent systems, both from the University of Verona.

 **Roberto Uribeetxeberria** is currently the head of research at the Faculty of Engineering, Mondragon University. He has participated in several European projects in the cyber-physical systems domain (eDIANA, pSHIELD, nSHIELD, ARROWHEAD, CITYFIED, DEWI, MANTIS [leader], MC-SUITE, PRODUCTIVE4.0). He has also participated in over 35 public-funded research projects and authored more than 30 publications. He has supervised three PhD theses, and he is currently supervising two PhD students. Dr. Uribeetxeberria obtained his PhD in Mobile Communications at Staffordshire University (UK) in 2001. Since then, he has combined lecturing and research at Mondragon University. He also directed the PhD program in New Information and Communication Technologies for several years and actively participated in the creation of the new Research Centre on Embedded Systems of the Faculty of Engineering, as well as designing the Masters in Embedded Systems. His research interests are in the fields of networking, information and network security, embedded system security, and data mining. He has represented Mondragon University in the ARTEMIS Industrial Association, the association for actors in Embedded Intelligent Systems within Europe, since 2007, and he is currently a member of the steering board of ARTEMIS-IA, representing Chamber B. He was also appointed to the presidium by the steering board, and thus, he has been vice-president of ARTEMIS-IA since March 2014.

# Contributors

**Antonio Abramo**
Dipartimento Politecnico di
    Ingegneria e Architettura
University of Udine
Udine, Italy

**Ignacio Arenaza-Nuño**
Electronics and Computing
    Department
Faculty of Engineering
Mondragon Unibertsitatea
Mondragon, Spain

**Paolo Azzoni**
Research & Development
    Department
Eurotech Group
Trento, Italy

**Marco Cesena**
Leonardo Company
Rome, Italy

**Cecilia Coveri**
Leonardo Company
Rome, Italy

**Kresimir Dabcevic**
Department of Electrical,
    Electronic, Telecommunications
    Engineering and Naval
    Architecture (DITEN)
Università di Genova
Genoa, Italy

**Javier Del Ser**
OPTIMA Research Area
TECNALIA
Zamudio, Spain

**Alessandro Di Giorgio**
Ingegneria Informatica Automatica
    e Gestionale
"Sapienza" Università di Roma
Rome, Italy

**Andrea Fiaschetti**
Ingegneria Informatica Automatica
    e Gestionale
"Sapienza" Università di Roma
Rome, Italy

**Konstantinos Fysarakis**
Institute of Computer Science
Foundation for Research and
    Technology - Hellas (FORTH)
Heraklion, Greece

**Iñaki Garitano**
Electronics and Computing
    Department
Faculty of Engineering
Mondragon Unibertsitatea
Mondragon, Spain

**Luca Geretti**
Dipartimento di Informatica
Università di Verona
Verona, Italy

**John Gialelis**
ATHENA/Industrial Systems
    Institute
Patra, Greece

**Stefano Gosetti**
Research and Development
    Department
Vigilate Vision
Brescia, Italy

**Marina Silvia Guzzetti**
Leonardo Company
Rome, Italy

**George Hatzivasilis**
School of Electrical & Computer
   Engineering
Technical University of Crete
Chania, Greece

**Christian Johansen**
Department of Informatics
University of Oslo
Oslo, Norway

**Francesco Liberati**
Ingegneria Informatica Automatica
   e Gestionale
"Sapienza" Università di Roma
Rome, Italy

**Charalampos Manifavas**
Rochester Institute of Technology
Rochester, Dubai

**Lucio Marcenaro**
Department of Electrical, Electronic,
   Telecommunications Engineering
   and Naval Architecture (DITEN)
Università di Genova
Genoa, Italy

**Silvano Mignanti**
"Sapienza" Università di Roma
Rome, Italy

**Andrea Morgagni**
Leonardo Company
Rome, Italy

**Josef Noll**
Department of Technology Systems
   (ITS)
University of Oslo
Oslo, Norway

**Martina Panfili**
Ingegneria Informatica Automatica
   e Gestionale
"Sapienza" Università di Roma
Rome, Italy

**Ioannis Papaefstathiou**
School of Electrical & Computer
   Engineering
Technical University of Crete
Chania, Greece

**Andreas Papalambrou**
ATHENA/Industrial Systems
   Institute
Patra, Greece

**Antonio Pietrabissa**
Ingegneria Informatica Automatica
   e Gestionale
"Sapienza" Università di Roma
Rome, Italy

**Francesco Delli Priscoli**
Ingegneria Informatica Automatica
   e Gestionale
"Sapienza" Università di Roma
Rome, Italy

**Konstantinos Rantos**
Computer and Informatics
   Engineering Department
Eastern Macedonia and Thrace
   Institute of Technology
Chania, Greece

**Carlo Regazzoni**
Department of Electrical,
   Electronic, Telecommunications
   Engineering and Naval
   Architecture (DITEN)
Università di Genova
Genoa, Italy

**Francesco Rogo**
Innovation and Technology
  Governance Unit
Leonardo Company
Rome, Italy

**Werner Rom**
Virtual Vehicle
Graz, Austria

**Dimitrios Serpanos**
ATHENA/Industrial Systems
  Institute
Patra, Greece

**Kyriakos Stefanidis**
ATHENA/Industrial Systems
  Institute
Patra, Greece

**Marco Steger**
Virtual Vehicle
Graz, Austria

**Vincenzo Suraci**
Università eCampus
Novedrate, Italy

**Nawaz Tassadaq**
Department of Electrical,
  Electronic, Telecommunications
  Engineering and Naval
  Architecture (DITEN)
Università di Genova
Genoa, Italy

**Andrea Toma**
Department of Electrical,
  Electronic, Telecommunications
  Engineering and Naval
  Architecture (DITEN)
Università di Genova
Genoa, Italy

**Massimo Traversone**
Leonardo Company
Rome, Italy

**Roberto Uribeetxeberria**
Electronics and Computing
  Department
Faculty of Engineering
Mondragon Unibertsitatea
Mondragon, Spain

# 1

## *Introduction*

**Andrea Fiaschetti, Josef Noll, Paolo Azzoni, and Roberto Uribeetxeberria**

### CONTENTS

In the new era of cyber-physical systems (CPSs) and Internet of Things (IoT), the driving force for new business opportunities is the dynamic interaction between the entities involved in the business. The Internet-based service world is currently based on collaborations between entities in order to optimize the delivery of goods or services to the customer as shown in Figure 1.1a. The evolution toward the dynamic interaction between entities, as indicated in Figure 1.1b, represents the ongoing evolution.

One of the real challenges to the way ahead is the disappearing borders between companies, and the automatic exchange of sensor- and process-based information between the entities. Given the second trend of dynamic modeling implies the creation of autonomous decisions, one of the big challenges is the lack of a measurable security when exchanging information. *"Is the information that your system receives from one of the suppliers (or competitors) reliable?"*: this is one of the key questions which you need to answer if your process or business model depends on those data.

Security has traditionally been a subject of intensive research in the area of computing and networking. However, security of CPSs is often ignored during the design and development period of the product, thus leaving many devices vulnerable to attacks. The growing number of CPSs today (mobile phones, pay-TV devices, household appliances, home automation products, industrial monitoring, control systems, etc.) are subjected to an increasing number of threats as the hacker community is already paying attention to these systems. On the other hand, the implementation of security measures is not easy due to the complexity and constraints on resources of these kinds of devices.

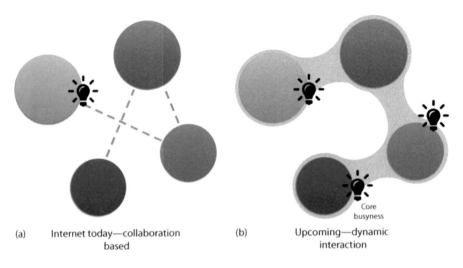

(a)    Internet today—collaboration          (b)       Upcoming—dynamic
                based                                              interaction

**FIGURE 1.1**
The upcoming business world of dynamic interaction between entities.

One of the biggest challenges in security today is related to the software of operating systems and applications. While traditional software avoid word breaking onto next line if poss have made (some) headway in developing more resilient applications, experts say embedded device and systems makers—from those who create implanted medical devices to industrial control systems—are way behind in secure system design and development maturity. There are a number of aspects that are different when it comes to embedded and industrial control system security. First, the consequences of poor system design can create substantially more risk to society than the risks created by insecure traditional software applications. Second, software being implemented on an embedded design will normally reside there for the lifetime of the device. Third, secure software on embedded devices is much more costly—if it is reasonably possible at all—if you also consider the need to update these systems (Hulme, 2012).

While computer software is undergoing version updates to react to malfunctions and new security threats, embedded devices like actuators, sensors, and gateways come with integrated software and, most of the time, non-upgradeable hardware. Maintenance costs for upgrading the software, and vulnerabilities during the upgrade process, make it practically impossible to upgrade an embedded device. Taking the example of Java, where computer systems were upgraded several times during the first months of 2013, no such upgrades were available for Java on embedded devices.

Thus, if the business depends on data and information originating from or going through CPSs, you should have an opinion on the quality of those data. Consider a simple example of a heating system: assume that it solely depends on temperature readings from an outdoor thermometer. Manipulating this outdoor temperature sensor, or malfunctions due to reflected sunlight, might

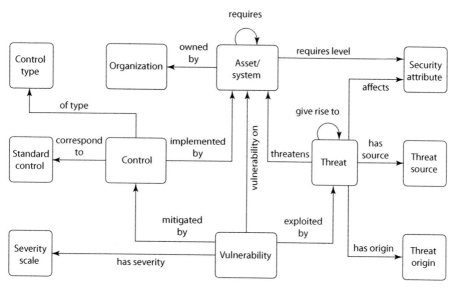

**FIGURE 1.2**
Typical view on security architecture.

alter the collected value of temperature, which can turn the heating system off and provide inconvenient indoor temperatures.

The traditional approach to security, as shown in Figure 1.2, is often addressed through an integrated approach. The integrated approach describes assets like the system and the organization, threats like the environment and attack scenarios, vulnerability, and control mechanisms. This traditional approach suffers mainly from two weaknesses:

1. Security issues in current systems are described through an integrated approach, and do not open for scalability.
2. Measurable security in terms of quantifiable results is not industry.

Security has impacted on the way of doing business. We expect that security will become a major factor for the ongoing business evolution cycle in several domains, including:

- *Avionics:* Where safety was the main driver for developing formalized methods and models for software developments, needed for the "fly-by-wire operation" of modern aircrafts.
- *Automotive:* Where the complexity of modern cars that integrate CPSs containing millions of lines of software codes, exposes cars to many new vulnerabilities and attacks.
- *e-Health:* Healthcare applications have a clear demand for privacy and reliability.

- *Energy*: Where renewables are the drivers of a smart electric grid that currently lacks security.
- *Control and automation industry*: Where current applications are mainly based on corporate implementations, in which only suppliers and users might share the sensor data.

This book will address the challenges of measurable security arising in the communication within and between enterprises, with the focus on information provided by sensor systems. We address the challenges of new infrastructures, new ways of communication, and new devices. The two dominant trends in this domain are (1) wireless sensors contributing to automated processes and (2) the migration of processing and control into mobile devices.

In addition to phones, tablets, and computers, sensors and embedded systems (ES) will deliver and request information too, contributing to automated processes. The traditional way of handling security, by securing the whole infrastructure of a company, leads to declining to allow employees to use their own devices. Control is based on an approach of integrating, managing, and securing all mobile devices in the organization in a traditional way and such a short-sighted approach, as suggested by leading IT companies, is deemed to fail. These strategies cannot be applied to sensors, sensor networks, or other kinds of embedded systems: security must be guaranteed by design as the integration of such devices must be seamless.

A paradigm shift in handling security is required, addressing the need for securing information instead of securing infrastructure. The paradigm shift includes the need for measurable security, privacy, and dependability (SPD), and is the core of this book. It addresses a metrics-based approach for a quantitative assessment of both the potential attack scenario and the security measures of the information, and will outline the methodology of measurable security for systems of embedded systems, or the so-called IoT.

Measurable SPD is often misinterpreted as a good risk analysis. When "banks are secure," it means that they have a decent risk analysis, calculating the loss against the costs of increased security. Hereby, loss does not only mean financial loss, but also loss of customers due to bad reputation or press releases. Likewise, costs of increased SPD are not only the costs of applying new security mechanisms, but also a loss of customers, as customers might find the additional security mechanisms too cumbersome.

---

## What Can Go Wrong?

Security has often been considered as mitigating the vulnerabilities of systems by introducing specifications on how to design, implement, and operate a system. A good example of such an approach was introduced by the

National Institute of Standards and Technology in their Special Publication 800-12, focusing on IT systems (Guttman and Roback, 2005).

Confidentiality, integrity, and availability compose the core principles of information security. These three security attributes or security goals are the basic building blocks of the rest of security goals. There are other classifications and the community debates about including other security objectives in the core group, but this is irrelevant for this book. We can summarize the security objectives as

- Confidentiality
- Integrity
- Availability
- Authenticity
- Accountability
- Non-repudiation

Some or all of these objectives must be fulfilled (depending on the specific application requirements) in order to offer security services such as access control, which can be divided in three steps: identification, authentication, and authorization or privacy. To achieve these goals, different techniques such as encryption and anonymization are used. For example, information security uses cryptography to transform usable information into a form that renders it unusable by anyone other than an authorized user; this process is called *encryption*. It is possible to classify the attacks based on their final goal and functional objective and on the method used to execute them. Our classification is based on Ravi et al.'s (2004) taxonomy of attacks.

At the top level, attacks are classified into four main categories based on the final goal of the attack (Grand, 2004): cloning, theft-of-service, spoofing, and feature unlocking. The second level of classification is the functional objective of the attack. Here, we would distinguish between attacks against privacy (the goal of these attacks will be to gain knowledge of sensitive information; manipulated, stored, or communicated by an ES); attacks against integrity (these attacks will try to change data or code within an ES); attacks against availability (a.k.a. "denial of service" attack, these attacks disrupt the normal operation of the system). The third level of classification is based on the method used to execute the attack. These methods are grouped into three categories: physical attacks, side-channel attacks, and software attacks.

## How Secure Are Cyber-Physical Systems?

Bruce Schneier (2014) in his blog talks about the Security Risks of Connected Embedded Systems, saying that

*The industries producing these devices (embedded systems connected to the internet) are even less capable of fixing the problem than the PC and software industries were. Typically, these systems are powered by specialized computer chips. These chips are cheap, and the profit margins slim. The system manufacturers choose a chip based on price and features, and then build a router, server, or whatever.*

*The problem with this process is that no one entity has any incentive, expertise, or even ability to patch the software once it is shipped. And the software is old, even when the device is new even if they may have had all the security patches applied, but most likely not. Some of the components are so old that they're no longer being patched. This patching is especially important because security vulnerabilities are found "more easily" as systems age.*

*The result is hundreds of millions of devices that have been sitting on the Internet, unpatched and insecure, for the last five to ten years. Hackers are starting to notice and this is only the beginning. All it will take is some easy-to-use hacker tools for the script kiddies to get into the game. The last time, the problem was computers, ones mostly not connected to the Internet, and slow-spreading viruses.*

*The scale is different today: more devices, more vulnerability, viruses spreading faster on the Internet, and less technical expertise on both the vendor and the user sides. Plus vulnerabilities that are impossible to patch. Paying the cost up front for better embedded systems is much cheaper than paying the costs of the resultant security disasters.*

## What Are the Consequences of This?

Taking all this into account, it is clear that poorly designed CPSs are a threat to the domain where they are being used. CPSs are used in medical devices, aircraft, and nuclear plants. Attacking any of these systems can have terrible consequences, including the loss of human lives. CPSs are usually interconnected as well as connected to the Internet. A vulnerable CPS from a third party included in a larger system can be the entry point to the whole system for hackers. The market will put pressure on CPS vendors to design their systems better and, if this is not taken into consideration, business will certainly be affected.

## What Can We Do About It? The SHIELD Approach

Still, there is very little work concerning the full integration of security and systems engineering from the earliest phases of software development. Although several approaches have been proposed for some integration of security, there is currently no comprehensive methodology to assist developers of security sensitive systems. All this becomes a special concern when considering complex security requirements (SECF, 2011). This book is a step forward in that direction, as SHIELD offers a methodology for building secure embedded systems.

Through SHIELD, we address measurable SPD, and introduce countable numbers for the security components of systems. Our suggested approach works toward measuring SPD in terms of cardinal numbers, representing the application of specific security methods as compared to the specific threat scenario. The approach is based on the semantic description of both a potential attack scenario, the security-related aspects of sensors/systems, and security policies that should be applied irrespective of the scenario.

We also address the scalability aspect by using composition techniques that are able to build a security representation of the composed system (system of systems) based on the individual security representations of each individual element. This eases the process of measuring the security of the composed system, and opens up for the opportunity to build the system in an incremental way.

Our approach is to specify security through the terms Security (S), Privacy (P) and Dependability (D). Thus, instead of assessing a system based on just one term, we introduce the triplet SPD for describing different and correlated aspects of a CPS.

The main goals of SHIELD architecture are:

- Be coherent, measurable, composable, and modular in order to allow for a flexible distribution of SPD information and functionalities between different embedded systems while supporting security and dependability characteristics.
- Be adaptable to each defined application scenario and cover most of their SPD requirements.
- Ensure the correct communication among the different SPD modules.
- To be interoperable with other scenarios (1) developing standard interfaces for easy adoption and (2) easy deployment of architecture in legacy systems.

The last three goals aim at making security interoperable among heterogeneous systems, the same way as standardized communication architectures (TCP/IP, OSI, etc.) aim at making communications among heterogeneous systems possible.

In order to be able to achieve those goals, a SHIELD system is organized according to the following layering:

- Node layer: This layer is composed of stand-alone and/or connected elements like sensors, actuators, or more sophisticated devices, which may perform smart transmission. Generally, it includes the hardware components that constitute the physical part of the SHIELD system.

- Network layer: This is a heterogeneous layer composed of a common set of protocols, procedures, algorithms, and communication technologies that allow the communication between two or more devices as well as with the external world.
- Middleware layer: This includes the software functionalities that enable:
  - Basic SHIELD services to utilize underlying networks of embedded systems (like service discovery and composition).
  - Basic SHIELD services necessary to guarantee SPD.
  - Execution of additional tasks assigned to the system (i.e., monitoring functionality).

  Middleware layer software is installed and runs on SHIELD nodes with high computing power.
- Overlay layer: This is a logical vertical layer that collects (directly or indirectly) semantic information coming from the node, network, and middleware layers and includes the "embedded intelligence" that drives the composition of the SHIELD components in order to meet the desired SPD level. It is composed of software routines running at the application level.

Figure 1.3 provides a conceptual picture of the four functional layers of the SHIELD system together with a number of important SPD properties that must be considered.

Each device in a SHIELD system is security interoperable both horizontally (at the same layer) and vertically (across layers). Thus:

- SHIELD allows for security interoperability between different components on the same layer by using common security information representations and communication protocols. It also employs a

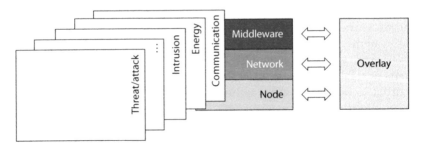

**FIGURE 1.3**
The four functional layers of a SHIELD system.

common semantic representation of the capabilities of each system with respect to their SPD functionalities.

- Security interoperability between different components across different layers is achieved by means of a common semantic framework that allows the security services registry and/or discovery, the exchange of SPD capabilities and metrics, and, finally, the composition and reconfiguration of components. This means that new components can be added to the SHIELD system with no need for existing architecture changes.

While the proposed SHIELD architecture substantially improves existing approaches to embedded systems security, it is clear that existing systems are not going to be replaced with SHIELD systems overnight. Thus, SHIELD architecture also includes a legacy systems gateway component to bridge the gap between both worlds and be able to migrate legacy systems to SHIELD systems gradually. Figure 1.4a and b describe how this migration can be achieved. Figure 1.4a shows the initial stage of a migration process, where the legacy system communicates through the gateway with no need for integrating new devices. Figure 1.4b shows a system that has partially replaced legacy components with SHIELD components but still needs some legacy compliancy.

Through the SHIELD methodology, we are able to measure security as well as to make possible the composition of security according to application requirements and/or attack scenarios. While some steps in the methodology focus on a mathematical treatment of SPD, others take into consideration system aspects or user awareness. Security assessment is on very different levels. While, for example, car and avionics industries, coming from the safety domain, can use their safety approaches and adapt them to the assessment of security, other industries like home/building industries have a less stringent understanding of security related to embedded systems.

The SHIELD methodology for measurable security is based on the system designer's awareness of the need of security and concludes with a successfully tested/verified secure CPS implementation. During the design process, the SHIELD framework guides and helps the designer to choose the appropriate security components according to its specific needs, the resources of the system, and the threats it is facing.

The approach adopted for the definition of SHIELD metrics considers three different theories internationally recognized and used. Among them, the most popular is the one introduced by the Common Criteria standard, an international standard adopted in the field of IT systems/products *security certification*. Common Criteria define increasing levels of assurance that can be targeted when performing a system/product security certification. These levels express the confidence that the security functionalities of an IT system/product as well as the assurance measures applied to it during its development meet the requirements stated in a document named Security Target.

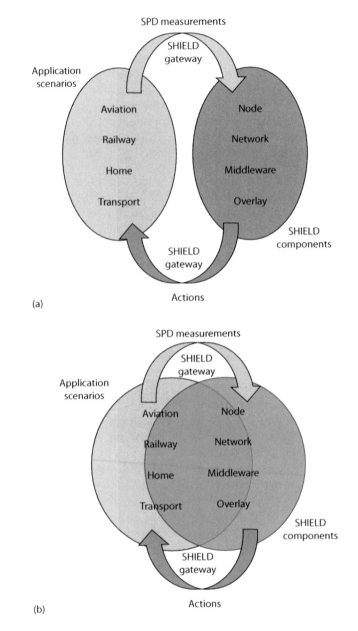

(a)

(b)

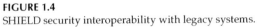

**FIGURE 1.4**
SHIELD security interoperability with legacy systems.

This document is the basis for performing a security certification against Common Criteria standard. The higher the level of assurance, the higher the confidence on the fulfillment of security requirements by the IT system/product under evaluation. Common Criteria standard is perfectly applicable to a generic SHIELD compliant system; however, while dealing with a generic system, it is not possible to describe its security functionalities and requirements through a Security Target: a Security Target is a document related to a specific system/product. As such, it should contain information highly dependent on implementation but, in this phase, we are dealing with the security features of a generic SHIELD compliant system. To address this type of situation, Common Criteria standard introduced the concept of Protection Profile, a document quite similar in its content to a Security Target but without implementation-specific characteristics. It could be written by a user group to specify their IT security needs and used as a guideline to assist them in procuring the right product or systems that best suits their environment. Basically, a Protection Profile is used to define the security features of a family/class of products. From this perspective, we considered particularly interesting the definition of a Protection Profile that is applicable to a generic system that aims to be SHIELD compliant. In the context of a Protection Profile, the class of products it is related to is named the target of evaluation (TOE). In the SHIELD context, the TOE is the generic system/product which aims to be SHIELD compliant and in our preliminary Protection Profile the TOE is the middleware.

The main security features contained in the Protection Profile defined for SHIELD middleware belong to the following class of security functions:

- Identification and authentication
- Auditing
- Data integrity
- Availability

These generic security functions can be mapped on SHIELD middleware TOE as follows:

- The orchestrator that improves services discovery/composition is able to identify and authenticate services/devices (discovered/composed)
- The middleware is able to record security-relevant events
- The middleware is able to verify the integrity of composition command definition
- The middleware is able to grant services availability

Of course, the defined Protection Profile is not meant to be a point of arrival, but a starting point that can be exploited to define Protection Profiles

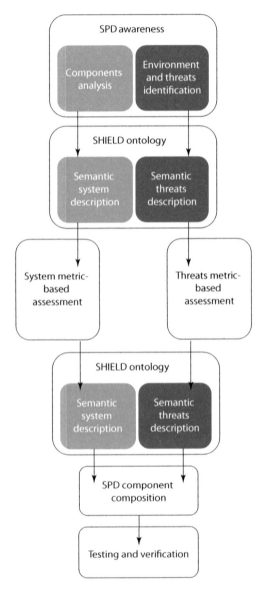

**FIGURE 1.5**
SHIELD framework for measurable SPD.

for all the other layers defined in SHIELD systems, up to hypothesizing a unique Protection Profile describing the security features of an entire system which aims to be SHIELD compliant (Figure 1.5).

The SHIELD methodology starts with *Step 1*, the *environment and threats identification*, where we study the location and access to the embedded systems (e.g., is the access to the environment where sensors are used restricted or open available? Are the sensors communication through wires or wireless? etc.).

*Step 2* is the *SPD assessment*, where application cases are used to visualize the potential of security failures.

*Step 3* is a *metrics implementation*, where we established a mathematical assessment based on the SPD guidelines.

*Step 4* is *SPD functionalities composition*, where the identified threats are balanced by activating proper countermeasures: the SPD functionalities, adequately described through the mentioned metrics and put together according to security standards (e.g., Common Criteria), domain policies, or specific composition algorithms.

The SHIELD methodology will be described in detail in Chapters 4 and 5.

---

## Summary

This chapter shows the benefits that the SHIELD approach provides to CPSs designers and integrators regarding the SPD functionalities that these systems must offer for a dynamic interaction between them.

The main advantages are:

- Measurable SPD: SHIELD defines a set of SPD metrics that determine the formal quantification of the SPD measurements. Current approaches use a qualitative security value at best, which makes it very difficult to compare different security strategies or to measure an increase/decrease in security.
- Scalability (by means of composition techniques):
  - At metrics evaluation level: Security metrics not only define the quantitative value of a given SPD functionality, but also how to compose the individual values of the components to obtain the value of the composed system.
  - At system configuration level: The ontological description of the SPD functionalities and metrics of each component allows the overlay layer to configure the individual subcomponents to reach the desired SPD level.
- Reduction of design complexity: As a consequence of the previous two advantages, the complexity of system design is greatly reduced. This is because we only need to deal with SPD requirements and threats for each component individually, and can

define the required SPD levels for the final system of systems independently.

- Interoperability: SHIELD allows for security interoperability between different components by using common security information representations and communication protocols, and a common semantic framework. This means that new components can be added to a SHIELD system with no need for existing architecture changes, even legacy systems.

- Testing and verification: The components that constitute the SHIELD platform are subject to a formal validation procedure that allows proving their correctness.

## References

Grand, J., Practical secure hardware design for embedded systems. In *Proceedings of the 2004 Embedded Systems Conference*, San Francisco, California, 2004.

Guttman, B. and E. Roback, *An Introduction to Computer Security: The NIST Handbook*, National institute of Standards and Technology, Special Publication 800–812, 2005. http://www.nist.org/nist_plugins/content/content.php?content.33; accessed: 25 September 2012.

Hulme, G. V., Embedded system security much more dangerous, costly than traditional software vulnerabilities, April 16, 2012. http://www.csoonline.com/article/704346/embedded-system-security-much-more-dangerous-costly-than-traditional-software-vulnerabilities; accessed: 11 February 2014.

Ravi, S. et al., Tamper resistance mechanisms for secure embedded systems. In *Proceedings of the International Conference of VLSI Design*, Mumbai, India, pp. 605–611, 2004.

Schneier, B., Security risks of embedded systems, January 9, 2014. https://www.schneier.com/blog/archives/2014/01/security_risks_9.html; accessed: 11 February 2014.

SECF, 2011. FP7-ICT-2009-5 European Project. Project ID: 256668. SecFutur: Design of secure and energy-efficient embedded systems for future internet applications. http://secfutur.eu/; accessed: 11 February 2014.

# Section I

# SHIELD Technologies and Methodology for Security, Privacy, and Dependability

# 2

## Security, Privacy, and Dependability Concepts

Andrea Fiaschetti, Paolo Azzoni, Josef Noll, Roberto Uribeetxeberria, John Gialelis, Kyriakos Stefanidis, Dimitrios Serpanos, and Andreas Papalambrou

### CONTENTS

### Introduction

Our society is built and driven by cyber-physical systems (CPSs), ranging from low-end embedded systems (ESs), such as smart cards, to high-end systems, like routers, smartphones, vehicle embedded control units, industrial controls, and so on. CPSs thus constitute one of the key elements of the Internet of Things (IoT) (Alam et al., 2011). The technological progress produced several effects, such as the power and performance boost of CPSs. Hence, their capabilities and services have raised, and in consequence, their usage has been substantially increased. Together, with the evolution of performance, energy consumption, and size, CPSs jump from isolated environments to interconnected domains (digitization). Although the evolution of connectivity enlarges the number of possible services, at the same time it increases the attackability of these kind of systems. When isolated, CPSs were hard to attack, since attackers need to have physical access. However, the open connection toward the Internet makes them vulnerable to remote attacks.

CPSs are used for multiple purposes, mainly to capture, store, and control data of a sensitive nature, for example, home, mobile, or industrial usage. Attackers could have different goals to compromising CPSs, from gathering sensitive data, thus compromising their privacy, to disruption of service by a denial of service (DoS) attack, exploiting their security and dependability. The consequences of a malicious and a successful attack could cause physical and economic losses, and thus it is important to keep them as secure, privacy-aware, and dependable as needed in a given situation.

In this chapter, we lay out the path from the traditional threat-based assessment to the security, privacy, and dependability (SPD) approach introduced through SHIELD. The chapter starts with a review on SPD threats, then provides a reference security taxonomy, and finally provides an industrial perspective on the impact of CPSs on industry.

## Security, Privacy, and Dependability Threats

In the traditional way of handling security, the attempt was to secure the whole infrastructure of a company. To follow the analogy, bring your own device (BYOD) is often seen as a threat, and answered in the classical way by preventing employees from using their devices, as security cannot be ensured. A second variant of counteracting classic threats as insufficient authentication and loss of devices is addressed through an approach of integrating, managing and securing mobile devices. But these strategies cannot be applied to sensors and other kind of CPSs. Companies cannot stop integrating ESs into their infrastructures, as their businesses and processes need them to remain competitive. So they need to be able to assess the security impact of integrating a new system into their infrastructure in a manageable way.

Such a short-sighted approach, as suggested by leading IT companies, is deemed to fail. A paradigm shift in handling security is required, addressing the need for securing information instead of securing infrastructure. The paradigm shift includes the need for measurable security, and is the core of this book. It addresses a metrics-based approach for a quantitative assessment of both the potential attack scenario and the security measures of the information, and will outline the methodology of measurable security for systems of CPS, or the so-called Internet of Things.

As anticipated, measurable security is often misinterpreted as a good risk, which allows to calculate the loss against the costs of increased security. Hereby, loss does not only mean financial loss, but also loss of customers due to bad reputation or press releases. Likewise, costs of increased security are not only the costs of applying new security mechanisms, but also a loss of customers, due to the perceived complexity of technology.

The SHIELD-suggested approach works toward measuring security in terms of cardinal numbers, representing the application of specific security methods as compared to the specific threat scenario. The approach is based on the semantic description of both a potential attack scenario, the security-related aspects of one sensors/systems, and security policies that should be applied irrespective of the scenario. The outcome is a methodology for measurable security that provides composable security for systems of CPS.

The threat-based approach (see Chapter 1), is still the dominating approach, but has a major challenge when it comes to scalability. Through the introduction of sensors, each system converts into a system of systems, which can't be economically assessed through the threat analysis. In addition, industrial systems are more and more software-driven or software-configurable (up to software-adaptive), and thus show a high degree of flexibility and extensibility. The SHIELD approach aims at decomposing a system of system along the line of components and functionalities, as shown in Figure 2.1.

These components and functionalities are then assessed with respect to their security, privacy, and dependability (SPD) behavior. As an example, a sensor has an *identity* component, and as functionality an *identity management*. The identity of the sensor can then be used to link the sensor data to the identity of the sensor, and thus communicate a traceable value. The threat taxonomy defined hereafter is used to set the basis for the SPD assessment described in the following chapters.

---

## The Fundamental Chain of Threats

Taxonomies are a tool for understanding the concepts of security, privacy, and dependability in the embedded systems environment. This first step has been taken into account in order to help designing the architecture and implement the prototypes of SHIELD. The following taxonomy, as derived by the relevant literature, is used as the basis for the design decisions behind the SHIELD architecture.

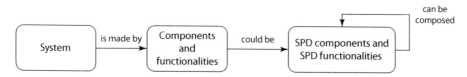

**FIGURE 2.1**
The SHIELD approach for composability of security, privacy, and dependability.

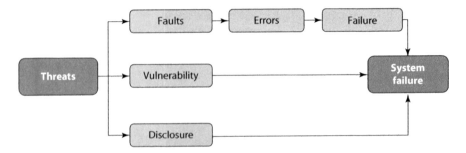

**FIGURE 2.2**
Threats taxonomy.

The importance of assessing SPD attributes is usually expressed in terms of the consequential damage resulting from the manifestation of threats. Threats are expressed in the taxonomy shown in Figure 2.2.

A fault is defined as the cause of an error and a failure is linked to the error that is outside of the error tolerance boundary and is caused by a fault. Threats are identified mainly through a risk analysis relevant to a scenario and are characterized in terms of a threat agent, a presumed attack method, any cause for the foundation of the attack, and the identification of the asset under attack. An assessment of risks to SPD qualify each threat with an assessment of the likelihood of such a threat developing into an actual attack, the likelihood of such an attack proving successful, and the consequences of any damage that may result.

To be more precise, we consider threats addressed against SPD attributes of assets requiring protection, that they can affect the system's normal function or cause faults, errors, failures from a dependability perspective, or exploit vulnerabilities from a security perspective, or cause disclosure from a privacy perspective. All these events lead to a system state that we can define as a system failure, that is, very briefly, what we want to avoid in designing and developing a secure system.

A system failure occurs when the delivered service deviates from fulfilling the system function, the latter being what the system is aimed at. An error is that part of the system state that is liable to lead to subsequent failure: an error affecting the service is an indication that a failure occurs or has occurred. The adjudged or hypothesized cause of an error is a fault. The creation and manifestation mechanisms of faults, errors, and failures may be summarized as follows:

- A fault is active when it produces an error. An active fault is either an internal fault that was previously dormant and which has been activated by the computation process, or an external fault. Most internal faults cycle between their dormant and active states. Physical faults can directly affect the hardware components only, whereas human-made faults may affect every component.

- An error may be latent or detected. An error is latent when it has not been recognized as such; an error is detected by a detection algorithm or mechanism. An error may disappear before being detected. An error may, and in general does, propagate; by propagating, an error creates other and new error(s). During operation, the presence of active faults is determined only by the detection of errors.

- A failure occurs when an error "passes through" the system–user interface and affects the service delivered by the system. A component failure results in a fault for the system which contains the component, and as viewed by the other component(s) with which it interacts; the failure modes of the failed component then become fault types for the components interacting with it.

These mechanisms enable the "fundamental chain," where a failure causes a component fault, which then signals an error. The error will cause other components to cause a failure, and repeatedly cause new and different errors.

---

## A Reference Security Taxonomy

The dependability concept in technical literature is introduced as a general concept including the attributes of reliability, availability, safety, integrity, maintainability, and so on. With ever-increasing malicious attacks, the need to incorporate security issues has arisen. Effort has been made to provide basic concepts and taxonomy to reflect this convergence, mainly because, traditionally there are two different communities separately working on the issues of dependability and security. One is the community of dependability that is more concerned with nonmalicious faults and more interested in the safety of systems. The other is the security community that is more concerned with malicious attacks or faults.

The analysis of a design concerning dependability usually focuses on accidental or random faults. Modeling these attributes can be performed easily; hence, the ability to optimize the design can be achieved. However, there is another fault cause that must be addressed, namely, the malicious or intentional ones. These types of faults are mainly associated with security concerns. The root causes of system failure in a dependability context (e.g., random accidental failures, system ageing, etc.) are fundamentally different from the root causes of security failure and privacy loss (e.g., intentional attacks).

In order to have a unified approach, we focused on threats. Having defined threats in the previous section, we can establish SHIELD SPD objective and from these derive the required SPD function, in other words the means to face the threats. When we are aware which kind of SPD function we need in our operational environment, we have a valid tool to contribute in the requirements definition useful to meet the security objectives to solve the SPD concerns.

In Figures 2.3 through 2.5, we present the categorization of SPD attributes and the more common means for their maintenance.

We have considered not only the traditional confidentiality, integrity, and availability (CIA) but even other attributes, to have more granularity in the definition (Figure 2.4).

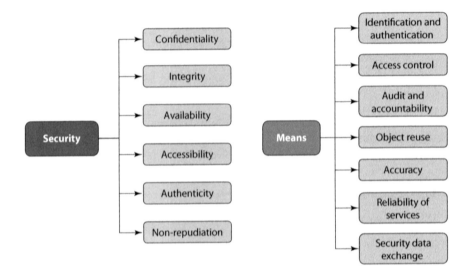

**FIGURE 2.3**
Security/means definition.

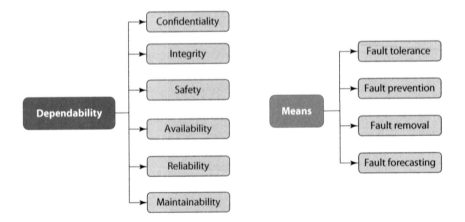

**FIGURE 2.4**
Dependability/means definition.

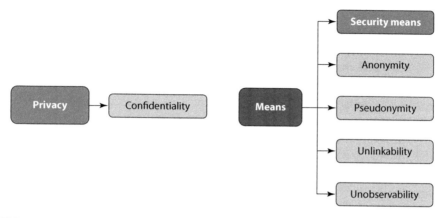

**FIGURE 2.5**
Privacy/means definition.

We have kept the traditional dependability taxonomy (Figure 2.5).

We have intended privacy as a reason for security rather than a kind of security. For example, a system that stores personal data needs to protect the data to prevent harm, embarrassment, inconvenience, or unfairness to any person about whom data are maintained, and for that reason, the system may need to provide a data confidentiality service, anyway, as indicated, we identified specific means that can be adopted to assure privacy.

## Taxonomy Definitions

Detailed descriptions of the terms that are present in the SHIELD reference taxonomy are given next in alphabetical order:

- *Access control*: It is the process of mediating every request of access to SHIELD assets determining whether the request should be granted or denied according to the security policies established.
- *Accessibility*: The ability to limit, control, and determine the level of access that entities have to a system and how much information they can receive.
- *Accountability*: The ability to track or audit what an individual or entity is doing on a system.
- *Anonymity*: The function that ensures that a user may use a resource or service without disclosing the user's identity.
- *Audit*: SPD auditing involves recognizing, recording, storing, and analyzing information related to SPD relevant activities. The resulting audit records can be examined to determine which SPD relevant activities took place.

- *Authentication*: Authentication is any process by which a system verifies the identity of a user who wishes to access it.
- *Authenticity*: The property of being able to verify the identity of a user, process, or device, often as a prerequisite to allowing access to resources in an information system.
- *Authorization*: Authorization is the process of giving someone permission to access a resource or some information.
- *Availability*: The readiness for correct service. It is the automatic invocation of a web service by a computer program or agent, given only a declarative description of that service, as opposed to when the agent has been preprogrammed to be able to call that particular service.
- *Composability*: Is the possibility to compose different (possibly heterogeneous) SPD functionalities (also referred to as SPD components) aiming at achieving in the considered system of embedded system devices a target SPD level that satisfies the requirements of the considered scenario.
- *Confidentiality*: Property that data or information are not made available to unauthorized persons or processes.
- *Configuration*: Configuration is the translation of logical configuration into a physical configuration.
- *Dependability*: There are a lot of definitions for dependability. Here are some of them:
  - Dependability is the ability of a system to deliver the required specific services that can "justifiably be trusted."
  - Dependability is the ability of a system to avoid failures that are more frequent or more severe than is acceptable to the users.
  - Dependability is a system property that prevents a system from failing in an unexpected or catastrophic way.
- *Error*: Deviation of system behavior/output from the desired trajectory.
- *Failure*: An event that occurs when system output deviates from the desired service and is beyond the error tolerance boundary.
- *Fault*: Normally the hypothesized cause of an error is called fault. It can be internal or external to a system. An error is defined as the part of the total state of a system that may lead to subsequent service failure. Observing that many errors do not reach a system's external state and cause a failure, Avizienis et al. have defined active faults that lead to error and dormant faults that are not manifested externally.
- *Fault avoidance (prevention)*: A technique used in an attempt to prevent the occurrence of faults.

- *Fault containment (isolation)*: The process of isolating a fault and preventing its effect from propagating.
- *Fault detection*: The process of recognizing that a fault has occurred.
- *Fault forecasting (prediction)*: The means used to estimate the present number, the future incidence, and the likely consequence of faults.
- *Fault location*: The process of determining where a fault has occurred so a recovery can be used.
- *Fault masking*: The process of preventing faults in a system from introducing errors into the informational structure of that system.
- *Fault removal*: The means used to reduce the number and severity of faults.
- *Fault restoration (recovery)*: The process of remaining operational or gaining operational status via a reconfiguration event in the presence of faults.
- *Fault tolerance*: Ability to continue the performance of tasks in the presence of faults.
- *Graceful degradation*: The ability of a system to automatically decrease its level of performance to compensate for hardware or software faults.
- *Hardware failures*: Hardware failures include transient faults due to radiation, overheating, or power spikes.
- *Identification*: Determining the identity of users.
- *Information*: Information is measured data, real-time streams from audio/video-surveillance devices, smart sensors, alarms, and so on.
- *Integrity*: Absence of malicious external disturbance that makes a system output off its desired service.
- *Maintainability*: Ability to undergo modifications and repairs.
- *Mean*: All the mechanisms that break the chains of errors and thereby increase the dependability of a system.
- *Non-repudiation*: Non-repudiation refers to the ability to ensure that a party to a contract or a communication cannot deny the authenticity of their signature on a document or the sending of a message that they originated.
- *SHIELD asset*: The SHIELD assets are information and services.
- *Performability*: The degree to which a system or component accomplishes its designated functions within given constraints, such as speed, accuracy, or memory usage. It is also defined as a measure of the likelihood that some subset of the functions is performed correctly.

- *Privacy*: The right of an entity (normally a person), acting on its own behalf, to determine the degree to which it will interact with its environment, including the degree to which the entity is willing to share information about itself with others.
- *Pseudonymity*: The function that ensures that a user may use a resource or service without disclosing its user identity, but can still be accountable for that use.
- *Reliability*: Reliability is continuity of correct service even under a disturbance. It is removal (fault) mechanism that permits the system to record failures and remove them via a maintenance cycle.
- *Safety*: The property that a system does not fail in a manner that causes catastrophic damage during a specified period of time.
- *Software failures*: Software failures include: bugs, crashes, data incompatibilities, computation errors, wrong sequence of actions, and so on.
- *Survivability*: Ability to fulfill its mission in a timely manner in the presence of attacks, failures, or accidents.
- *Testability*: The degree to which a system or component facilitates the establishment of test criteria and the performance of tests to determine whether those criteria have been met.
- *Traceability*: The ability to verify the history, location, or application of an item by means of documented recorded identification.
- *Unlinkability*: The function that ensures that a user may make multiple uses of resources or services without others being able to link these uses together.
- *Unobservability*: The function that ensures that a user may use a resource or service without others, especially third parties, being able to observe that the resource or service is being used.

---

## Next Steps

The following chapters will address the challenges of measurable security from a goal-based approach, with the focus on information provided by sensor systems. We address the challenges of new infrastructures, new ways of communication, and new devices. A detailed introduction of the novel SHIELD approaches for measurable security is further outlined in Chapters 4 and 5, followed by their applicability in several use cases in Chapters 6 through 9, technology demonstrators (Chapter 10), and industrial perspective and new research initiatives (Chapter 11).

# References

Alam, S., M. M. R. Chowdhury and J. Noll, Interoperability of security-enabled internet of things. *Wireless Personal Communications*, 61(3), 567–586, 2011.

Avizienis, A., J.-C. Laprie, and B. Randell, Dependability of computer systems: Fundamental concepts, terminology, and examples. Technical Report, LAAS-CNRS, October 2000.

Avizienis, A., J.-C. Laprie, B. Randell, and C. E. Landwehr, Basic concepts and taxonomy of dependable and secure computing. *IEEE Transactions on Dependable and Secure Computing*, 1(1), 11–33, 2004.

# 3

# Security, Privacy, and Dependability Technologies

Paolo Azzoni, Luca Geretti, Antonio Abramo, Kyriakos Stefanidis,
John Gialelis, Andreas Papalambrou, Dimitrios Serpanos,
Konstantinos Rantos, Andrea Toma, Nawaz Tassadaq, Kresimir
Dabcevic, Carlo Regazzoni, Lucio Marcenaro, Massimo
Traversone, Marco Cesena, and Silvano Mignanti

## CONTENTS

Introduction .................................................................................................. 30
Node-Level Hardware and Software Technologies ................................. 31
   Smart Card–Based Security Services ...................................................... 31
      Smart Card Security Services .............................................................. 32
      Implementation Overview .................................................................. 35
      SHIELD Integration .............................................................................. 37
      Application of SPD Metrics ................................................................. 38
   Secure and Dependable Biometrics ........................................................ 38
      Biometric Face Recognition ................................................................ 39
      Face Recognition Software .................................................................. 40
      The HW Platform ................................................................................. 47
      SHIELD Integration .............................................................................. 49
   The SHIELD Gateway ............................................................................... 50
      The S-Gateway Architecture ............................................................... 52
      SHIELD Integration .............................................................................. 54
      Application of SPD Metrics ................................................................. 55
   Open Mission Network Integrated Architecture .................................. 57
      The OMNIA Architecture .................................................................... 58
      Hardware Architecture ........................................................................ 59
      Software Architecture .......................................................................... 60
Network Technologies ................................................................................ 61
Distributed Self-x Models ......................................................................... 61
Recognizing and Modeling of DoS Attacks ............................................ 61
   Categories .................................................................................................... 62
   Detection ...................................................................................................... 62

## Introduction

A SHIELD ecosystem is based on a set of heterogeneous security, privacy, and dependability (SPD) technologies and solutions in the so-called secure service chains (SSC). The concept is to provide system functionalities, from the SPD perspectives, in a tightly integrated way at node, networks, and middleware/overlay layer. Starting from SPD state of the art in embedded systems (ESs), this chapter describes new technologies and solutions that constitute a solid basement for the SHIELD framework and for the creation of "SPD-ready" cyber-physical systems (CPS). Three different levels are covered:

- The node level consists of CPSs that natively provide SPD functionalities and that can seamlessly interact through appropriate service-oriented interfaces.
- The network level is focused on ensuring secure communications and service dependability.
- The middleware/overlay level provides the software components that compose the SHIELD framework and that allow the orchestration of a SHIELD ecosystem in terms of SPD.

This baseline of SPD technologies and solutions, and the resulting SHIELD framework, has been adopted in several vertical domains and technology demonstrators that are described in the second part of the book.

## Node-Level Hardware and Software Technologies

### Smart Card–Based Security Services

Smart cards are technologically mature devices resilient to physical attacks and to unauthorized data manipulation, and are therefore utilized in environments where application security is of major importance, such as payment applications, healthcare, identity management, and physical access control. These applications need to rely on a robust system that safely stores and handles critical information and sensitive data, such as cryptographic keys. Smart cards are a strong candidate due to their tamper resistance.

Smart cards can provide multiple security levels for data and applications stored on them. For instance, information stored on the smart card can be accessible to authorized entities only, can be marked as read-only, can be used only within the smart card, or shared among applications that reside on the card. One of the main advantages of deploying a smart card–based solution is that all the sensitive operations can be accomplished on the card rather than the terminal or application, which in many cases is not considered trustworthy. Smart cards can provide, among others, the following security services:

- Message authentication code (MAC)
- Encryption
- Digital signatures
- Hash functions
- Secure key management

Additionally, smart cards can communicate through standardized interfaces, such as USB or wirelessly, and are not bound to specific platforms. Their small size and low power consumption make them a strong candidate to deploy as a complementary device to adequately protect nodes and provide security to their stored data and communications. The SHIELD architecture can take advantage of their security properties in providing the required SPD functionalities.

### Smart Card Security Services

In this section, the use of two smart card–based services, namely secure messaging and mutual authentication, is analyzed in the context of the SHIELD architecture.

#### Secure Messaging

In the SHIELD architecture, participating nodes interact with each other and with remote nodes outside the SHIELD networks. The aim of deploying a secure messaging service is to provide SHIELD nodes with the means to establish a secure channel and exchange information while ensuring confidentiality, integrity, and message origin authentication. Moreover, nodes might have to prove their identities prior to joining the SHIELD system. In the SHIELD terminology, this means that the overlay layer has to mutually authenticate with the node, with the latter using the services provided by the smart card, in order to register the node and its services as legitimate SHIELD components. The provided services can therefore be used to build trust among the participating nodes while mitigating the risk of the node and its data being compromised by unauthorized entities.

The establishment of a secure channel relies on a robust key management scheme where cryptographic keys are used for nodes' mutual authentication and the derivation of session keys. The required protocols might depend on public-key cryptography or on symmetric cryptosystems. While the former is generally more flexible, they require a supporting infrastructure, which will be used for the issuance of digital certificates and their management. Key management based on symmetric cryptosystems was chosen instead mainly for its simplicity in closed environments and the low requirements in terms of resources.

Prior to the smart card being issued and deployed in the node's environment, it must be personalized for the specific system. One of the data items that must be provided to the personalization system is the node's unique serial number. This serial number will be included in the card together with additional information, so that the personalized smart card will have the following data stored in it:

- Node's serial number, which also serves as the node's ID
- Node's master encryption key
- Node's authentication key

Each smart card will bear two keys: one used for encryption and one for authentication purposes. Nodes' encryption keys are derived from the system's master key—SMK (encryption or authentication) and the node's serial number. The generation of nodes' secret keys is therefore derived using the following mechanism:

$$\text{Node Master Encryption} - \text{Key (NMEK)}$$
$$= \text{AES}_{\text{SMK-E}}\left(\text{Node's Serial Number}\right)$$

$$\text{Node Master Auth} - \text{Key (NMAK)}$$
$$= \text{AES}_{\text{SMK-A}}\left(\text{Node's Serial Number}\right)$$

where SMK-E is the system's master key used for encryption, SMK-A is the system's master key used for message authentication, and AESSMK-x(y) denotes that encryption of data y using the Advanced Encryption Standard algorithm and the key SMK-x. Master keys typically reside in a hardware security module (HSM) while card personalization should take place in a safe environment. The node's serial number must be padded to the left with zeroes to produce a 16-byte long number.

Following the smart card's personalization, it can be deployed with the node and used for the establishment of a secure channel between two parties, such as the node and the central authority, which includes mutual authentication and the secure channel establishment of session keys. Messages exchanged using the secure channel are protected against unauthorized disclosure and modification, while their origin can be authenticated.

There are two options regarding the adoption of a secure channel depending on its usage. In the first one, encryption takes place on card. This option is only applicable in situations where small messages are exchanged between the node and the central authority and a higher level of security is a strong requirement. Given that some of these nodes might have limited resources, including power, we want to restrict the number of exchanged messages. Therefore, the most efficient way to secure messages is by the use of an implicit secure channel, that is, a secure channel that does not need extra messages to be exchanged between the two entities for its establishment. The secure channel is initiated implicitly when a cryptographically protected message is received by either entity. The cryptographic keys never leave the card and this constitutes an advantage for this method. The following steps have to take place:

If the smart card–enabled node wants to send a secure message

- The node sends the data to the smart card with a request to appropriately encrypt them.
- The smart card generates a random number (RN), a session encryption key (SEK), and a session authentication key (SAK) using NMEK and NMAK, respectively, as follows:
  - $\text{SEK} = \text{AES}_{\text{NMEK}}\left(\text{RN}\right)$
  - $\text{SAK} = \text{AES}_{\text{NMAK}}\left(\text{RN}\right)$

- The card generates a hash-based message authentication code (HMAC) on the data and appends it to the data. Following that, it encrypts the data together with the HMAC and passes to the node the RN and the result of the encryption process:
  - Card's response $= RN \mid\mid AES_{SEK}$ (Data $\mid\mid HMAC_{SAK}$ (Data))

  If encryption is not necessary yet integrity protection remains a strong requirement, there is no encryption taking place and only the HMAC is computed on data (the card only needs to compute SAK). In that case, the card's response becomes
  - Card's response $= RN \mid\mid HMAC_{SAK}$ (Data)
- Upon receiving the message, the remote party uses the RN to calculate the session keys, decrypt the message, and verify the MAC for message origin authentication. Note that this is a one-pass protocol that can be used to send encrypted and integrity-protected messages to the remote party.

If a remote entity wants to initiate an encrypted communication with the smart card–enabled node:

- It generates the session keys, using the same method described above, calculates the HMAC on the data, encrypts the data, and constructs the message.
- The node passes the encrypted message to the card where the card generates the same session keys, decrypts the messages, verifies the MAC, and if successful, it returns the decrypted message to the node.

The other option addresses the need to be able to exchange large amounts of data in a protected manner and typically refers to nodes with more resources, such as power nodes that can support the exchange of these volumes of data. In this case, the card can only be utilized for session key establishment. The node requests for a session key from the card, which is handed over to the node for the encryption of the exchanged data. The key is derived using the same method described previously. As a result, the following steps take place:

- The node requests for session keys from the card.
- The card generates an RN, and generates an SEK and an SAK using the aforementioned method.
- The card returns the keys to the node.

Note that a session key does not have to be regenerated for every message the two entities exchange, saving valuable computing resources. It can be used for a session, as the name suggests or for multiple sessions, depending on the system's policy. Note that although the node's ID is critical for the proper establishment of the required keys, it is intentionally omitted from the above analysis since it will typically be part of the message that the node will prepare for the central authority.

*Node Authentication*

One of the building blocks of the SHIELD system is trust built among participating nodes and system components. This is typically achieved following proper successful authentication that will provide assurances regarding the identities of the peers. Smart cards can provide the means to safely store the appropriate credentials needed for this purpose, such as identities and cryptographic keys.

Mutual authentication between Nodes A and B includes the following steps (note that we assume that both entities make use of smart cards attached to the nodes), while Node A is considered a central node that has the capacity to securely store and manage SMK:

- Node A, generates an RN, $R_A$, either on-card or off-card, and sends it to remote Node B. Note that this might be the result of a request sent by Node B to be authenticated by a central node upon Node B joining the network.
  - Node A → Node B: $R_A$
- Upon receiving $R_A$, Node B generates a new RN, $R_B$, and forwards both random numbers to the card, which uses the node's authentication key to compute HMAC on the two values. Note that the RN $R_B$ can be generated on-card. It then returns HMAC to the node, which forwards it to Node A together with the random value $R_B$ and the node's serial number.
  - Node B → Node A = $R_B$ || $HMAC_{NMAK}$ ($R_A$ || $R_B$) || Serial Number
- Node A, using the serial number and the SMK, computes the shared NMAK. It then verifies the received HMAC value. If the verification succeeds, Node A computes a new HMAC using the same key as follows and sends it to Node B.
  - Node A → Node B: $HMAC_{NMAK}$ ($R_B$ || $R_A$)
- Upon receiving this value, Node B verifies that the HMAC is the one expected to be sent by Node A.

After the successful exchange of the aforementioned messages, both entities have assurances regarding the identity of the remote party. The central authority for example can start using the services of the node while the node can proceed with the exchange of additional information, configuration data, and services.

## Implementation Overview

The aforementioned protocols can be implemented on any smart card platform assuming the host, that is, SHIELD node, can support the necessary interface to communicate with this external additional component, that is, the smart card. A strong candidate is a Java-based smart card solution, which is a mature platform that has been extensively used for various applications and environments where security is of primary concern.

To support the functionality of node authentication and secure communication described in the preceding section, we implemented the architecture illustrated in Figure 3.1.

The prototype consists of two main modules:

- The applet that resides on the card and acts as a server waiting for the client's commands.
- The client that resides on the node and issues commands to the card, necessary for performing the authentication and security services provided by the card. The commands might be the result of a communication taking place between the overlay layer and the node.

An appropriate set of commands has been defined to support the secure channel establishment and messaging functionality. The commands and responses typically follow the ISO 7816 standardized form of commands, that is, application protocol data unit (APDU), comprising the fields shown in Table 3.1.

The following four commands and equal number of responses are necessary for the previous prototype functionality: two commands for secure channel/messaging and two for authentication.

*Secure Channel/Messaging*

Regarding secure channel/messaging between a node and the overlay layer, two commands are required, one of them used for securing the message (Table 3.2) and the other one for message decryption and verification (Table 3.3).

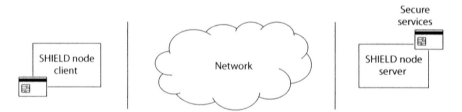

**FIGURE 3.1**

High-level components of the proposed implementation.

**TABLE 3.1**

Command Fields

| CLA | INS | P1 | P2 | Lc Field | Data Field | Le Field |
|---|---|---|---|---|---|---|
| Class of instruction | Instruction code | Instruction parameter 1 | Instruction parameter 2 | Number of bytes present in the data field of the command | String of bytes sent in the data field of the command | Maximum number of bytes expected in the response |
| *(1 byte)* | *(1 byte)* | *(1 byte)* | *(1 byte)* | *(1 or 3 bytes)* | | |

**TABLE 3.2**

Message Protection

| CLA | INS | P1 | P2 | Lc Field | Data Field | Le Field |
|---|---|---|---|---|---|---|
| $0 \times 80$ | $0 \times 74$ | $0 \times 02$ or $0 \times 04$ | $0 \times 00$ | Lc | Data that needs protection | Le |

Upon receiving the Message Protection command (Table 3.2), the card generates either both session keys (P1 = $0 \times 02$) or only the authentication session key (P1 = $0 \times 04$), it then generates the HMAC, and either encrypts the message as defined previously and sends it back to the host (P1 = $0 \times 02$) or simply responds with the computed HMAC (P1 = $0 \times 04$).

Upon receiving the Message Decryption and Verification command (Table 3.3), the card proceeds with the decryption and HMAC verification (P1 = $0 \times 02$), or only with the HMAC verification (P1 = $0 \times 04$). If the verification succeeds, the card returns the decrypted data (P1 = $0 \times 02$) or a command-successful response code SW1-SW2 = "9000."

*Authentication*

Regarding authentication, two commands are required, one of them being HMAC generation and the other one being HMAC verification. For HMAC generation, the standardized Internal Authenticate command can be used, while for HMAC verification we propose a variant of standardized external authentication command.

### SHIELD Integration

Smart card services provide the means to authenticate SHIELD components and secure their communications. They are mapped to the node layer of the SHIELD architecture while they provide services to the overlay layer. Moreover, they enable the secure authentication of a SHIELD node by the overlay layer, thus facilitating a node's integration to the SHIELD system.

The level of security services provided by the prototype can be adapted accordingly by the overlay layer considering the corresponding SPD metrics. The commands issued to the card have to be based on the required security levels (Figure 3.2).

**TABLE 3.3**

Message Decryption and Verification

| CLA | INS | P1 | P2 | Lc Field | Data Field | Le Field |
|---|---|---|---|---|---|---|
| $0 \times 80$ | $0 \times 78$ | $0 \times 02$ or $0 \times 04$ | $0 \times 00$ | Lc | Protected data concatenated with random number | Le |

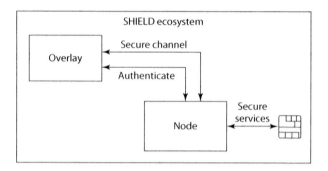

**FIGURE 3.2**
Smart card security services in the SHIELD architecture.

### Application of SPD Metrics

The level of protection provided by the smart card on the exchanged data can be controlled by the issuance of the appropriate commands. The prototype, as previously mentioned, provides authentication, message integrity, and confidentiality. The provided functionality can be combined with any kind of communication required among participating components. As such, the SPD levels for this prototype are defined in Table 3.4.

### Secure and Dependable Biometrics

Surveillance emerging trends frequently consider SPD application scenarios where the detection and tracking of objects (devices, cars, goods, etc.) play a crucial role. One of the most important objectives of this trend is to increase the intrinsic SPD of the scenario, providing added value services that improve our life (automatic tolling payment, navigation, traceability, and logistics). Frequently, these services and functionalities are based on the identification of a device while we are using it, or directly on the identification of people: currently, the real-time recognition, monitoring, and traceability of people is emerging as a crucial service in several application domains. From a technical point of view, these services introduce new challenges derived from the combination of the "real-time" requirements and the necessity to adopt ESs to provide recognition, monitoring, and tracking services. The SHIELD methodology, combined with the SPD hardware

**TABLE 3.4**

Smart Card Security Services SPD Levels

| SPD Level | Functionality |
| --- | --- |
| 0 (low) | No authentication/no message protection (the service is not called) |
| 1 (medium) | Message authentication only |
| 2 (high) | Message authentication and encryption |

infrastructure and software layers, represents the correct answer to these important challenges.

### *Biometric Face Recognition*

The adoption of biometric technologies in ESs is currently becoming a real opportunity in the surveillance market due to the convergence of the intelligent functionalities of recognition algorithms and the increasing hardware capabilities of ESs.

Several new face recognition techniques have been proposed recently, including recognition from three-dimensional (3D) scans, recognition from high-resolution still images, recognition from multiple still images, multimodal face recognition, multi-algorithms, and preprocessing algorithms to correct illumination and pose variations. The evolution of recognition technologies offers new potential toward improving the performance of automatic people identification. The design and development of a solution that supports the new recognition methods presents many challenges that can be tackled with an assessment and evaluation environment based on a significant data set, a challenging problem that allows the evaluation of the performance improvement of a selected algorithm, and an infrastructure that supports an objective comparison among the different recognition approaches.

The embedded face recognition system (EFRS) has been conceived to address all these requirements. The EFRS data corpus must contain at least 50,000 records divided into training and validation partitions. The data corpus contains high-resolution still images, taken under controlled lighting conditions and with unstructured illumination, 3D scans, and contemporaneously still images collected from cameras.

The identification of a challenging problem for the evaluation ensures that the test is performed on sufficiently reasonable, complex, and large problems, and that the results obtained are valuable when compared between different recognition approaches. The challenging problem identified to evaluate the EFRS consists of six experiments. The experiments measure the performance

1. On still images taken with controlled lighting and background
2. Uncontrolled lighting and background
3. 3D imagery
4. Multi-still imagery
5. Between 3D and still images

The evaluation infrastructure ensures that the results from different algorithms are computed on the same data sets and that performance scores are generated by the same protocol. The approach adopted for the improvements introduced by the EFRS has been inspired by the Face Recognition Vendor Test (FRVT) 2013 [1].

At the end of the evaluation, the face recognition method that has been selected as the baseline solution for the EFRS is the eigenface method [2]. This method is based on the extraction of the basic face features in order to reduce the problem to a simplified dimension. The principal component analysis (PCA [3]) is the selected method (also known in the pattern recognition application as the Karhunen–Loève [KL] transform) to extract the principal components of the faces distribution. These eigenvectors are computed from the covariance matrix of the face pictures set (faces to recognize); every single eigenvector represents the feature set of the differences among the face picture set. The graphical representations of the eigenvectors are also similar to faces: for this reason, they are called *eigenfaces*.

The eigenfaces set defines the so-called face space. In the recognition phase, the unknown face pictures are projected on the face space to compute the distance from the reference faces.

Each unknown face is represented (reducing the dimensionality of the problem) by encoding the differences from a selection of the reference face pictures. The unknown face approximation operation considers only the eigenfaces providing higher eigenvalues (variance index in the face space). In other words, during face recognition the unknown face is projected on the face space to compute a set of weights of differences with the reference eigenvalues. This operation first allows the ability to recognize if the picture is a face (known or not) and if its projection is close enough to the reference face space. In this case, the face is classified using the computed weights, deciding for a known or unknown face. A recurring unknown face can be added to the reference known face set, recalculating the whole face space. The best matching of the face projection with the faces contained in the reference faces set allows the identification of the person.

### Face Recognition Software

The core technology of the application for face recognition on ESs is the extraction of the principal components of the faces distribution, which is performed using the PCA method (also known as the KL transform). The principal components of the faces are eigenvectors and can be computed from the covariance matrix of the face pictures set (faces to recognize). Every single eigenvector represents the feature set of the differences among the face picture set. The graphical representations of the eigenvectors are also similar to real faces (also called eigenfaces). The PCA method is autonomous and therefore it is particularly suggested for unsupervised and automatic face recognition systems. The recognition process is based on the following functional steps (Figure 3.3):

- Face detection
- Face normalization

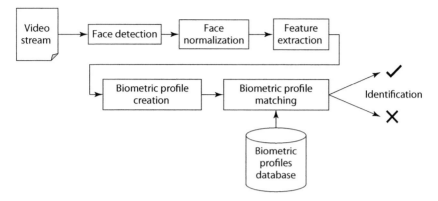

**FIGURE 3.3**
The face recognition process.

- Face features extraction
- Biometric profile creation
- Biometric profile matching
- Identification

The face recognition application is composed of a set of functional components, which implement each of the logical steps of the face recognition process.

- The *face detection* component finds the position and the size of each visible face present in the acquired image. To reach this important goal, the component is based on two pipelined algorithms for high-speed face localization, which are based on the training of a chain of Haar features, acting as weak recognizers. This solution is specialized for real-time face detection in video streams and provides:
  - Fast localization of image areas that are likely to contain faces.
  - Fast multiscale and multilocation search.
  - Accurate localization of facial features (eyes, mouth, eyebrows, etc.), robust against unfavorable illumination conditions, and against occlusions (dark sunglasses, scarves, etc.).

  The training procedure has been implemented using a proprietary implementation of the algorithm known as AdaBoost.
- The *face normalization* component identifies the features of morphological interest throughout the face area (eyes, mouth, and eyebrows), in order to properly rotate and scale the image. This operation moves all such points to predefined locations. As in the previous component, the algorithm for the localization of characteristic traits adopts

a chain of weak recognizers. This component has been designed, privileging the requirement of operational efficiency, to be easily portable across embedded hardware and to limit the impact on hardware performance that, in the classic surveillance solutions, is in charge of people identification. A second solution, for face localization, is an algorithm based on statistical models of appearance, which allows the alignment of the face even in the presence of severe occlusions of some characteristic points (sunglasses, scarves, etc.). This solution contributes to improving the SPD level of the identification process in the presence of noncollaborative users.

- The *feature extraction* component selects the relevant features from the normalized image in order to maximize the robustness against environmental disturbance and noise, non-optimal pose, non-neutral facial expressions, and variable illumination conditions. In the implementation, to maintain high-quality results in the presence of non-optimal environmental conditions, two specific solutions have been adopted: the advanced techniques for illumination compensation and the appropriate nonlinear differential filtering that create features inherently insensitive to global illumination and are not very sensitive to inaccuracies of alignment.

- The component in charge of *biometric profile creation* has been implemented by a feature vector template: the extracted features vector is processed by a trained statistical engine and is reduced to a smaller one that optimally describes the user's identity with a biometric profile.

- The component for *biometric profile matching* is responsible for person identification: the face recognition and the face verification processes are based on measurements of differences between biometric profiles. The normalized profile distance is the similarity value of the compared identities. This component is responsible for the creation of the normalized template distance.

- Finally, a *classification component* has been developed to provide the previous components with a horizontal tool capable of offering the classification features required during the various steps of the recognition process. The classification component is based on methods for the statistical classification of models and is able to manage information partially compromised by noise or occlusions. The classification component is based on an optimized version of the Bayesian classifier, because the problem of biometric recognition is inherently a problem of classification between two classes only. This component is used during face detection, during the relevant face features identification, and during the final profiles matching.

The classification component is based on statistical pattern analysis and addresses the classification problem providing an inexpensive training process suitable for ESs, and ensuring accurate and fast classification. Regarding the classification problem, given a tagged sample database (positives and negatives), to obtain the correct classification on unknown samples it is necessary to

- Select the optimal features extractor
- Train a statistical engine (classification algorithm)

Starting from these two objectives, the classification component has been developed using a recursive approach based on training and progressive adaptation. The entire classification process is described in the block diagram in Figure 3.4.

The initial phase of recursive training can be split into the following steps:

- During the *extractor selection* step an algorithm for feature extraction is selected in the database.
- A *feature extractor* test is performed applying the selected extractor to all positive (e.g., images with a face) and negatives results (e.g., images without faces), obtaining, for each image, a vector of real numbers as a result of the test (the filter is applied in all possible positions inside the images).
- The *statistical engine training* prepares the recognition system for the performance evaluation: on the vectors of numbers belonging to the positive and negative classes, an algorithm applies a statistical classification algorithm (it is based on support vector machines, neural

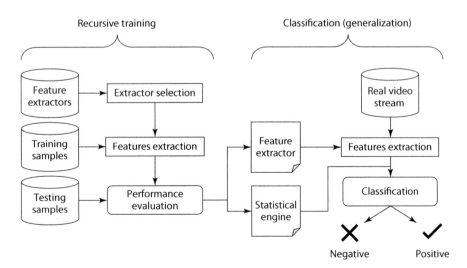

**FIGURE 3.4**
The classification process.

networks, etc.). After this step, the statistical engine should be able to correctly recognize a face.

- *Performance test*: This step checks whether the training produced generalized specimens not seen during training. This control is performed applying it to positive and negative images (obviously not used during training). The evaluation of the results is based on the International Civil Aviation Organization (ICAO) [4–6] standard and at least 90% of identification success is required. If the results don't satisfy this requirement the recursive training procedure can increase the dimension of the training set, or change the parameters of the extractor and the statistical engine, or restart the entire procedure with a different extractor.

The feature extractor algorithm, with its own configuration, and the statistical engine configuration that are selected during the training phase are finally adopted by the classification component for real-time people identification.

The classification phase consists conceptually of the generalization of the identification capabilities of the system from the training set to real-life unknown examples. A database of real examples is analyzed by the feature extractor and the statistical engine tries to identify the face recognized in the real images. If the result still satisfies the ICAO standards, the selected setup can be adopted on real cases.

During the feature extraction, the training and identification process algorithm uses a biometric profile template. The information in this template is organized by tag and is structured as follows:

- Position of the face mask points
- *Photographic set*: Focus, illumination, background, etc.
- *Subject*: Gender, age, ethnic group, etc.
- *Face*: Pose, eye expression, gaze, mouth expression, etc.
- *Morphology*: Eye type, lip type, nose type, mouth type, etc.

Figure 3.5 describes an example of the result of the matching and identification phase. The screenshot illustrates the face features identified in the photo and all the information extracted from the database after the matching and identification process.

The functional components previously described have been implemented in three software modules:

- The face finder (FF) module
- The ICAO module
- The face recognition (FR) module

**FIGURE 3.5**
Example of the recognition and identification process.

These modules cooperate to implement the recognition and identification procedure illustrated in Figure 3.6

The face recognition application can operate in two different modes: "enroll" mode and "transit" mode. The enroll mode is used to populate the database with the biometric profiles of the people who will be accepted by the system. The transit mode is used to recognize and identify the people who pass ("transit") in front of the camera. Figures 3.7 and 3.8 illustrate the operations performed in these two working modes.

The FF module is responsible for the acquisition of the video stream, for the analysis of the video stream itself, and for the generation of the output messages when a human face is found in the input stream. Every time a face is detected in the video stream, the extracted features are passed to the ICAO module.

The ICAO module is responsible for the selection of the best image in the set of images identified by the FF module. Once the module identifies a face with an ICAO score that allows identification, the face information is passed to the FR module.

The FR module is responsible for the extraction of the biometric profile and for the identification of a matching profile in the database. This module has two different behaviors depending on the software recognition mode (enroll or transit). In "enroll" mode, the detected biometric profile and personal information of a person are stored in the database. In "transit" mode,

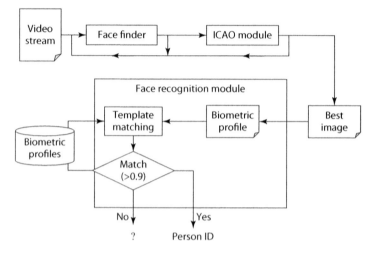

**FIGURE 3.6**
The people identification process.

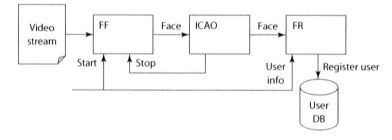
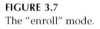

**FIGURE 3.7**
The "enroll" mode.

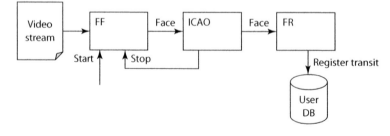

**FIGURE 3.8**
The "transit" mode.

the module searches the database for the biometric profile matching the one extracted from the video stream.

### The HW Platform

The face recognition software has been conceived specifically for ESs that play a crucial role in the new generation of distributed surveillance solutions. During the initial phase of study and design of the face recognition software, a general-purpose hardware platform has been designed to support the development process and simplify the test and evaluation activities. This initial hardware platform for face recognition is an all-in-one self-contained device capable of showing and demonstrating, step by step, all the phases of the identification process. The prototype provides a touch screen and a simple interface that allows the user to interact with the recognition system during the entire recognition process, from the acquisition of the video stream, the selection of the best photo of the face, to the final results of the identification. The prototype is composed of the following key parts: a main board, a USB camera, a set of white light illuminators, a touch screen, and a custom case (Figure 3.9).

The prototype has been developed to provide a standard software development platform that also includes the identification components of a surveillance application, mainly for demonstration purposes, focusing on the possibility to show the quality of the results of the recognition process in an effective way. The prototype is based on ANTARES i5 1 GHz from Eurotech (Figure 3.10).

The ANTARES main board is a 5.25″ single board computer based on the Mobile Intel QM57 Express Chipset. Designed to offer high performance with low power dissipation, the ANTARES is available with Intel Core i7 or i5 processor options and is ideal for use in compact spaces with restricted ventilation. This board is a good choice also as a development platform for

**FIGURE 3.9**
The face recognition development platform.

**FIGURE 3.10**
Eurotech ANTARES i5 1 GHz.

ESs because it provides features such as Watchdog, GPIO, four serial ports, and an integrated SD/MMC Flash port for accessing storage cards and Mini PCIe expansion for wireless modules such as Wi-Fi, Bluetooth, and cellular modems. The development platform includes a USB high-definition (HD) camera, integrated in the enclosure, and a set of illuminators. The light conditions represent an important factor in face recognition, significantly influencing the results and the quality of the identification process: vertical light, coming from ceiling lights, create shadows on the face, especially under the eyes and the nose. These shadows disturb the correct face analysis and for this reason should be eliminated during the recognition process. To solve this issue, the prototype is equipped with two illuminators, composed of six LEDs that illuminate the front of the face, eliminating shadows.

The evaluation activities performed by the face recognition development platform helped to identify the requirements and specifications of an intelligent camera that was specifically conceived for face recognition. The intelligent camera is based on a custom board designed around the DM816x DaVinci system on a chip manufactured by Texas Instruments and based on ARM architecture. The board has been conceived specifically for video processing and provides native hardware video encoders and a digital signal processing. The key elements of the DM816x DaVinci are three HD video and imaging coprocessors (HDVICP2). Each coprocessor can perform a single 1080p60 H.264 encode or decode, or multiple lower resolution frame rate encodes and decodes. It provides TI C674x VLIW floating-point DSP core, and HD video and imaging coprocessors and multichannel HD-to-HD or HD-to-SD transcoding along with multicoding is also possible. This solution has been conceived specifically for demanding HD video applications (Figure 3.11).

**FIGURE 3.11**
The intelligent camera custom embedded board based on TI SoC.

### *SHIELD Integration*

The face recognition application is one of the components of a "people identification" system that is orchestrated by the SHIELD middleware, providing native SPD support (see Chapter 8). In this context, the face recognition application runs on the intelligent embedded camera that, with the help of a smart card security service (see the first section of this chapter), manages access to a large private infrastructure by means of automatic people identification at the turnstiles. The people identification system is a distributed system, with a client–server architecture, based on SHIELD embedded nodes and a central server. The components of the system are connected through a local area network via cable or Wi-Fi.

The intelligent embedded camera is responsible for video stream acquisition, analysis, and face recognition. A smart card reader, integrated in the turnstile, collects the biometric profile from the smart card of the person. With this approach, a double check of the person's identity is performed comparing the profile extracted from the video stream with the one read from the smart card. An embedded PC is "paired" with different turnstiles and executes some modules of the face recognition application. The turnstile infrastructure is finally connected to a central server that remotely monitors the entrances, the ongoing identification processes, and the SPD levels. A framework for distributed execution (see the related section) manages face recognition and the SHIELD middleware orchestrate all the system components, monitoring the SPD levels and providing feedback and countermeasures when SPD changes.

Figure 3.12 illustrates the architecture of the people identification system.

The components of the people identification system have been analyzed to understand their SPD behavior following the SHIELD approach described in Chapters 4 and 5; details of the SPD-level calculation and a test-use case are illustrated in Chapter 8.

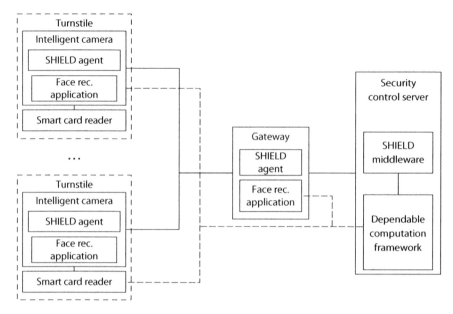

**FIGURE 3.12**
The architecture of the people identification system.

## The SHIELD Gateway

Modern systems of systems (SOS) are very frequently composed of a large variety of ESs. The SHIELD framework orchestrates this heterogeneity with a special purpose gateway (S-Gateway) that operates as a bridge among the SHIELD components of the SOS. The S-Gateway plays an important role also when interconnecting SHIELD components with legacy ESs, either with proprietary physical or logical communication capabilities. It performs a proxy-like behavior for legacy nodes; by translating the SHIELD middleware messages into legacy messages and vice versa, and adapting the proprietary interfaces, so that legacy component can interact with the SHIELD network. This approach allows non-SHIELD-compliant devices to inherit the communication capabilities necessary to make their services available to the SHIELD middleware. Besides this proxy role, the S-Gateway prototype performs a series of supporting tasks to ensure real-time and safe behaviors for internal data management of the exchanged data. Leveraging on the S-Gateway is possible to extend the SHIELD layered architecture to a wide range of devices that can't natively implement it, either because of limited computational resources or because of non-upgradeability due to cost and time constraints. Furthermore, the S-Gateway can be integrated during the design process of a new SHIELD-compliant ESs, reducing the design costs by providing an already functioning interface with the SHIELD architecture.

This SHIELD native node is endowed by a rich set of interfaces and protocols to simplify the integration of subsystems. Furthermore, the S-Gateway performs complex operations such as data encryption and smart message forwarding. The S-Gateway is a pivotal SHIELD framework element, providing several SHIELD services and functionalities, and has been designed to mainly address aspects such as flexibility and adaptability. It is based on the Zynq chip and is characterized by a sophisticated HW architecture that allows it to dynamically adapt resource usage, ensuring real-time signal processing in every working condition and reducing the power consumption. Notable fault detection hardware blocks have been integrated into the field-programmable gate array (FPGA) section to promptly detect, isolate, and recover a fault. Furthermore, the internal architecture is fully modular to support future evolutions. Aggregating hardware and software components make it possible to balance, opportunely, any computational requests, encryption, fault detection, reconfiguration, decryption, and so on (Figure 3.13).

The development of the gateway prototype has been performed on the ZedBoard development board from Digilent, which is based on the Zynq-7000 MP SoC. This processor-centric all-programmable platform from Xilinx combines the powerful and flexible Dual-core ARM A9 processor and the high-performance 7-series Xilinx FPGA programmable logic subsystem. The

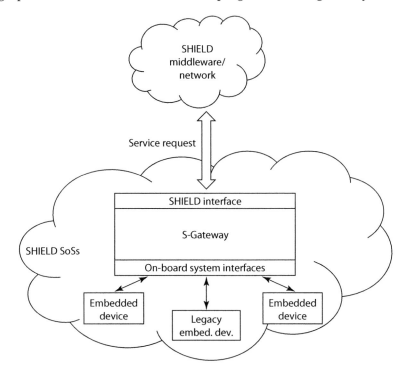

**FIGURE 3.13**
The S-Gateway in the SHIELD ecosystem.

choice of this architecture is motivated by its high customizability and by the availability of a series of prebuilt interfaces (SPI, I2C, CAN, Eth, and others), which can be used to interface the gateway to a large variety of devices. Furthermore, this architecture has other important advantages, such as the presence of a high-speed internal interface (AMBA AXI4 Interface) between the processor and the programmable logic, and a powerful development ecosystem that greatly simplifies the development process and reduces the time to market. Another important element to consider is the presence of a hard-coded dual processor, usable in asymmetric configuration, which allows obtaining certifications necessary for the integration of the prototype in demanding domains. This is an important enabling factor that allowed the adoption of the S-Gateway in the avionic domain (see Chapter 6).

Exploiting the PL Zynq section can considerably increase the computational power of the S-Gateway, for example, executing into the PL tasks atomic operations, encryption/decryption, CRC check/generation, and so on. This approach introduces a dual benefit: on one side, it allows the reduction of the CPU workload and, on the other side, it allows the optimization of the power consumption. In fact, the FPGA in some specific operations can achieve a rate of MSample/joule even 20 times higher than a CPU, making the system extremely efficient in terms of energy consumption.

### The S-Gateway Architecture

The entire system has been developed according to a modular approach, by implementing its various functionalities both in hardware and in software, in a way ensuring good balancing between the high customizability of the software modules and the high performance and reliability of the FPGA-based solutions. Figure 3.14 presents the S-Gateway internal architecture.

The architecture offers the ability to extend the connectivity capabilities of the S-Gateway by means of the programmable logic subsection. Thanks to this feature, it is possible to interface the processors to any custom I/O controller core through the high-speed AXI4 interconnection, enabling the interfacing of the gateway to any kind of ESs endowed with communication capabilities.

Cryptography plays a vital role in securing information exchange. Cryptographic algorithms impose tremendous processing power demands that can represent a bottleneck in high-speed and real-time networks. The implementation of a cryptographic algorithm must achieve high processing rates to fully utilize the available network bandwidth. For all these reasons, an FPGA-based HW accelerator is used to provide high-performance data encryption based on elliptic curve cryptography, also leveraging on a parallelized architecture.

To evaluate the correctness of data exchanged with other nodes, CRC code check is used. To perform its high-speed calculation, and to avoid the waste of processor computing resources, a specific and highly parallelized FPGA

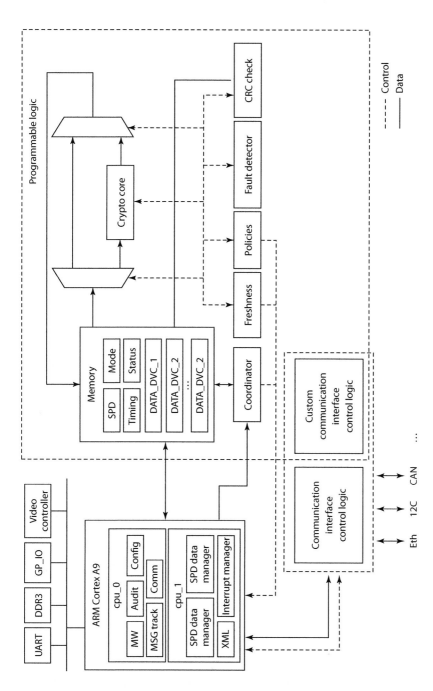

**FIGURE 3.14**
The internal architecture of the S-Gateway.

solution has been employed to calculate the CRC codes and to compare it with the one provided with the data under analysis.

To grant the correctness and freshness of all stored and processed data, the embedded freshness and policies save blocks are used for operation time definition. These blocks are directly controlled by the system coordinator and produce interrupt events that are quickly served by the processors. Eventual issues faced during the resolution of these interruptions, together with errors on CRC code check, are identified by the fault detector core and communicated to the system's coordinator, which acts as configured to limit the bad consequences due to this bad functioning, such as making the timing constraints less stringent or inducing the retransmission of corrupt data.

The dual-core processor is used in Asymmetric Multiprocessor configuration, so that the master processor (CPU_0) runs a Linux-based distribution compiled for ARM processors, while the second processor is used to execute a bare metal program as a parallel task. This configuration takes advantage of both solutions; in fact, the multi-thread native characteristics of the Linux operating system can be adopted to execute various tasks simultaneously, and the already existing drivers to interface it with some peripherals, such as Ethernet or USB. Furthermore, this solution can take advantage of the bare metal software to get directly interfaced with the custom PL cores thanks to the memory addressable nature of the Zynq architecture and of the AXI4 interconnect, without the necessity of specific drivers and consequently with improved performances.

To allow information sharing between the two processors and the various PL cores, a distributed Block RAM is instantiated on the PL, on which a strong partitioning of its address space allows fast and direct access to the desired information stored in it, both regarding the gateway's configuration parameters and the data exchanged with the devices interfaced to it.

### SHIELD Integration

The S-Gateway belongs to the SHIELD power node category and offers in one self-contained board, high performance computing, data storage, and networking capabilities. In the SHIELD ecosystem, the S-Gateway can execute at least the following operations:

- Acquire SPD information from the legacy nodes and calculate the related SPD metrics.
- Register the services provided by the legacy nodes to the SHIELD middleware, using an appropriate protocol.
- Discover the commands provided by the SHIELD middleware and directed to its cluster.
- Translate the commands coming from the SHIELD middleware into a language comprehensible to the legacy nodes.

In addition to these basic functionalities, the S-Gateway also offers more advanced features

- It is endowed with a communication policies advisor that regulates the interrogations (interval, priority, broadcast, unicast, etc.), type of messages to be dispatched and/or to be accepted, and the operational modes. These policies are not upgradeable, avoiding, by design, the risk of malicious attacks.
- In accordance to the defined policies, the S-Gateway evaluates the SPD level provided by the middleware and changes its operational mode in accordance to it:
  - Increasing/decreasing the rate of the messages exchanged.
  - Enabling/disabling cryptographic modules for the encryption of internal data.
  - Increasing/decreasing the writing of the log file concerning the audit function.
- To optimize S-Gateway performances, a coordination module provides a service balancing, according to the SPD level. A module that provides fault detection and SPD evaluation encompasses the SHIELD algorithms. It consists of several software modules and some FPGA-based Internet Protocol (IP) processing blocks.

The integration between the S-Gateway and the SHIELD middleware, to perform service registration and discovery, is based on the service location protocol (SLP), which is defined in RFC 2608 and RFC 3224 and relies on the exchange of user datagram protocol (UDP) packets for connectionless communication. Furthermore, in the context of the avionic use case described in Chapter 6, the gateway prototype is interfaced with the prototype that implements the software-defined radio (SDR) for smart SPD transmission, which implements reliable and efficient communications even in critical (physical) channel conditions. The physical interface towards this prototype is the standard Ethernet 10/100 Mbps, while on the applicative layer the simple object access protocol (SOAP) is adopted, which relies on the exchange of UDP packets with an XML payload formatted with the configuration Head-Body, in which the Header contains information about routing, security, and parameters for the orchestration, while the Body segment transports the informational content.

### Application of SPD Metrics

The S-Gateway is able to change its operative behavior according to the SPD level obtained from the SHIELD middleware. According to this value, the coordinator core can adapt the rate of message exchanges with other devices. In particular, the message throughput can be increased in case of SPD-level

**TABLE 3.5**

Gateway Configuration Parameters

| SPD Ranges | [0, SPD_LOW] | [SPD_LOW, SPD_MED] | [SPD_MED, SPD_HIGH] | [SPD_HIGH, 100] |
|---|---|---|---|---|
| SPD refresh timing | SPD_timing_1 | SPD_timing_2 | SPD_timing_3 | SPD_timing_4 |
| Cluster's data refresh timing | REF_timing_1 | REF_timing_2 | REF_timing_3 | REF_timing_4 |
| Data logging activity timing | LOG_timing_1 | LOG_timing_2 | LOG_timing_3 | LOG_timing_4 |
| | | Increasing speed | | |

reduction, to allow a higher check rate. The system coordinator, depending on the SPD level, activates cryptography in order to hide the information relating to the occurred fault within the cluster and to the solutions adopted by the SHIELD middleware. To achieve this functionality, the gateway features four possible states, and the transitions among them are performed by the system's coordinator comparing the SPD level with three configurable thresholds.

Furthermore, it is possible to enable or disable the CRC code generation/ check by configuration, according to the needs of the devices that compose the cluster. When this feature is enabled, the activation of the CRC is performed by the system's coordinator according to the SPD level. In this way, the gateway ensures that, when the correctness of the exchanged data becomes crucial to restore the correct functioning of the cluster, the received data are indeed correct thanks to the decisions taken by the SHIELD middleware (Table 3.5).

The system configuration parameters are

- Normal state
  - Cryptography: off
  - CRC check: off
  - SPD refresh timing: SPD_timing_1
  - Cluster's data refresh timing: REF_timing_1
  - Data logging activity timing: LOG_timing_1
- Moderate state (CRC off by configuration)
  - Cryptography: off
  - CRC check: off (by configuration)
  - SPD refresh timing: SPD_timing_2

- Cluster's data refresh timing: REF_timing_2
- Data logging activity timing: LOG_timing_2
- Alarming state (CRC off by configuration)
  - Cryptography: on
  - CRC check: off (by configuration)
  - SPD refresh timing: SPD_timing_3
  - Cluster's data refresh timing: REF_timing_3
  - Data logging activity timing: LOG_timing_3
- Critical state (CRC off by configuration)
  - Cryptography: on
  - CRC check: off (by configuration)
  - SPD refresh timing: SPD_timing_3
  - Cluster's data refresh timing: REF_timing_3
  - Data logging activity timing: LOG_timing_3
- Moderate state (CRC on by configuration)
  - Cryptography: off
  - CRC check: on
  - SPD refresh timing: SPD_timing_2
  - Cluster's data refresh timing: REF_timing_2
  - Data logging activity timing: LOG_timing_2
- Alarming state (CRC on by configuration)
  - Cryptography: on
  - CRC check: on
  - SPD refresh timing: SPD_timing_3
  - Cluster's data refresh timing: REF_timing_3
  - Data logging activity timing: LOG_timing_3
- Critical state (CRC on by configuration)
  - Cryptography: on
  - CRC check: on
  - SPD refresh timing: SPD_timing_3
  - Cluster's data refresh timing: REF_timing_3
  - Data logging activity timing: LOG_timing_3

## Open Mission Network Integrated Architecture

Open Mission Network Integrated Architecture (OMNIA) is a dependable avionic open system HW/SW architecture that has been built on

SHIELD concepts and on an innovative avionic standard, following the rules of the integrated modular avionic (IMA) to implement the IMA2G architecture.

Avionic systems must ensure all three aspects on which SHIELD is focused: security, privacy, and dependability. The trend for modern civilian aircraft is to support the aircraft application with an IMA platform. OMNIA is a SHIELD native power node conceived to provide a scalable solution, to define a minimal set of modules, to increase the number of supported functions, and to demonstrate fault tolerance and reconfiguration and, consequently, the high dependability of the SHIELD solution in avionics.

The approach adopted to design a "new" avionics architecture is based on the definition, development, and validation of the "avionics module" that provides native SPD features and functionalities and becomes the component of a dependable avionic SoS.

The OMNIA system is based on an IMA platform that collects, analyses, and distributes the data acquired from the aircraft sensors. It provides a fault recovery engine that is able to dynamically replace faulty components with spare ones, if available.

Interoperability is provided through the adoption of "system level" middleware services. Among them, the data-distribution service (DDS) has been implemented to support data and events exchange among applications. The HW/SW units developed in accordance to the IMA are connected by a high-speed serial network based on the ARINC 664p7 standard.

The OMNIA system has been adopted in the dependable avionics scenario (see Chapter 6).

### The OMNIA Architecture

The OMNIA architecture is composed of a network of a reduced aircraft and mission management computers (AMMC) and related remote interface units (RIU) connected through a high-speed deterministic serial line. Each "unit" is connected to the A/C (AirCraft) sensors. The system/network, by means of an additional middleware layer built around the real-time publisher–subscriber (RTPS) architecture, virtualizes the connection of a related sensor with all "computer units" that will be present in the system, enabling fault tolerance functionalities. More specifically

- The main unit AMMC is used as the IMA central unit.
- The other "units" (referenced as RIU) are mainly used as sensor interfaces.
- Both the IMA central unit and the RIU-IMA are "computer units."
- All the "computer units" are connected via Ethernet (rate constraint or best-effort methodology). Each "computer unit" can implement the interface with the avionic sensor.

The RTPS functionality has been integrated, as a library, in the equipment software (EQSW) environment. This library, according to the IMA concept, is segregated in a partition to increase the reliability and flexibility of the system. The RTPS represents the communication pattern used by the IMA and the RIU to exchange data. In particular, the publish–subscribe architecture is designed to simplify one-to-many data-distribution requirements. In this model, an application "publishes" data and "subscribes" to data. Publishers and subscribers are decoupled from each other too. Real-time applications require more functionalities than those provided by the traditional publish–subscribe semantics. The RTPS protocol adds publication and subscription timing parameters and properties so that the application developer can control different types of data flows and therefore the application's performance and reliability goals can be achieved. The RTPS has been implemented on top of the UDP/IP and tested on Ethernet; the determinism of the Ethernet is provided by the Avionics Full-Duplex Switched Ethernet (AFDX) network. A logical view of the OMNIA architecture is presented in Figure 3.15.

Data exchanged between all the IMA "computer units" include

- Discrete signals
- Bus1553 data words
- ARINC 429 data words
- General I/O signals

Data are managed by the RIU and sent/received by the IMA central unit.

### Hardware Architecture

The hardware components adopted for the prototype consists of two N-AMMC reduced equipment units connected to each other via AFDX, where:

- The IMA central unit is constituted mainly by a rack with

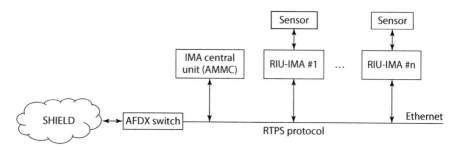

**FIGURE 3.15**
OMNIA architecture.

- A processor module based on the PPC microprocessor (APM460), Ethernet 10/100/1000 BASE-T link, RS232 serial I/F (monitor–debug channel)
- An AFDX End System Interface (mezzanine PCI card on CPU)
- The RIU is constituted by a rack with
  - A processor module based on the PPC microprocessor (APM460), Ethernet 10/100/1000 BASE-T link, RS232 serial I/F (monitor–debug channel)
  - An AFDX mezzanine card
  - Two I/O boards (RS422, Arinc429, discrete, and analog) (DASIO)
  - A 1553 mezzanine card
- The high-speed deterministic Ethernet switch is the AFDX-Switch 3U VPX Rugged and enables critical network-centric applications in harsh environments. The avionic full-duplex switched Ethernet interface provides a reliable deterministic connection to other subsystems.

Being a modular architecture, every component has been developed according to the actual avionic standards in terms of processing cycles, data buses, signal types, memory use, and so on.

### Software Architecture

The OMNIA software main blocks span across different layers (Figure 3.16):

- SHIELD application layer
- Middleware layer
- Module support layer

The SHIELD application layer provides all the OMNIA functionalities of communication with the SHIELD prototypes, collecting the flight and plant

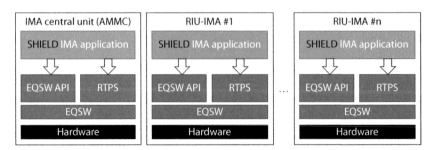

**FIGURE 3.16**
OMNIA software architecture.

information, computing the aircraft and OMNIA health status, managing and assuring the equipment redundancy.

The SHIELD Application SW on the Main IMA is responsible for the communications of all the computing units connected to the network. It interfaces with RTPS API to subscribe data and adopts an UDP/IP library interface to accomplish communication via Ethernet.

The SHIELD Application SW running on RIUs is responsible for the acquisition of data from sensors, for detecting and managing errors occurred in the acquisition, and for the interaction with other support processing units (e.g., a flight simulator). It interfaces with RTPS API to publish data and uses a UDP/IP library interface to accomplish communication via Ethernet. Finally, it uses an EQSW library interface to acquire data from the sensors.

The middleware layer is constituted by the EQSW (API and virtual device drivers) and the RTPS software.

The RTPS software provides the management of the RTPS protocol and interface to the application software layer (RTPS API). These components run on the main processor. A space partitioning policy has been implemented and each application SW operates in one partition, while the RTPS SW operates in a different partition to maintain the safety of the whole system in case of any failure. The publish–subscribe architecture is designed to simplify one-to-many data-distribution requirements.

## Network Technologies

### Distributed Self-x Models

Development of the distributed self-x models technologies have focused on design of distributed self-management and self-coordination schemes for unmanaged and hybrid managed/unmanaged networks. With the aim to reduce the vulnerability to attacks depleting communication resources and node energy. Recognizing denial of service (DoS) attacks as well as dependable distributed computation framework modules will be described.

## Recognizing and Modeling of DoS Attacks

DoS attacks are attacks that attempt to diminish a system's or network's capacity and resources and thus render it unable to perform its function. Resources that can be targeted by attacks typically include network capacity, processing resources, energy, and memory. Usually, a combination of these resources is affected as they are related, for instance, increased network traffic will result in increased processing requirements and therefore energy.

## Categories

There are several categories of DoS attacks. The most important ones, related to the area of secure ESs, are Jamming, Tampering, Collision, Exhaustion, Unfairness, Neglect and Greed, Homing, Misdirection, Egress filtering, Flooding, and Desynchronization. Some of these concern physical implementations, known as side-channel attacks, or affect very specific protocol mechanisms. DoS attacks can often appear in the form of distributed attacks (known as DDoS) where the attacker is not a single node but rather a smaller or larger number of nodes aimed at disarming defense mechanisms capable of blocking attacks originating from single points.

## Detection

In general, DoS attacks can be detected with methods based on the concept of faults and abnormal behavior. Faults in a system are events that are clearly related to a malfunction of the system or a particular subsystem. Detected faults can be normal, part of routine malfunctions, or abnormal. These detected faults need to be distinguished against normal faults, especially when a large number of nodes exist. Normal behavior is defined as the set of expected operation status of the system and the subsystems. Behavior needs to be monitored in order to be judged as normal or abnormal. Abnormal behavior can include increased or decreased usage of resources (network, hardware, etc.), unexpected patterns of events, and more. Abnormal behavior is distinct to a fault in that the system continues to be operational, regardless of its possible inability to fulfill the requested services due to exceeding of its capacity. DoS attack detection is a field partially overlapping with intrusion detection in methods and techniques. It is, however, different in that a DoS attack can take place without any sort of intrusion or compromise.

## Countermeasures

DoS attacks are in general complicated attacks regarding appropriate defense methods because there is usually no control over the source of the attack. Assuming successful detection of an attack, various countermeasures of different aggressiveness can be taken, ranging from passive response, such as blocking network traffic and waiting, in order to save system resources, to active responses, such as throttling traffic or tracing the source of the attack in order to maintain operation of the network.

## Network Architecture Issues

DoS attacks can take place toward all layers including the physical, MAC, network, transfer, and application. As a result, successful mechanisms against DoS attacks need to be implemented to various layers where protection is needed. Moreover, cross-layer mechanisms and inter-layer communication are necessary in order to detect and defend. When systems under attack are distributed rather than centralized, a distributed mechanism is needed in order to detect attacks.

## Module Description

The adopted solution was of a similar approach to the one commonly used on the Internet, based on IP marking schemes. However, in order for a complete approach to be applied to ESs, an additional level of DoS attack detection has been added. As a result, it was necessary for the right parameters to be identified that would be used to monitor a system. These parameters needed to be simple and not require increased computational resources, so that the approach could be used even in scenarios where nodes have limited resources. The goal was to use these identified parameters and—through a DoS detection algorithm—to be able to detect when an attack is taking place and if possible to issue reconfiguration commands.

### Architecture

The scheme can be seen as an algorithmic operation that is fed with inputs from various components and provides a set of results relating to the identification of a DoS attack that is underway. As is evident from Figure 3.17, the algorithmic operations are fed data input that comes from different components, that are described in "Components." The data input is analyzed and used to provide the result of the algorithm that leads to attack identification and the issue of reconfiguration commands.

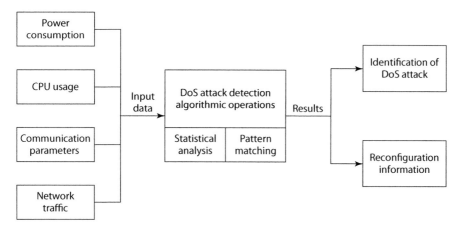

**FIGURE 3.17**
DoS atack detection algorithm.

## Components

A proper organization of components leads to efficient and simple operation. The components identified in the system are organized in the following two types:

*Input Modules*

- *Power unit monitoring module*: This is a process that runs in the power unit module and provides information related to power consumption.
- *CPU monitoring module*: This is a process that provides CPU usage information and can typically be considered a service of the operating system.
- *Communication monitoring module*: This is a process that provides information from the physical transmission, such as signal strength.
- *Network traffic module*: This module has access to the data packets exchanged and can sample either complete packets or specific parts in the content or header.

*Algorithmic Modules*

There are two algorithmic software processes in the system, the statistical analysis algorithm and the pattern matching algorithm.

- *Statistical analysis algorithm*: This software process communicates with all four input modules it reads and processes that information.
- *Pattern matching algorithm*: This software process communicates with the network traffic module. Its task is to sample network packets and compare them with the signature database.

## Interface between Modules

Communication between the input modules and the algorithmic modules is made with the exchange of messages. As Figure 3.18 depicts, a typical message is composed of the following three fields: Node ID, timestamp, and payload. The Node ID identifies the node that is reporting the information and the timestamp identifies the exact timing of the sampled information. These two fields are common for all messages originating from all modules. The third field, called the payload, contains the actual information being reported and is different for each of the modules.

**FIGURE 3.18**
DoS attacks—fields composing the message.

The payload is defined as follows for the four different types of messages:

- *Power consumption payload*: Contains the power consumption wattage at the timestamp sampling time in mW.
- *CPU usage payload*: Contains the current CPU usage percentage and number of running processes at the timestamp sampling time.
- *Communication parameters payload*: Contains the incoming and outgoing bitrate, the number of established connections as well as the number of other nodes to which there are established connections at the timestamp sampling time.
- *Network traffic payload*: Contains complete data packets (typically TCP/IP) or headers of the packets, depending on mode of operation, which can be fed to the pattern matching algorithm.

## Algorithmic Operation

The statistical analysis algorithm receives messages from input modules and correlates inputs in order to detect whether an anomaly is taking place. The algorithm can produce an output by examining any number of input message types, from one to three. The algorithm is based on simple but multi-parameter statistical analysis.

Typical communication patterns produce a highly predictable hardware and software response. For example, a typical communication scenario of a certain number of interacting nodes and the amount of exchanged traffic corresponds to specific amounts of CPU usage and power consumption. If there is a deviation from this correlation, a DoS attack may be under way.

A score is calculated by performing all the correlations, after which the scores of all the correlations are summed up. If the total score is above the predetermined threshold, then a DoS attack warning is issued.

The pattern matching algorithm receives messages from the network traffic module, which contains either a header of data packages or complete data packages. The algorithm analyzes the data and tries to detect patterns in the provided traffic according to a predetermined attack pattern database. The database is composed of known risks such as malicious software or known network protocol vulnerabilities. Whether the operation is in a header or full payload analysis depends on the mode of operation, which in turn depends on the available node resources.

### DDoS Attack Mitigation

For embedded devices that have the capabilities to offer services and resources to their network of peers or other client applications and users, being a potential target of DDoS attacks is a credible risk. Those devices belong to the same environment as conventional service providers and are targets to the same kinds of attacks.

In this section, we present the design of a packet marking scheme that, when used together with an intelligent filtering and traceback mechanism, can effectively mitigate ongoing DDoS attacks. The architecture relies on a highly configurable filtering mechanism that ensures the service of legitimate users while under an ongoing DDoS attack for different attack scenarios. The configurable features of this filtering mechanism allow an organization to provide both effective and flexible protection against DDoS attacks depending on the organization's network traffic characteristics.

The IP specification allows any system to provide an arbitrary source IP address for each outgoing packet regardless of the true source of the packet. This is exploited by the systems that participate in a DDoS attack by adding either random or foreign IP addresses into the packets that they produce, effectively making traceback of such attacks a very hard task. This technique is called IP spoofing and is one of the main reasons why DDoS attacks are still considered a credible threat against networked systems that need to provide services. Any attempts to filter incoming traffic that is part of a DDoS attack contains the risk of filtering out traffic from legitimate users or systems thus strengthening the effect of the attack itself.

A requirement to defending against DDoS attacks that use IP spoofing is to be able to identify the true source of the incoming (attack) packets. Even an indication that will allow us to classify incoming packets based on their true sources can be used to implement an effective filtering mechanism. In this chapter, we will present two schemes for classifying packets. Both schemes are based on the packet marking technique. Given those schemes, real-time filtering of DDoS attack packets and traceback of those packets is possible. The architecture of such a filtering and traceback mechanism will also be described.

### General Marking Scheme Description

As described in "DDoS Attack Mitigation," in order to be able to effectively stop an ongoing DDoS attack, we need to be able to classify all incoming packets as part of an attack or not. In order to do this effectively, the classification and filtering needs to be real time. There are a number of known mechanisms that identify if a set of packets are part of a DDoS attack. Given those mechanisms, and since the source IP address is not a reliable indicator of the true source of an incoming packet, extra information inside the packet (packet markings) is used to filter out all incoming packets that arrive from sources that are part of the attack.

The basic notion of the packet marking scheme is that the routers along the packets path inject some additional information that denotes the true path of the packet. This information can be used instead of the source IP address to determine the true source of each packet or at least classify the incoming packets based on the networks that they originate from. The filtering procedure, based on that scheme, can be performed in real time since, unlike existing packet marking schemes, it relies only on the information that exists inside each individual packet without needing to perform any kind of correlation between incoming packets.

For a more detailed description of the marking scheme, we first need to define the following key terms:

- *Router signature*: Part of a router IP address that is injected into the packet's header.
- *Distance field*: A counter, measuring the hops along the packet's path, that is injected into the packet's header.

The first router along the packet's path injects its router signature into the packet's header (by overwriting parts of the IP header as we will describe in the following sections) and initializes the distance field to zero. Each subsequent router along the path increases the distance field by one and <xor>'s the existing router signature with its own router signature.

The final result is that when the packets arrive at their destination the router signature/distance field tuple acts as a distinctive marking that denotes the true source, as well as the true path, of the incoming packet. The mechanisms and modules that are tasked with filtering and performing traceback can use this marking instead of the source IP address for their operations. The filtering operation can be performed in real time since all the required information about the true origin of the packet is contained inside each packet.

One inherent limitation of the methodology is that it is only capable of providing information up to the nearest router to the source. All the hosts that are behind the edge router are treated identically hence if one of those hosts is part of a DDoS attack, all the hosts behind that edge router will be filtered out as attacking hosts.

While filtering can be performed in real time, traceback requires more elaborate correlation and also a priori information. First, the victim needs to have an updated map of all upstream routers from his or her location and outwards. Such a map can be quite easily constructed using standard traceroute tools. Given such a map, the victim gathers the packet markings of the packets that are part of a DDoS attack and essentially follows the reverse marking procedure for each packet marking. The result is a collection of networks that participated in the DDoS attack.

The traceback procedure is a forensics mechanism that can be used after an attack since its computationally intensive and not suitable for real-time results.

Lastly, this marking procedure does not introduce any kind of additional bandwidth requirements. No additional control traffic is generated during the marking procedure or exchanged during the filtering and traceback procedure. While the computational overhead for the routers is not zero, it is limited to an <xor> operation. There are no additional memory require-ments for the routers also.

### Individual Schemes and Encoding Details

Depending on what IP header fields we choose to overload for the packet markings, we identify two distinct marking schemes. The trade-off between those two marking schemes is the false positive probability versus the prob-ability of receiving a corrupted IP packet.

In the first marking scheme, we choose to overload the "Identification" field and one bit of the unused "Flags" field. This gives us 17 bits from which we use 12 bits for the "Router Signature" and 5 bits for the "Distance" fields.

The "false positive" probability is the probability to classify a host or net-work that is not part of a DDoS attack as an attacking host or network. The "false negative" probability is the probability to identify a host or network that is part of a DDoS attack as legitimate host or network.

As we will describe in the next section, one of the filtering mechanism's modules is a DoS detection module. This module is in charge of detecting if the packets that are originating from a specific network are part of a DDoS attack. With spoofed source IP addresses, such a module could not reliably correlate the packets from distinct networks and thus could not classify them as part of the attack or not. This module now uses the packet markings instead of the source IP address. By design, all the packets that originate from a specific network have the same packet marking. In the not so com-mon exception that the route changes, the origin host will appear as two or more hosts (part of the hosts traffic will take a different route). In either case, the marking scheme itself does not produce any false negatives and the false negative probability depends only on the DoS detection module.

The false positive probability, on the other hand, is not zero and it depends on the number of attacking hosts, the total number of edge routers, and the length of the marking field. The false positive probability can be calculated as a special case of the occupancy problem and the calculations show that given the length of the marking field for the first scheme and for 100–150 k edge routers and 0–50 k attackers, the false positive probability is between 0% and 9% depending on the number of attackers. An interesting effect is that when the number of attackers rises significantly (more than 30 k in our case), so do the number of collisions between sources that actually belong to the attack, which in turn lowers the false positive probability back to 0%.

In the second marking scheme, we choose to overload the "Identification" field, the "Flags" field, and the "Fragment offset" field. This gives us 32 bits, we use 27 of them for the "Router signature" and 5 bits for the "Distance" fields.

In order to not corrupt the IP packets due to the header field overload, the router closest to the destination (the one that collects the final packet markings) sets the flags and offset fields back to zero. All the IP packets remain uncorrupted except the actual fragmented packets.

Due to the high number of bits reserved for the "Router signature" field, the false positive probability is practically zero as long as the routers use the 27 most significant bits of their IP address as a "Router signature," the false negative probability is the same as the previous marking scheme.

### Filtering and Traceback Mechanism

The aforementioned packet marking schemes can be used in order to design a defense mechanism that is able to provide real-time filtering of ongoing DDoS attacks, as well as traceback of those attacks to their true origin. A detailed description of the architectural and functional design of this mechanism can be found in Chapter 10.

### Application of SPD Metrics

The SPD level of the core of the prototype, the filtering, and traceback mechanism are defined by the existence of the traceback procedure and the use of the network repository for network prioritization during both the filtering and traceback procedure.

The traceback procedure is not necessary for the successful mitigation of a DDoS attack and mainly serves the purpose of identifying and taking legal or other measures against the systems that participated in the attack. Therefore, its utilization depends on the amount of the SPD level required for the affected system.

In a similar fashion, the affected system can be configured to treat all potential networks with the same priority, in terms of filtering, or identify the interesting ones and treat them with a lower probability of filtering as described in the preceding sections.

Therefore, we can identify three SPD levels as depicted in Table 3.6.

### Smart SPD-Driven Transmission

Smart SPD-driven transmission refers to a set of services deployed at the network level designed for SDR-capable power nodes, whose goal is ensuring smart and secure data transmission in critical channel conditions. To achieve

**TABLE 3.6**

SPD Levels

| SPD Level | Functionality |
| --- | --- |
| 1 (low) | Only filtering, no known networks |
| 2 (medium) | Only filtering, network prioritization |
| 4 (high) | Filtering and traceback, network prioritization |

this, it utilizes the reconfigurability properties of the SDR technology as well as the learning and self-adaptive capabilities characteristic for cognitive radio (CR) technology. Developed functionalities are demonstrated by means of an SPD-driven smart transmission layer. These functionalities include remote control of the transmission-related parameters, automatic self-reconfigurability, interference mitigation techniques, energy detection spectrum sensing, and spectrum intelligence based on the feature detection spectrum sensing.

### Starting Point and Objectives

The emerging communication paradigms of SDR and CR technologies may be used to provide enhanced SPD features compared to the "legacy" communication technologies. For this reason, it was decided to go on with exploring the potentials of the aforementioned paradigms, and put together an SDR/CR test bed architecture useful for developing, deploying, and testing various SPD relevant algorithms. Even though SDR and CR technologies inherently bring their own set of security issues, stemming mainly from dependence on reliable firmware/software, on-the-fly reconfigurability, and potential problems associated with adaptive/cognitive algorithms based on machine learning techniques, the advantages of proffering safer and more reliable communication outweigh the mentioned problems. SPD-driven smart transmission technologies have been adopted in the Airborne use case (see Chapter 6) and in a specific technology demonstrator (see Chapter 10).

### Prototype Description

The proposed smart transmission layer SDR/CR test bed prototype (Dabcevic et al., 2014b) consists of three secure wideband multi-role–single-channel handheld radios (SWAVE HHs), each interconnected with the OMBRA v2 multiprocessor embedded platform (power node).

SWAVE HH (from now on referred to as HH) is a fully operational SDR radio terminal capable of hosting a multitude of wideband and narrowband waveforms.

The maximum transmission power of HH is 5 W, with the harmonics suppression at the transmit side over −50 dBc. A super heterodyne receiver has specified image rejection better than −58 dBc. The receiver is fully digital; in VHF, 12-bit 250 MHz analog-to-digital converters (ADCs) perform the conversion directly at RF, while in UHF, ADC is performed at intermediate frequency (IF). No selective filtering is applied before ADC. Broadband digitized signal is then issued to the FPGA, where it undergoes digital down conversion, matched filtering, and demodulation.

HH has an integrated commercial global positioning system (GPS) receiver, and provides the interface for the external GPS receiver. GPS data is available in National Marine Electronics Association (NMEA) format and may be output to the Ethernet port.

The radio is powered by Li-ion rechargeable batteries; however, it may also be externally powered through a 12.6V direct current (dc) source. Relatively small physical dimensions ($80 \times 220 \times 50$ mm), long battery life (8 h at the maximum transmission power for a standard 8:1:1 duty cycle), and acceptable weight (960 g with battery) allow for portability and untethered mobile operation of the device.

Hypertach expansion at the bottom of HH provides several interfaces, namely: 10/100 Ethernet; USB 2.0; RS-485 serial; dc power interface (max 12.7 V), and PTT.

The radio provides operability in both very high frequency (VHF) (30–88 MHz) and ultrahigh frequency (UHF) (225–512 MHz) bands. The software architecture of the radio is compliant with the software communications architecture (SCA) 2.2.2 standard. Following that, HH provides support for both legacy and new waveform types. Currently, two functional waveforms are installed on the radio: SelfNET soldier broadband waveform (SBW) and VHF/UHF line of sight (VULOS), as well as the waveform providing support for the IP communication in accordance with MIL-STD-188-220C specification.

The considered power node—OMBRA v2 platform—is composed of a small form factor system-on-module (SOM) with high computational power and the corresponding carrier board. It is based on an ARM Cortex A8 processor running at 1 GHz, encompassed with powerful programmable Xilinx Spartan 6 FPGA and Texas Instruments TMS320C64+DSP. It can be embodied with up to 1 GB LPDDR RAM, has support for microSD card up to 32 GB, and provides interfaces for different RF front ends. Support for IEEE 802.11 b/g/n and ANT protocol standards are proffered. Furthermore, several other external interfaces are provided, namely, 16-bit VGA interface; mic-in, line-in, and line-out audio interfaces; USB 2.0; Ethernet; and RS-232 serial. The node is dc-powered, and has Windows CE and Linux distribution running on it. System architecture of the power node is shown in Figure 3.19.

Connection to HH is achieved through Ethernet, as well as serial port. Ethernet is used for the remote control of the HH, using SNMP. For the serial connection, due to different serial interfaces—RS-232 and RS-485—an RS-232-to-RS-485 converter is needed. Serial connection is used for transferring the spectrum snapshots from HH to power node.

The described test bed allows for testing the cognitive functionalities of the system. Talking about cognitive functionalities, the following ones were developed:

- *Self-awareness*: The network learns the current topology, the number, and the identity of the participants and their position, and reacts to variations of their interconnection.

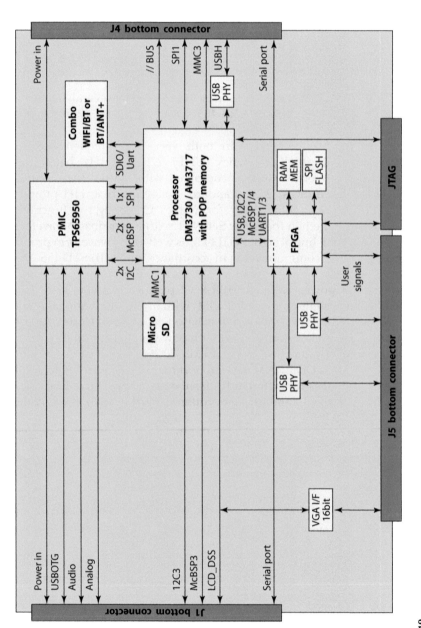

**FIGURE 3.19**

STL—OMBRA v2 power node—system architecture.

- *Spectrum awareness*: The node collects information of the spectrum occupancy, either by spectrum sensing or indirectly by using geo-location/database method, and shares the data with other nodes in order to create a map of existing sources.
- *Spectrum intelligence*: The node analyzes the spectrum information focusing on specific sources, trying to identify them by extracting significant signal parameters.
- *Jamming detection and counteraction*: The nodes recognize the presence of hostile signals and inform the rest of the network's nodes; the nodes cooperate in order to come up with the optimal strategy for avoiding disruption of the network.
- *Self-protection*: The nodes cooperate to detect and prevent the association of rogue devices and to reject nodes that exhibit illegal or suspicious behavior.

Because of the high output power of the radios, programmable attenuators had to be included in the coaxial path, and were programmed to their maximum attenuation value of 30 dB. An Agilent 778D 100 MHz to 2 GHz dual directional coupler with 20 dB nominal coupling was placed between the attenuators, allowing sampling and monitoring of the signal of interest. An Agilent E4438C vector signal generator was connected to the incident port of the coupler, for the purpose of injecting a noise/interference signal in to the network. An Agilent E4440A spectrum analyzer was connected to the coupler's reflected port, facilitating the possibility of monitoring the radio-frequency (RF) activity.

Figure 3.20 shows the list of functionalities with which the smart SPD-driven transmission prototype is embodied.

The main functionalities that could be adopted for the SPD level control are the following (see Chapter 10):

- *Energy detector spectrum sensing*: Obtaining information on the current spectrum occupancy is paramount for the CRs to be able to opportunistically access the spectrum, but it may also aid them in recognizing anomalous or malicious activity by comparing the current state to those stored in their databases. There are three established methods for CRs to acquire knowledge of the spectrum occupancy: spectrum sensing, geolocation/database, and beacon transmission. HH has the capability of performing energy detection spectrum sensing: the HH's FPGA provides a functionality that allows transmission of 8192 samples every 3 s from the ADC over the RS-485 port. Currently, there is not a synchronization pattern; however, the idle interval between the two transmissions may be used, for example, to perform analysis of the received data.

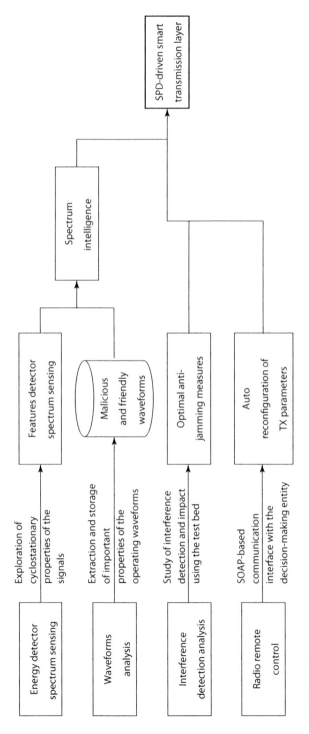

**FIGURE 3.20**
Smart SPD-driven transmission functionalities.

- *Interference influence analysis*: Various DoS attacks, especially jamming attacks, have for a long time been posing, and continue to pose, significant security threats to radio networks. RF jamming attacks refer to the illicit transmissions of RF signals with the intention of disrupting normal communication on the targeted channels. SDR and CRs allow further improvements to the jamming capabilities of malicious users; however, they also offer the possibility of developing advanced protection methods and countermeasures. One of the main focuses of the SPD-driven smart transmission layer is providing precise, safe, and reliable communication in jamming-polluted environments. For that, a detailed study of various jamming attack strategies and the development of appropriate security solutions needs to be done.

- *Waveform analysis*: To alleviate the problems analyzed in the previous point, theoretical analysis of electronic warfare was done, and basic anti-jamming principles based on spread spectrum techniques were studied for future improvement of the prototype, which for the time being is focused on the analysis of alternative techniques namely: retroactive frequency hopping (channel surfing), altering transmission power, and modulation (waveform) altering. Channel surfing refers to the process of changing the operating frequency of the radio in the presence of a strong interfering signal. Increasing transmission power and/or gain may further improve the probability of detection and reconstruction of a signal. Modulation altering may play an important role in alleviating RF jamming, provided that the radios are equipped with feature detection possibilities, allowing them to recognize the features of jamming waveforms.

---

## Middleware Technologies and Software Platforms

### Dependable Distributed Computation Framework

Dependability in terms of computation means the enforcement of two major properties on the application along with the running system:

- The application is defined in a way that some guarantees over its execution can be introduced.
- The application is able to tolerate faults in the system that runs it.

In particular, the guarantees of an application usually fall into the following categories:

1. *Data integrity*: Received data are still correct after being transmitted over a medium.
2. *Code correctness*: Data conversions and code semantics operate as expected or at least the designer is able to observe the behavior of the application both at the debug and production phases.
3. *Code authentication/authorization*: Code can be identified as coming from a trusted source, with an unambiguous versioning scheme that prevents an incorrect deployment.

The Dependable Distributed Computation Framework (DDCF) is designed to provide such guarantees by means of a distributed middleware that handles deployment, scheduling, computation, and communication between a set of nodes. The DDCF has been adopted in the biometric security scenario (see Chapter 9) to manage the dependability of the people identification system components.

### Overview

To achieve the mentioned result, the application is written according to a dataflow metamodel that enhances dependability in both the development and production phases (Figure 3.21).

Given an existing application, we can partially rewrite its behavior in order to express it as a graph, where vertices are code portions and edges are data dependencies. Then, if the hardware platform offers some redundancy, DDCF is able to guarantee that node faults do not prevent the application from running to completion (or to provide a service in a continuous way). This feature is obtained by changing the execution node for a code fragment, under the assumption that some redundancy is available. If this is not the case, the framework still helps during the deployment phase, since

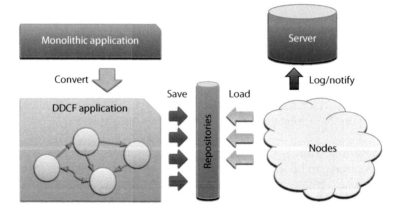

**FIGURE 3.21**
Overview of the DDCF flow.

code is automatically distributed from repositories to the nodes that participate in a computation task. In particular, the signature of the repository can be verified for security purposes. In addition, the application metamodel allows the definition of specific graph edges where data can be encrypted: this approach reduces the use of secure/private communication to the actual minimum required by the application. It also makes explicit where the most critical communication operations take place.

*Framework Architecture*

DDCF is designed to enhance both the development and production phases of an application. The development phase is concerned with dependable design, while the production phase with dependable execution. Here, we mainly focus our attention on the production phase, but many concepts (like the application metamodel) also directly impact the development phase.

We envision a generic distributed application as being run on a set of worker nodes (here called "wnodes") that are embedded and capable of accepting workload. Not all wnodes necessarily perform computation at a given moment, some of them being redundant resources that may be called upon when a computing wnode fails.

In addition to wnodes, other sets of nodes that are relevant within DDCF are the following:

- *Synchronization nodes ("snodes")*: Nodes that have the task of synchronizing communication between worker nodes; their primary role is that of triggering an event to all its listeners as soon as new data are available.
- *Persistence nodes ("pnodes")*: Nodes that persist data; they are nothing more than a (distributed) database that wnodes can use to load/store data.
- *Client nodes ("cnodes")*: Nodes that can inject or extract data for the application by leveraging the pnodes; they represent the external interface to the application, so they also allow composition of distinct applications.
- *Directory nodes ("dnodes")*: Nodes that handle the participation of other nodes in a specific application, also performing authentication and possibly authorization; they represent the authority within the framework, by grouping nodes and handling security.
- *Repository nodes ("rnodes")*: Nodes that provide (parts of) the application to be deployed on the wnodes; they hold versioned repositories, that is, subversion, git, or mercurial.

It must be noted here that an embedded node can participate on multiple sets and consequently can take on multiple roles within this software architecture. Clearly, multiple independent devices for each role are envisioned to improve the robustness of both the execution of the code and the consumption of the computed data. Consequently, the roles as previously defined must be considered logical components of the runtime, rather than disjointed physical embedded platforms.

Multiple wnodes and snodes are particularly important to the runtime, in order to run to completion even in the presence of faults. Secondarily, multiple pnodes are preferable especially when data must be collected for later analysis. Multiple dnodes are not as important, since their concerns are mainly restricted to the setup phase of the execution of an application. However, their impact increases if the set of wnodes is dynamic, as is the case with mobile wireless networks. Multiple repository nodes make sense mostly due to the potentially significant data transfer required: if we offer mirrors, we can better balance the communication load, and account for server downtimes. Finally, multiple cnodes instead are present only when multiple users interact with an application, hence they are not actually a replicated resource.

In Figure 3.22, the architecture of the runtime is shown, where a letter stands for a role (e.g., C for cnode), and an arrow starts from the initiator of a transaction and ends with the target of such transaction. We identify two important groups of roles that are involved in the processing and setup parts of the runtime:

1. The processing part includes wnodes, snodes, pnodes, and cnodes and it is rather straightforward: each wnode independently elaborates a code section (i.e., a vertex of the application graph), while getting/posting data (i.e., edges of the application graph) to the set of pnodes. The snodes are used to synchronize the wnodes by means of events. The cnodes may be used to inject or extract data at runtime by a user.

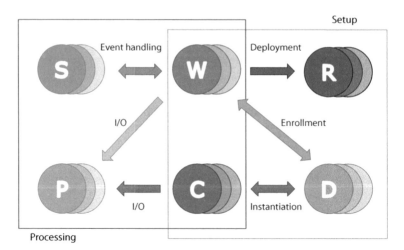

**FIGURE 3.22**
Architecture of the DDCF runtime.

2. The setup part is more complex: First, we need to instruct a distributed platform to actually perform the processing of a given application. This operation is clearly controlled by the user from a cnode, who "instantiates" an application by submitting a job to the directory nodes. Then, depending on the policy used, wnodes enroll in a job or are asked to enroll by the dnodes. Such enrollment also informs wnodes/cnodes on which snodes/pnodes to use; in fact, snodes and pnodes, being server-like resources, are semi-statically chosen by the dnodes as services to the DDCF runtime. The rnodes instead are completely static: the corresponding resources are encoded in the application metamodel itself. Wnodes are initially given a pointer to a repository that holds the highest-level information on the application. Then, a wnode can pull the information it actually needs for execution, thus performing deployment.

### Application Metamodel

The architecture discussed in the preceding section clearly needs a dedicated metamodel for the application.

While a monolithic application is certainly reasonable, on the other hand a composable application can exploit the distributed nature of the embedded network in terms of multiple resources that operate concurrently. Consequently, the application metamodel relies on a model-based approach: the application is made of "blocks" that can contain either pure code or other blocks connected with each other. A concrete model, based on such a metamodel, is concerned with specifying the how "implementations" interface between each other using given "data types." We call implementations and data types "artifacts" of the application model, since they are concrete objects that are created by the designer to fully describe the interaction between parts of the application.

Due to the splitting of the application code in separate sections, we are able to distribute the code to multiple nodes. A monolithic application is still possible, but in that case only a "migration" of the whole application from one node to another is feasible.

To accommodate this metamodel, state and time models have been designed that are able to unambiguously determine data dependencies in the more general case of directed cyclic graphs. This feature is necessary to perform a dataflow-like execution of the application, but it is also useful during the design phase to observe the application evolution. In particular, it is feasible to "rewind" the execution to any previous state and change blocks on demand, to observe the impact of improved/debugged portions of code on partial execution. While these aids do not concern the execution in the "production" phase, they aim to provide more guarantees related to the correctness of execution by enhancing the "development" phase.

The application graph is, in general, a cyclic graph. This generality is necessary to account for complex iterative subroutines that are common, for example, within algorithms based on optimization. For this reason, the runtime is designed to address cyclicity by enriching data with temporal properties. These temporal properties allow unambiguous determining of the order of data even when code portions are run asynchronously, as it is the case in a distributed computation scenario.

Another important aspect of the metamodel is that all data types and code portions are hosted at external repositories whose authenticity can be verified by the wnode that requests the sub graph of the application to execute. In this way, a secure deployment is guaranteed as long as the repositories are not breached. The use of common repositories also enables a safe collaborative development, since data type specification is defined globally, possibly with explicit semantics defined using a natural language.

When an application must be designed as distributed from the ground up, DDCF provides several advantages. The first one is that both top-down and bottom-up approaches can be used, as shown in Figure 3.23. Here, we identify three layers: the bottom layer is concerned with the actual code and does not have any notion of communication or data conversion. The top layer is concerned with representations in terms of building blocks called "structures" connected using specified composite "abstract data types." In the intermediate layer, we concern ourselves with specifying how to make the other two layers communicate, which requires specifying the "concrete data types" used and wrapping the code sections into "behaviors."

The metamodel easily allows testing of a sub graph of the application and to refine it as soon as a more detailed implementation is needed. Other advantages stem from the methodology required to design applications in DDCF. Since all artifacts (implementations and data types) must be versioned and

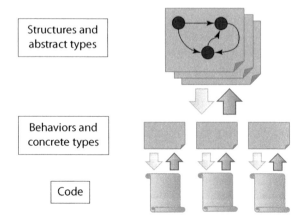

**FIGURE 3.23**
Top-down and bottom-up approaches in the application metamodel definition.

outputs are organized into pnodes, a designer can always trace back his or her simulation results to the specific application that generated them. This is in contrast with the too-common development scenario where simulation/testing is performed on unversioned (or monolithically versioned) software that produces unorganized, often overlapping, data. It can be seen that this holistic approach to development improves the quality of the software and reduces issues in the production phase, thus adding to the overall dependability of execution.

*Artifact Descriptors*

An application in DDCF is written using the dataflow paradigm, expressed as a graph comprising edges and vertices: edges represent data, and vertices represent implementations. Implementations may be just code sections to be executed, or they may represent a sub graph. Therefore, we can see that the description of an application may become rather complex. To tackle such complexity and also, to simplify abstraction/refinement development approaches, the descriptor of an application is hierarchical. This means that we do not store the entire application graph within one descriptor file, but we rather let descriptors reference other descriptors.

We generically consider an artifact any entity with information useful to define a DDCF application. Artifacts are declared using descriptor files, which provide the basic information that DDCF needs to deploy, build, and run applications. The two most important artifacts are data types and vertex implementations, as explained before. For example, the descriptor file of a vertex implementation may specify the script that must be run to build the library.

In practice, artifact descriptors are text files that use the YAML language. YAML has been chosen because it offers a natural syntax with almost no overhead, since it relies on indentation: this is important to keep descriptor files readable, compared for example to XML or even JSON. Readability is not to be taken lightly, if we consider that designers must write their own descriptors to create a DDCF-compliant distributed application; while IDE tools may mitigate the complexity of maintaining descriptors, manual editing always offers the maximum insight and control over the design process. Software libraries exist for several programming languages to automatically parse and produce YAML documents (Figure 3.24).

In Program listing 1 (Figure 3.24), an example of a fragment of a descriptor is shown, which only lists the repositories element for an artifact. This example also illustrates our concept of repository, that is, a remote location where we can find a versioned repository containing an artifact. The fragment shows the coordinates of the repository, that includes a version control system (DDCF supports git, subversion, and mercurial repositories), an entry URL, a path within the repository, and the version of the content. In addition, optional mirror entries are provided for redundancy purposes.

```
repositories
- name: myartifactname
  control: git
  entry: 'git@mysite.com:myself/myproject.git'
  path: myproject/path/to/myartifactname
  version: 477125dbe78fe0a51be2486d8902b49ec2161450
  mirrors:
  - 'git@thirdsite.com:myname/myproject.git'
  - 'git@fourthsite.com:othername/someproject.git'
```

**FIGURE 3.24**
Program listing 1. DDCF—repositories fragment of a YAML artifact descriptor.

By design, a repository identifies one specific artifact; the descriptor for the remote artifact can be found as "artifact.yaml" at the root of the repository path. In other terms, it is possible to have one versioned repository with all the required artifacts organized into directories, and declare repository entries for each remote artifact of interest.

*Data Types*

As explained before, the two main artifacts in DDCF are data types and vertex implementations, since their composition allows the definition of a data-flow application graph.

Data types are the most delicate aspect of a distributed computation framework, since they represent the "glue" that connects the different parts of the application. The final goal is to define complex types through which implementations can communicate in a safe and practical way (Figure 3.25).

An artifact definition of a data type is provided in Program listing 2 (Figure 3.25). Here, we see that there exist some built-in types, namely, boolean, byte, real, and text; these types are sufficient to describe simple types, but to do more we must be able to compose them. Consequently, the *town*

```
artifact: datatype
name: town
fields:
- name: name
  type: text
- name: loc
  fields:
  - {name: province, type: text}
  - {name: region, type: text}
  - {name: state, type: text}
  - {name: coords, type: real, array: '2:3'}
- name: props
  type: text
  array: ':,2'
```

**FIGURE 3.25**
Program listing 2. DDCF—YAML artifact descriptor of a type.

```
repositories:
- name: 2Dcoordinates
  control: svn
  entry: 'http://mysite.com/someproject'
  path: types/xy
  version: 19

types:
- name: 3Dcoordinates
  fields:
  - {name: xy, type: 2Dcoordinates}
  - {name: z, type: real}
```

**FIGURE 3.26**
Program listing 3. DDCF—types and repositories fragment of a YAML artifact descriptor.

data type is a composite type featuring both scalars and multidimensional arrays. More than that, it is hierarchically composite: the *loc* field is itself a composite type and it is declared inline.

When we reference a repository pointing to this artifact as in Program listing 1 (Figure 3.24), that is, when we have an external declaration of a type, we provide a custom name that may override the name of the artifact. In fact, the *name* field in the artifact descriptor is optional: this choice is due to the fact that externally declared types are reusable and may have a different name in different domains; also, it is preferable to have all type names explicit within a descriptor. If we consider the descriptor fragment of Program listing 3 (Figure 3.26), where the *xy* field references a custom type, the type name corresponds to a repository name within the repositories element of the same descriptor file. The *types* element is where internal declarations of types are provided: when the validity of a type is restricted to the artifact declaring it, we can save ourselves the burden of setting up and referencing a repository (Figures 3.26 and 3.27).

This simple example suggests that we can distribute artifacts in different repositories, thus offering the maximum flexibility, especially when the application is built from contributions coming from independent developers.

### *Runtime*

Even if different interacting roles exist, the actual runtime is distributed on worker nodes. Figure 3.28 shows a structural view of a worker, where it is apparent that the DDCF software module is composed of multiple behaviors that are managed by the runtime.

```
artifact: datatype
fields:
- {name: x, type: real}
- {name: y, type: real}
```

**FIGURE 3.27**
Program listing 4. DDCF—YAML artifact descriptor of a type for an x-y couple of reals.

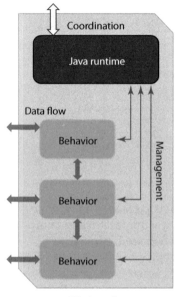

**FIGURE 3.28**
Structural view of a DDCF worker.

*Data Conversion*

Remember, DDCF is a Java framework that supports applications written in different languages. This generality implies the following: interprocess communication is required to transfer data between the middleware and any process that runs the code for a vertex. Also, since vertex implementations run on platforms with different data type precision, even the "same" data types may have different representations among the wnodes. While a common representation is available at middleware level due to the platform-independent type system of Java, it is still necessary to account for the specific platform.

To account for these problems, DDCF employs Apache Thrift, a remote procedure call framework for heterogeneous services. Given a generic representation of a service and the data types of its arguments/return, Apache Thrift produces source files for client and server implementations. These source files can be used to perform cross-language remote communication.

Our current use of Apache Thrift within DDCF is for interprocess communication: all the vertexes that are executed on a wnode have their own process, and each process exchanges data with the DDCF middleware using Thrift. It must be noted that we can extend this approach for inter-wnode data communication with no effort. This would allow the exchange of application data between wnodes directly, bypassing pnodes. While such a choice would decrease the communication latency, we would break our guarantees of data persistence against node failures.

*Data Communication*

As already explained, the DDCF architecture consists of six different kinds of nodes, also called roles. Those roles that directly involve the communication of application data are worker nodes (wnodes), synchronization nodes (snodes), and persistence nodes (pnodes).

The array of snodes is currently implemented using Apache ZooKeeper. ZooKeeper is a replicated synchronization service with eventual consistency. It is robust, since the persisted data are distributed between multiple nodes (this set of nodes is called an "ensemble") and one client connects to any of them (i.e., a specific "server"), migrating if one node fails. As long as a strict majority of nodes are working, the ensemble of ZooKeeper nodes is alive.

In detail, a master node is dynamically chosen by consensus within the ensemble; if the master node fails, the role of master migrates to another node. The master is the authority for writes: in this way writes can be guaranteed to be persisted in order, that is, writes are linear. Each time a client writes to the ensemble, a majority of nodes persist the information: these nodes include the server for the client and, obviously, the master. This means that each write makes the server up to date with the master. It also means, however, that you cannot have concurrent writes. As for reads, they are concurrent since they are handled by the specific server, hence the eventual consistency: the "view" of a client is outdated, since the master updates the corresponding server with a bounded but undefined delay.

The guarantee of linear writes is the reason that ZooKeeper does not perform well for write-dominant workloads. In particular, it should not be used for the interchange of large data, such as media. The advantage that ZooKeeper brings to DDCF is the ability to robustly listen for events and to issue them to all listeners. In addition, it can be used as an arbiter for consensus algorithms, like those related to scheduling and load balancing. For example, a very important event to listen to is the completion of a transaction to the pnodes, signaling the availability of data returned by a vertex implementation.

While a ZooKeeper ensemble could also be used as a set of pnodes, linear writes would be detrimental for heavy data streams. For the maximum generality, we consequently choose pnodes to be nodes of a distributed (replicated) database. The current DBMS of choice is MySQL Cluster, which offers both SQL and "NoSQL" APIs; the second one is particularly useful for the persistence of custom data structures such as all DDCF data types. It is important to say that the framework may accommodate a different persistence technology with a rather simple change of driver library. On the contrary, ZooKeeper is to be considered a consolidated choice that will be replaced only if a more efficient and versatile solution is identified. Both snodes and pnodes are not by themselves "DDCF-aware," in the sense that they operate as generic services with which wnodes can interact; again, all the logic required to interface with snodes/pnodes resides within wnodes.

In terms of reliability, we purposefully avoided any single-point-of-failure situation that may arise in a distributed environment. Wnodes "push" the data they produce to the pnodes, and "pull" the data they need to consume from the pnodes. This decoupling shields from node failures and allows the saving of data as soon as it is produced. Compare this approach to the opposite one, where data are simply transferred from producers to consumers: as soon as one link of the chain breaks, the application needs to be restarted from scratch since we have not saved any "snapshot" of the application state.

It could be argued that this methodology introduces inefficiencies, and it is certainly such a case. However, we believe that a paradigm shift is necessary to be able to address the intrinsic dependability problems in a distributed environment. This is especially true for mobile ad hoc networks, where connectivity and autonomy are relevant obstacles to collaboration.

## Service Discovery Framework

Service discovery is the functionality allowing any SHIELD middleware adapter to discover the available SPD functionalities and services within heterogeneous environments, networks, and technologies reachable by the SHIELD embedded service device (ESD). The SHIELD secure service discovery component is compatible with a number of discovery protocols (such as SLP [7], SSDP [8], NDP [9], DNS [10], SDP [11], and UDDI [12]), in order to maximize the possibilities to discover the interconnected ESDs and their available SPD services, functionalities, resources, and information. This can lead to a higher number of services/functionalities that can be composed to improve the SPD level of the whole system.

The base of any discovery process is a secure and dependable service registration/publication that, in turn, requires a maximally precise service description and proper service filtering capabilities (to improve the precision of the discovery results). The service registration consists in a proper advertising (doing which depends on the selected discovery protocol), in a secure and trusted manner, the available SPD services: each service is represented by its formal description, known in literature with the term service description.

Within the discovery phase, the registered services are discovered (i.e., returned as a result of the discovery itself) whenever their description matches with the query associated to the discovery process: this process is also known as service filtering.

Once discovered, the available SPD services must be properly prepared to be executed, in particular assuring that their dependencies (if present) and preconditions are fulfilled: assuring this is the core functionality of the service composition process and components.

Figure 3.29 depicts the discovery engine of the SHIELD framework, with the details of its components (called "bundles," considering they were implemented as OSGI bundles):

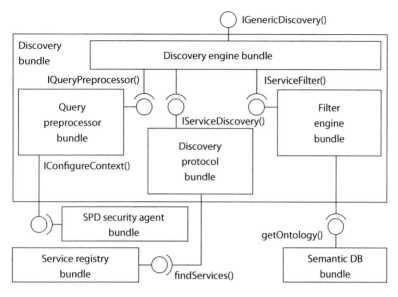

**FIGURE 3.29**
Discovery engine structure.

- *Discovery engine bundle*: It is in charge of handing the queries coming from the *IGenericDiscovery()* interface, managing the whole discovery process, activating the different functionalities of the discovery service; in particular calling the *IQueryPreprocessor()* interface to add semantic and contextual information to the query. This last, in turn, is passed to the various discovery protocol bundles, by means of the *IServiceDiscovery()* interface, to harvest over the interconnected systems all the available SPD components by using the various discovery standards. After these phases the list of discovered services is passed to the filter engine bundle, by the *IServicesFilter()* interface, in order to further filter the discovered services by discarding the ones not matching with the enriched query.

- *Query preprocessor bundle*: It is in charge of enriching the query passed by the discovery engine with semantic and context-related information; it could be configured by the SPD security agent to take care of the context information by using the *IConfigureContext()* interface.

- *Discovery protocol bundles*: It is a bundle able to securely discover all the available SPD components registered within the service registry bundle, by using the *findServices()* interface.

- *Filter engine bundle*: This bundle performs the semantic filtering of the discovered services, by matching the semantically enriched query with the (semantic) descriptions of the discovered SPD components. In order to do so, it can retrieve the semantic information, associated to the SPD components, from the semantic DB, by using the *getOntology()* interface.

### Service Discovery in SHIELD

One of the major achievements of SHIELD was to make the discovery process "secure and dependable." This goal has been achieved by means of guaranteeing both secure registration (whenever possible) and secure discovery.

Furthermore, a certain effort has been invested into the definition of a "lightweight version" of the discovery protocol, in order to make it more suitable to be run in the ESs domain.

The service discovery architecture was based on the SLP which has been enhanced, without introducing any modification to the SLP standard, by securing the SLP signaling thanks to the usage of messages' signatures.

### Service Location Protocol

The service location protocol (SLP) is a protocol created by the service location protocol working group (SVRLOC) of Internet engineering task force (IETF).

It is simple, light, scalable, decentralized, and designed for TCP/IP networks; furthermore, it is independent from HW, SW, and the language programs.

*Services* have some *properties*, such us the *URL* and a list of attribute-values couples. The service URL is a string with a specific form, and it specifies the general category of the service that it describes. For example, a service URL may be

service:content-adaptation:sip://can1@daidalos.org

it says that the current service is a content-adaptation service that may be reached, with session initiation protocol (SIP), at the SIP address can1@daidalos.org. The "service:" is a fixed string that says only that following is the URL of a service.

Each service, beyond a service URL, has a *list of attribute-values couples*. Each attribute is a property of the service, and it is indicated by a name. This property, typically, has one or more values: so, a couple can be "Supported_Resolutions = 640 40480,800 80600,1024 10768." This attribute indicates that that service (that can be a monitor or a projector) has an attribute, named "Supported_Resolutions" (i.e., auto-explicative) that can assume three possible values: 640 va480, 800 80600 and 1024 00768.

So, suppose describing a projector as a service on a LAN, its service URL can be something like "service: video-output device: projector://p1"

(supposing that a protocol exists that uses the projector://p1 as a way to indicate an address of a projector) with the "Supported_Resolution" attribute mentioned in the last paragraph.

*The SLP architecture* is based on three kinds of entities: user agents (UAs), service agents (SAs) and directory agents (DAs).

- The UA is the client that interrogates the DA to find a specific service of a SA.

- The SA is the service supplier: it advertises the location and characteristics of services, and has to register its services on the DA.

- The DA is the "core" of the architecture, because it registers all the services that are offered by a network: it collects service addresses and information received from SAs in their database and responds to service requests from UAs.

On the DA, for each service beyond service URL and its attributes (with their values), there is stored a service scope (i.e., "visibility area"), a service type (i.e., indicated service URL) and a service life time (i.e., a time that, after expired, causes the service to be unregistered). This additional piece of information is used to simplify the discovery process and to make the entire architecture failure-safe: indeed, if a service crashes it can't renew its registration on the DA, the life time expires and it is removed; so no one DA can find that service further.

Figure 3.30 shows the interactions between the three agents. When a new service connects to a network, the SA contacts the DA to advertise its existence (service registration). When a user needs a certain service, the UA queries the available services in the network from the DA (service request). After

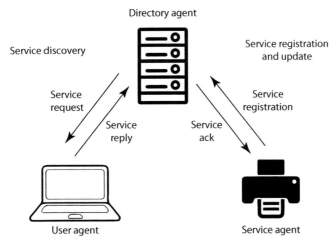

**FIGURE 3.30**
SLP agents and their transactions for service discovery and registration.

receiving the address and characteristics of the desired service, the user may finally utilize the service. Before a client (UA or SA) is able to contact the DA, it must discover the existence of the DA.

There are three different methods for DA discovery: static, active, and passive.

With *static discovery*, SLP agents obtain the address of the DA through dynamic host configuration protocol (DHCP). DHCP servers distribute the addresses of DAs to hosts that request them. In *active discovery*, UAs and SAs send service requests to the SLP multicast group address (239.255.255.253). A DA listening on this address will eventually receive a service request and respond directly (via unicast) to the requesting agent. In case of *passive discovery*, DAs periodically send out multicast advertisements for their services. UAs and SAs learn the DA address from the received advertisements and are now able to contact the DA themselves via unicast.

It is important to note that the *DA is not mandatory*. In fact, it is used especially in large networks with many services, since it allows to categorize services into different groups (scopes).

In smaller networks (e.g., home or car networks) it is more effective to deploy SLP without a DA. SLP has therefore two operational modes, depending on whether a DA is present or not.

If a DA exists on the network (as shown in Figure 3.30), it will collect all service information advertised by SAs. UAs will send their service requests to the DA and receive the desired service information.

If there is no DA (see Figure 3.31), UAs repeatedly send out their service request to the SLP multicast address. All SAs listen for these multicast requests and, if they advertise the requested service, they will send unicast responses to the UA. Furthermore, SAs multicast an announcement of their existence periodically, so that UAs can learn about the existence of new services.

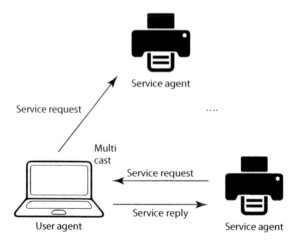

**FIGURE 3.31**
Service discovery without DA.

Services are advertised using a "service URL" and a "service template." The Service URL contains the IP address of the service, the port number, and path. Service templates specify the attributes that characterize the service and their default values. A service template associated with a network printer could look like the following:

service:printer://lj4050.tum.de:1020/queue1 scopes = tum, bmw, administrator printer-name = lj4050 printer-model = HP LJ4050 N printer-location = Room 0409 color-supported = false pages-per-minute = 9 sides-supported = one-sided, two-sided.

To summarize, the service discovery is done as follows. The UA sends a message to find a service. This message can be either in unicast mode to a specific DA, or in multicast: in this last case, each DA or SA that is on the network receives that message. If there are only SAs, the ones that satisfy the UA requests, respond to it; otherwise, Each SA that responds to the UA requests, as well as each DA where a service that satisfies the UA requests is registered, responds to the UA. So, SLP can work either in a centralized or in a decentralized way! The UA message contains a visibility area, a service type, and a list of capabilities that the service has to comply. All messages exchanged by entities are extensively defined in SLP specifics: these messages implement a robust asynchronous communication protocol.

Figure 3.32 explains the SLP architecture in the most general case.

The branch where SLP excels (on other protocols above and below reported) is in its capability to perform "complex" queries on the attributes' values: it

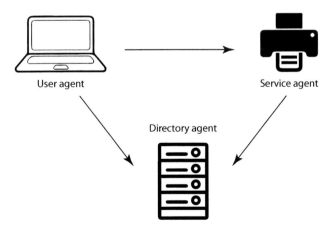

**FIGURE 3.32**
Overall SLP architecture.

can use Boolean operators (AND, OR, NOT), comparators (<, >, ≤, ≥), and functions of string matching.

Some SLP advantages are

- It has the best query language and capability.
- It is highly scalable: with service scopes and the possibility to work either in a centralized or decentralized way, it can perfectly work either in small LANs and on the Internet with the same simplicity.
- It supports the possibility of using more DAs, hierarchically interconnected: with SLP it is possible to create a service discovery architecture like the DNS one.
- It uses a moderate number of multicast messages and has very low traffic overhead.

### Enhancing the SLP Standard to Support Secure Signaling

As previously indicated, the SLP was enhanced in the SHIELD framework by the introduction of messages' signatures; the idea underlying the solution for the signature of the SLP messages (both of the discovery and registration ones, at client side) and for the validation of the signatures and relative certificates (at server side) is depicted in Figure 3.33.

As it is possible to see, it uses the perfectly functioning and well tested sign/verify method adopted for any kind of digital document, which was adapted to the SLP signaling using a public-key infrastructure, based on asymmetric keys.

The SLP messages (both registration and discovery) contain the so-called *<Authentication Block>* which, according to the SLP standard, "are returned with certain SLP messages to verify that the contents have not been modified, and have been transmitted by an authorized agent."

As usual, the verification process requires a couple of keys: the first, the private one, is used to compute the digital signature of the message to be

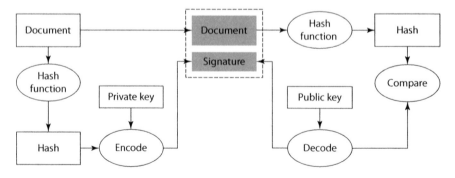

**FIGURE 3.33**
SLP flowchart.

sent (at sender side); the other is public and it is necessary to verify the message (at receiver side). The SLP standard requires that DAs and SAs exchange their keys during the configuration phase, which means that the trusted relation among the entities involved in the discovery process can only occur during the configuration, that is, there could not be a secure message exchange among new, unknown, nodes (DAs and SAs). Furthermore, this solution presents two additional problems:

- The first is that it is not possible to guarantee, in any manner, that only authorized UAs can have access (and discover) potentially critical registered services (the SLP standard requires DAs to respond to any SLP request, not considering any signature, authentication, or whatever kind of security issue).
- The second concerns the fact that UAs do not have any possibility to ascertain that a certain DA is effectively trusted.

Considering it is not possible to handle the first problem without modifying the SLP standard, it was decided to not include such a functionality within the SHIELD discovery architecture: the authentication of the discovery messages is let to the authentication mechanisms of other protocols (e.g., TCP/IP) or systems (e.g., firewalls). The second problem, instead, can be resolved simply guaranteeing that the public key is securely exchanged with the DA, as better explained in the following paragraphs.

*Authentication Block*

The structure of the SLP *<Authentication Block>* is shown in Figure 3.34.

The *<Authentication Block>* is used both to guarantee the integrity of the message and that the source is a trusted one. Excluding the obvious usage of the length fields, the *<Authentication Block>* is composed of four fields:

- *Block Structure Description (BSD)*: Used to identify the algorithm used to compute the digital signature. The standard indicates as default

```
0                   1                   2                   3
0 1 2 3 4 5 6 7 8 9 0 1 2 3 4 5 6 7 8 9 0 1 2 3 4 5 6 7 8 9 0 1
+-+-+-+-+-+-+-+-+-+-+-+-+-+-+-+-+-+-+-+-+-+-+-+-+-+-+-+-+-+-+-+-+
|  Block structure descriptor   |  Authentication block length  |
+-+-+-+-+-+-+-+-+-+-+-+-+-+-+-+-+-+-+-+-+-+-+-+-+-+-+-+-+-+-+-+-+
|                           Timestamp                           |
+-+-+-+-+-+-+-+-+-+-+-+-+-+-+-+-+-+-+-+-+-+-+-+-+-+-+-+-+-+-+-+-+
|     SLP SPI string length     |         SLP SPI string        |
+-+-+-+-+-+-+-+-+-+-+-+-+-+-+-+-+-+-+-+-+-+-+-+-+-+-+-+-+-+-+-+-+
|                       Digital signature                       |
+-+-+-+-+-+-+-+-+-+-+-+-+-+-+-+-+-+-+-+-+-+-+-+-+-+-+-+-+-+-+-+-+
```

**FIGURE 3.34**
Structure of the SLP authentication block.

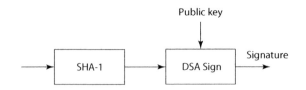

**FIGURE 3.35**
Signature generation process (according with DSA + SHA-1 standards).

choice the usage of the DSA algorithm, paired with the SHA-1 hash function, but different solutions could be used as well.

- *Timestamp*: Indicating the expiration time of the signature.
- *Security Parameter Index (SPI)*: Containing parameter useful to identify and to verify the digital signature (e.g., length of the key, the BSD parameters, or the public key).
- *Digital signature*: The actual digital signature.

*Signature Generation Process*

The generation process of the signature consists of two different phases: first of all, the hash function is used to calculate the digest of the message, then the cryptographic algorithm is used over the digest to generate the signature, exactly as for the digital signature of any usual document. The overall process is depicted in Figure 3.35.

*Signature Verification*

When receiving a message including the Authentication Block, the receiver must extract the BSD and the SPIs in order to identify the algorithm used to sign the message and related parameters as well as the public key. Once identified the parameters of the signature, the receiver decodes the messages and the signature. After that, using the indicated hash function (the same as the sender), calculates the digest of the messages and compares it with the digest obtained by decodification of the signature. If these match, the identity of the sender and the integrity of the message are both verified and guaranteed; otherwise, the message should be discarded. The overall process is depicted in Figure 3.36.

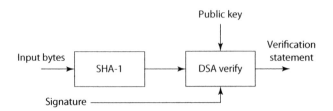

**FIGURE 3.36**
Signature verification (according with DSA + SHA-1 standards).

### Enhancing the SLP Standard for Lightweight Implementation: NanoSLP (nSLP)

The SLP is a well-established standard and there exist a plethora of implementations of such a protocol in a number of languages, most of which are open source, including Java and C implementations. Excluding Java implementation, which is obviously not well suited for low power nodes, C implementations are usually simple enough for quite low-powered devices; unfortunately, some very small devices, for example wireless sensor nodes, are so thin and extreme that they are not able to support even the smallest standard SLP C implementations.

In order to also include SLP implementation into the set of (extremely simple) devices that can be controlled and discovered by the SHIELD framework, a sort of lightweight version of the SLP was developed and implemented, the so-called "nanoSLP," specifically designed for low-power/low-capacity embedded devices, with a particular focus on the wireless sensor networks. The rationale for this choice is that, for their particular nature, such devices, in the near/mid-term, are not expected to have their processing power increased so much as to support an entire JVM or complex C programs and/or services.

The nSLP is a protocol that could be considered a sort of a particularization, a branch, of the standard SLP: the main aim of the introduced changes was to make the overall protocol less "wordy," minimizing both the size and the quantity of the exchanged messages. This permits to reduce the resources necessary to have a SLP-like communication among the nodes within a WNS; details of the nSLP are provided in the following paragraph.

#### Secure Service Discovery with nSLP

The particularization of the SLP developed within the SHIELD framework, called nSLP, inherits the architecture, the functions, and the protocol structure of the standard SLP.

The SLP agents (DAs, SAs, and UAs) are properly mapped in the nSLP ones:

- *nService Agent (nSA)*, a network device able to offer one or more services. The nSA is able to:
  - Discover one or more nano directory agents.
  - Publish its own services.
  - Receive, process and, in certain cases, respond, to nSLP messages.
- *nUser Agent (nUA)*, is the network device initializing the service discovery.
- *nDirectory Agent (sDA)*, a network device able to store the services provided by the various nSAs.

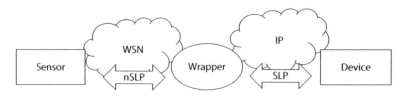

**FIGURE 3.37**
System architecture.

In order to integrate nSLP devices within a traditional network with standard SLP ones, a specific "wrapper" is needed, that is able to connect the nSLP subnetwork (which usually is not IP-based) with the SLP IP-based one (Figure 3.37).

Within this solution, the wrapper provides network address translation (NAT) functionalities, allowing the communication among devices located in different networks; for example, if the WSN network is sort of a MAC-based one, the nSLP signaling uses MAC addresses instead of the IP addresses used by usual SLP messages.

*Structure of the nSLP Messages*

nSLP maintains the same messages structure as the SLP standard, which is based on different message types to optimize the information exchange and uses unicast and broadcast transmission mechanisms. Due to the nature of the WSNs, having limited bandwidth and low computational capabilities, it was mandatory to introduce, in nSLP, some modification to the original SLP messages, they are as follows:

- nSLP uses a subset of the SLP signaling.
- The size of the nSLP messages is reduced, only maintaining the strictly necessary fields.
- The multicast transmission system is replaced by a broadcast one.

For details on the various SLP (and corresponding nSLP) fields, please refer to the SLP standard.

*nSLP Header*   Following all the considerations from the previous section, the header size was reduced, only leaving the strictly necessary fields. In particular, the following fields:

- *Version* was maintained in order to ease the detection of the version of the used nSLP.
- *Length* was shred from 24 to 8-bit, more than sufficient to describe the dimension of nSLP messages.

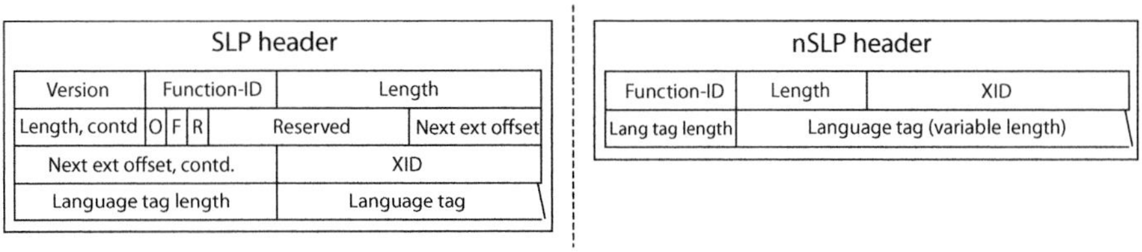

| SLP header | | | |
|---|---|---|---|
| Version | Function-ID | Length | |
| Length, contd | O F R | Reserved | Next ext offset |
| Next ext offset, contd. | | XID | |
| Language tag length | | Language tag | |

| nSLP header | | |
|---|---|---|
| Function-ID | Length | XID |
| Lang tag length | Language tag (variable length) | |

**FIGURE 3.38**
From SLP to nSLP header.

- *Flags O, F, and R* were removed because fragmentation of the messages is not allowed in nSLP (flag O), there are not multicast messages (flag R), and every message is considered as new (flag F).
- *Reserved* was removed due to its uselessness.
- *Next-ext-offset* was removed because nSLP does not support message extensions.
- *Language-tag-length* was reduced from 16 to 8-bit (to save space) (Figure 3.38).

*URL Entry* This field is used in SLP (and consequently also in nSLP) to represent the service URL: even in this case, the useless fields were removed or reduced. In particular the fields:

- *Reserved* was removed due to its uselessness.
- *URL length* was reduced from 16 to 8-bit.
- *Authentication blocks* and the relative authentication fields were removed (Figure 3.39).

*Service Request (SrvReq)* This message is used to discover services; considering the memory limitations of a typical nSA, the nSLP does not allow the caching of the already calculated responses to previous requests. The field

- *PRList* was removed due to its uselessness.

| SLP URL entry | | |
|---|---|---|
| Reserved | Lifetime | URL length |
| URL Len, contd | URL (variable length) | |
| # of URL auths | Auth. blocks (if any) | |

| nSLP URL entry | | |
|---|---|---|
| Lifetime | URL length | URL |

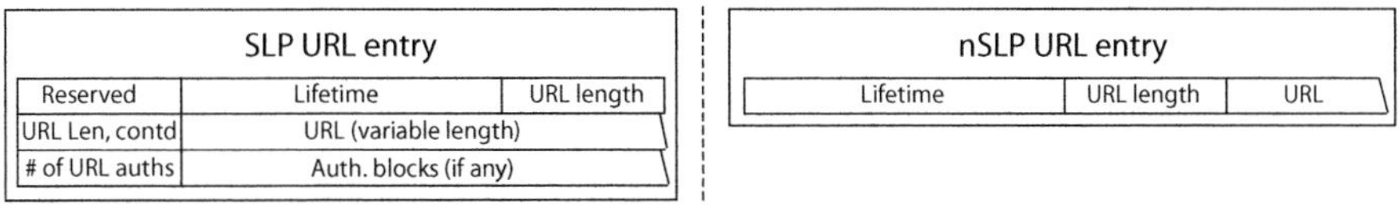

**FIGURE 3.39**
From SLP to nSLP URL entry.

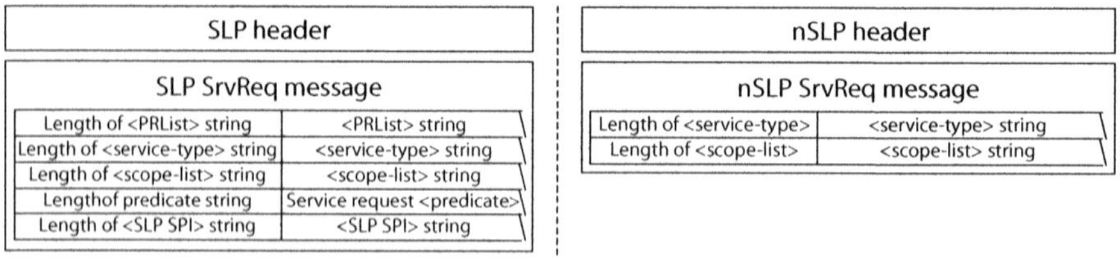

| SLP header |  |  | nSLP header |  |
|---|---|---|---|---|

| SLP SrvReq message |  |  | nSLP SrvReq message |  |
|---|---|---|---|---|

| Length of <PRList> string | <PRList> string |
|---|---|
| Length of <service-type> string | <service-type> string |
| Length of <scope-list> string | <scope-list> string |
| Lengthof predicate string | Service request <predicate> |
| Length of <SLP SPI> string | <SLP SPI> string |

| Length of <service-type> | <service-type> string |
|---|---|
| Length of <scope-list> | <scope-list> string |

**FIGURE 3.40**
From SLP to nSLP SrvReq message.

- *Predicate* was removed considering in nSLP it is not possible to spec-ify predicates for the attributes.
- *SLP SPI*, required for authentication, was removed (the hypothesis is to trust the nodes of a WSN).
- *All the relative length fields* were removed as well (Figure 3.40).

*Service Replay (SrvRply)*   The SLP standard permits the SrvRply message to encapsulate an undefined number of URL entries; within nSLP, such a mes-sage can only contain a URL entry. The structure of the nSLP service reply is shown in Figure 3.41: the field "error code" was removed, considering nSAs will ignore messages if errors occur.

*Service Registration (SrvReg)*   Service registration is essential in nSLP, allow-ing the registration of the services of the nSAs in the nDAs; the only differ-ence among nSLP messages and the standard SLP ones is the lack of the authentication field (Figure 3.42).

*Service Acknowledge (SrvAck)*   In order to limit the number of messages exchanged in the network, SrvAck is not contemplated in nSLP.

*Directory Agent Advertisement (DAAdvert)*   The aim of this message (in SLP) is to communicate the presence of a certain DA to all the agents within the same network, with a certain periodicity. Trying to pursue the minimization of the messages exchanged within the network, in nSLP this message is sent

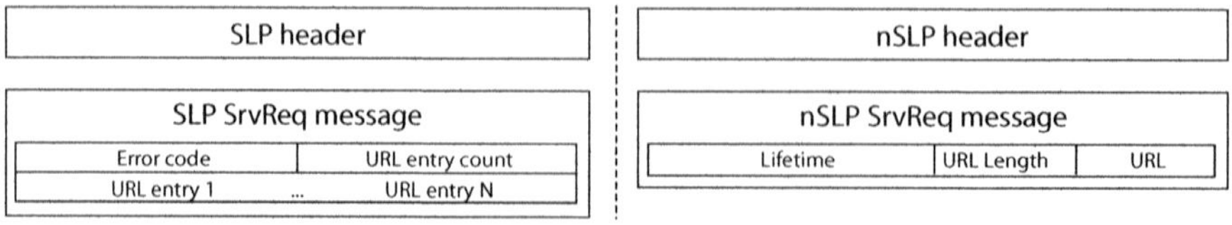

| SLP header |  |  | nSLP header |  |  |
|---|---|---|---|---|---|

| SLP SrvReq message |  |  | nSLP SrvReq message |  |  |
|---|---|---|---|---|---|

| Error code | URL entry count |
|---|---|
| URL entry 1 ... | URL entry N |

| Lifetime | URL Length | URL |
|---|---|---|

**FIGURE 3.41**
From SLP to nSLP SrvRply message.

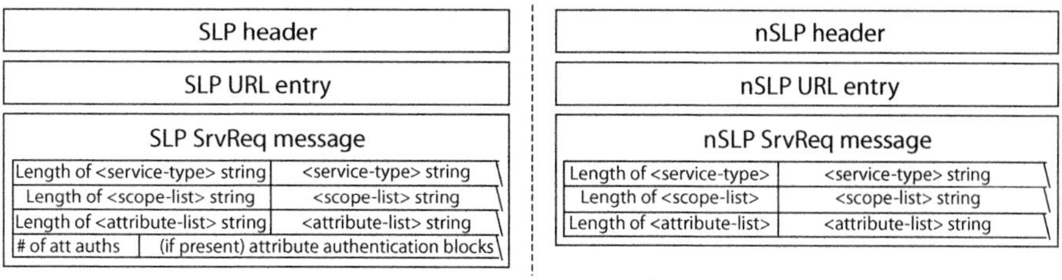

| SLP header |
|---|
| SLP URL entry |

| SLP SrvReq message | |
|---|---|
| Length of <service-type> string | <service-type> string |
| Length of <scope-list> string | <scope-list> string |
| Length of <attribute-list> string | <attribute-list> string |
| # of att auths | (if present) attribute authentication blocks |

| nSLP header |
|---|
| nSLP URL entry |

| nSLP SrvReq message | |
|---|---|
| Length of <service-type> | <service-type> string |
| Length of <scope-list> | <scope-list> string |
| Length of <attribute-list> | <attribute-list> string |

**FIGURE 3.42**
From SLP to nSLP SrvReg message.

by DAs only after a proper request by an nSA; in the new message only the URL remains, due to the fact it is essential to communicate with the nDA (Figure 3.43).

*Attribute Request (AttrReq)*    This message is as essential as the service request, permitting an agent to retrieve the attributes of the services. Following the same considerations made so far, in nSLP the fields:

- *PRList* was removed due to its uselessness.
- *Length of URL* was reduced from 16 to 8-bit.
- *Length of scope-list* was reduced from 16 to 8-bit.
- *Length of tag-list* was reduced from 16 to 8-bit.
- *SLP SPI* was removed (as every authentication field) (Figure 3.44)

*Attribute Reply (AttrRply)*    This message is dual to the previous one and, likewise, the fields:

- *Error code* was removed.

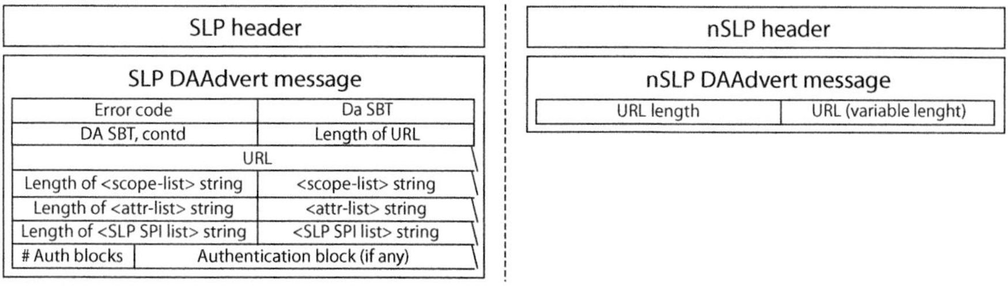

| SLP header |
|---|

| SLP DAAdvert message | |
|---|---|
| Error code | Da SBT |
| DA SBT, contd | Length of URL |
| URL | |
| Length of <scope-list> string | <scope-list> string |
| Length of <attr-list> string | <attr-list> string |
| Length of <SLP SPI list> string | <SLP SPI list> string |
| # Auth blocks | Authentication block (if any) |

| nSLP header |
|---|

| nSLP DAAdvert message | |
|---|---|
| URL length | URL (variable lenght) |

**FIGURE 3.43**
From SLP to nSLP DAAdvert message.

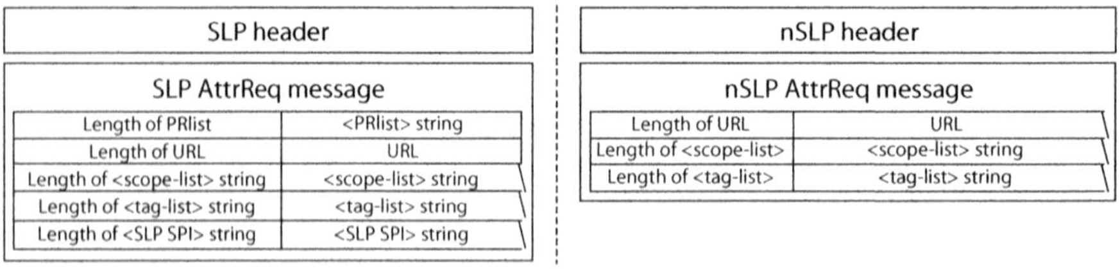

| SLP header | |
|---|---|
| SLP AttrReq message | |
| Length of PRlist | <PRlist> string |
| Length of URL | URL |
| Length of <scope-list> string | <scope-list> string |
| Length of <tag-list> string | <tag-list> string |
| Length of <SLP SPI> string | <SLP SPI> string |

| nSLP header | |
|---|---|
| nSLP AttrReq message | |
| Length of URL | URL |
| Length of <scope-list> | <scope-list> string |
| Length of <tag-list> | <tag-list> string |

**FIGURE 3.44**
From SLP to nSLP AttrReq message.

- *Length of attr-list* was reduced from 16 to 8-bit.
- *Authentication blocks* were removed (Figure 3.45).

## Service Composition and Orchestration Framework

Service composition is the SHIELD functionality in charge of the selection of the atomic SPD services that, once composed, can provide a complex and integrated SPD functionality essential to guarantee the required SPD level. The service composition is one of the SHIELD middleware adapter functionalities, cooperating with the SHIELD security gent to implement the configuration strategy decided by the SHIELD control algorithms (see Chapter 5). In order to compose SPD functionalities in the best possible configuration, the service composition components have to take into account the technology details at node, network, and middleware layers. In few words, control algorithms could drive composition as technology independent thanks to the abstraction introduced by the composition modules, which, instead, are technology dependent. Whenever the control algorithms calculate that a specific SPD configuration, of the available SPD services, must be enforced (taking into account the services' descriptions, capabilities, and requirements), the service composition process has to guarantee that all the dependencies, configuration, and preconditions associated with the composed service are met. This is mandatory to guarantee all the atomic SPD services will be able to work properly once composed.

The composition ecosystem of the SHIELD framework is illustrated in Figure 3.46.

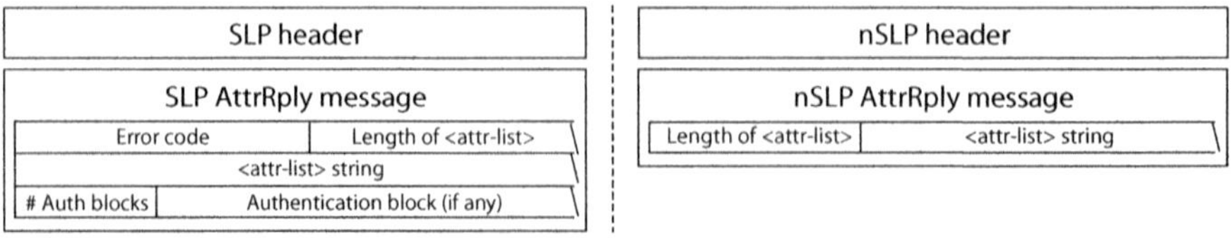

| SLP header | |
|---|---|
| SLP AttrRply message | |
| Error code | Length of <attr-list> |
| <attr-list> string | |
| # Auth blocks | Authentication block (if any) |

| nSLP header | |
|---|---|
| nSLP AttrRply message | |
| Length of <attr-list> | <attr-list> string |

**FIGURE 3.45**
From SLP to nSLP AttrRply message.

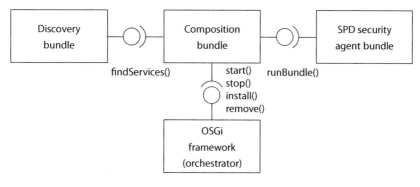

**FIGURE 3.46**
Composition framework.

### Composition Engine Bundle

This module actually performs the composition of the available SPD services according to the composition rules and constrains determined by the security agent. If the security agent indicates (by using the *runBundle()* interface) it is required to be run a certain composed functionality, the composition engine uses the *findServices()* interface to discover, through the service discovery framework, any suitable SPD component or service to be composed to the overall required service. After the discovery of all the required components, the composition engine composes them (also taking care of the inter-services dependencies) and uses the *start()*, *stop()*, *install()*, and *remove()* interfaces provided by the orchestrator (i.e., the OSGi framework itself) to properly manage them.

### Service Orchestration Framework

The role of the service orchestration framework, part of the SHIELD middleware, is to deploy, execute, and continuously monitor the SPD services that have been discovered and composed. The main difference among the service composition component and the orchestrator is that the first works "off-line," being triggered by a certain event or by the SHIELD overlay, while the second works "on-line," continuously monitoring the SPD status of the running services.

#### Implementation of the Service Orchestration: The OSGi Framework

Achieving the complex and challenging functionalities required for the service orchestration framework from scratch would have been an overwhelming effort for the SHIELD project, also considering there have already been, during the execution phase of the project, some well-functioning and quite diffused frameworks to this aim. Among all the possibly available open

Service-Oriented Architecture (SOA) solutions, it was decided to adopt the OSGi framework as the reference service orchestration platform to develop the proof-of-concept demonstrator.

The rationale behind this choice has the following reasons:

- OSGi is an open standard.
- There exists a number of open source implementations for the OSGi specifications (Equinox, Oscar, Knopflerfish, JBoss) and some of them are also included in widely adopted software (e.g., Eclipse).
- Despite its complexity, some of the OSGi implementations can be executed even over lightweight nodes (e.g., on EDS).
- There exists OSGi implementation for a number of programming languages (e.g., Java, C, C#).
- The selected Java implementations of the OSGi specification (namely, the Knopflerfish framework) is very fast to be deployed and it is very easy to be used.
- A plethora of plugins, developed taking into account, the OSGi directives, are available for a number of IDE tools (i.e., Eclipse, Visual Studio, etc.).
- Java-based OSGi implementations are OS-independent (as well as Java) and can be easily deployed in Windows, Linux, Macintosh, and Android.

Following these considerations we decided to adopt the open source, Java-based, OSGi-compliant Knopflerfish (KF) service platform. KF is a component-based framework for Java able to run a number of atomic services (called "bundes") that can be installed, run, stopped, composed, and updated. Each bundle can export its own services and/or run processes, while, if required, using the services of other ones; the KF is in charge of managing the dependencies of each bundle, in particular supplying a method to specify, for each bundle, the requirements. Considering the KF strictly follows the OSGi specifications (and, quite important, the KF implementation are usually well aligned with any news within the OSGi specification) any valid KF bundle can be installed in any valid OSGi container (Oscar, Equinox, JBoss, etc.).

Not all the OSGi implementations are really easy to use (e.g., JBoss has a certain complexity and is quite resource demanding), but the usage of KF is extremely easy. The OSGI KF implementation can found at http://www.knopflerfish.org/; and the typical KF interface is shown in Figure 3.47.

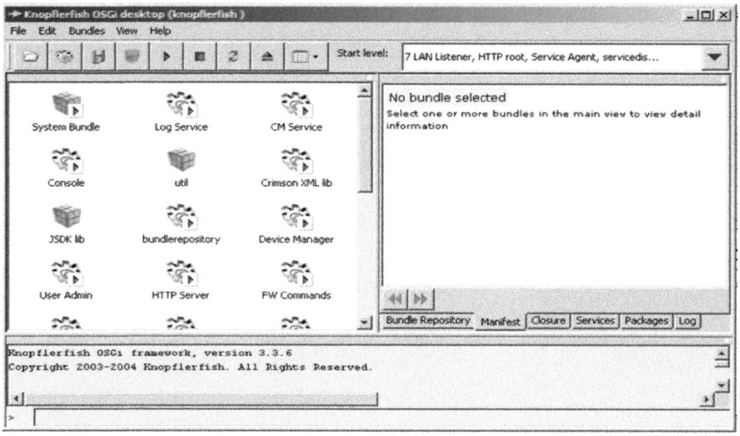

**FIGURE 3.47**
Service orchestration engine: The Knopflerfish start-up environment.

## References

1. NIST, Face Recognition Vendor Test (FRVT) 2013. https://www.nist.gov/itl/iad/image-group/face-recognition-vendor-test-frvt-2013.
2. H. Kyong, I. Chang, K. W. Bowyer, and P. J. Flynn, An evaluation of multi-modal 2d+3d face biometrics, *The IEEE Transactions on Pattern Analysis and Machine Intelligence*, 27 (4), pp. 619–624, 2005.
3. H. Moon and P. J. Phillips, Computational and performance aspects of PCA-based face-recognition algorithms, *Perception*, 30 (3), pp. 303–321, 2001.
4. ETSI TS 33.401 V12.10.0 (2013–2012): 3rd Generation Partnership Project; Technical Specification Group Services and System Aspects; 3GPP System Architecture Evolution (SAE); Security architecture (Release 12).
5. 3GPP TS 33.203 V12.4.0 (2013–2012): 3rd Generation Partnership Project; Technical Specification Group Services and System Aspects; 3G security; Access security for IP-based services (Release 12).
6. 3GPP TS 33.210 V12.2.0 (2012–2012): 3rd Generation Partnership Project; Technical Specification Group Services and System Aspects; 3G security; Network Domain Security (NDS); IP network layer security; (Release 12).
7. IETF Service Location Protocol V2. http://www.ietf.org/rfc/rfc2608.txt.
8. UPnP Simple Service Discovery Protocol. http://upnp.org/sdcps-and-certification/standards/.
9. IETF Neighbor Discovery Protocol. http://tools.ietf.org/html/rfc4861.
10. IETF Domain Name Specification. http://www.ietf.org/rfc/rfc1035.txt.
11. Bluetooth Service Discovery Protocol. https://www.bluetooth.com/specifications/assigned-numbers/service-discovery.
12. OASIS Universal Description Discovery and Integration. http://www.uddi.org/pubs/uddi_v3.htm.

# 4

## *The SHIELD Approach*

Andrea Fiaschetti, Paolo Azzoni, Josef Noll, Roberto
Uribeetxeberria, Antonio Pietrabissa, Francesco Delli Priscoli,
Vincenzo Suraci, Silvano Mignanti, Francesco Liberati, Martina
Panfili, Alessandro Di Giorgio, and Andrea Morgagni

### CONTENTS

## Introduction

The SHIELD approach can be summarized in two words: *measurable* and *composable* security. *Measurable* means that security must be quantifiable, because measuring a property is the basic condition to control it; *composable* means that the SHIELD framework is able to put together different elements (i.e., security, privacy, and dependability [SPD] functionalities), creating an overall, end-to-end behavior that ensures the desired SPD level. The SHIELD approach can then be described as a logical chain of actions, as indicated in Figure 4.1 and further detailed in Table 4.1.

### SPD Awareness

Security awareness can be defined as the knowledge and attitude that members of an organization possess regarding the protection of the physical and especially information assets of that organization. It is common that staff members do not understand the security risks relevant to their duties, even if they handle sensitive data as a part of their everyday routine. In conventional information and communication technology (ICT) systems, human actions can result in the loss of intellectual property and exposure of customer data. The consequences of failure to properly protect information include potential loss of employment, economic consequences to the firm, damage to individuals whose private records are divulged, and possible civil and criminal penalties.

For a cyber-physical systems (CPSs) developer company, the lack of security awareness can be even more dangerous. CPSs are being deployed in a wide range of application areas, ranging from control of safety-critical systems to data collection in hostile environments. These systems are inherently vulnerable to many operational problems and intentional attacks due to their embedded nature. Network connectivity opens even more ways for remote exploits. Therefore, it is clear that successful attacks against those systems may cause fatal consequences. If an embedded systems designer is not aware of the implications of not implementing SPD features in the design, the consequences may have a great cost: economical, environmental, or even loss of human lives. Unfortunately, it is not uncommon that embedded systems designers dedicate their effort to the development of functional aspects, not considering nonfunctional aspects, such as SPD, especially when time to market is critical for the business.

Figure 4.2 represents the first building block of the SHIELD framework: SPD awareness. The first and most important step is to be aware of the security risks. Even if the problems may be similar between traditional software

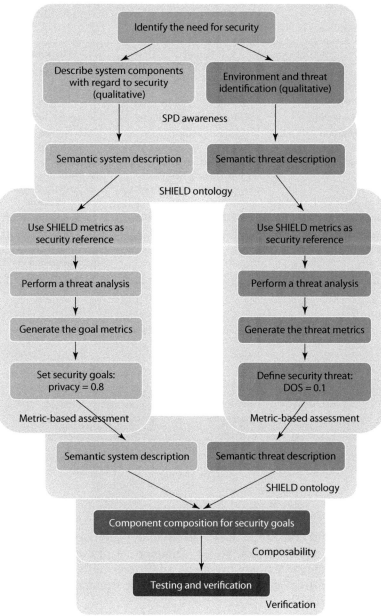

**FIGURE 4.1**
SHIELD framework for measurable and composable security.

**TABLE 4.1**

Run-Through Steps of the SHIELD Framework

| Action | Input | Outcome |
|---|---|---|
| Environment and threats identification | | Awareness |
| SPD assessment | Application case | SPD guidelines |
| Metrics implementation | SPD guidelines and tools | Metrics (security measure) |
| Ontology definition | Application case, tools | Web ontology language (OWL) |
| Technological injection | Software module, intellectual property (IP), template, trusted runtime environment | Software- or firmware-customized modules (SPD security framework [SF] module) |
| Integration | SPD SF modules + design files | Validation report |
| Validation/verification | SPD validation tools | Validation report |
| Deployment | | Embedded systems physical implementation + end user application note |

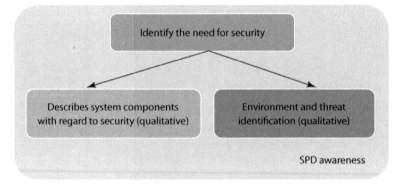

**FIGURE 4.2**
SPD awareness.

development and embedded device security, the engineering teams do not take them into consideration so far. Therefore, a preliminary qualitative description of the system, environment, and threats will serve as input for the next, more formal steps: semantic description, threat modeling, and risk analysis.

Being security aware means you understand that there is the potential for some people to deliberately or accidentally steal, damage, or misuse the data that are used by your embedded systems or system of systems. Therefore, it would be prudent to support the secure design of such systems.

The basic action for SPD awareness is "identify the need for security" of the system. It should serve to define the overall security objectives to be fulfilled by the system. Then, this will derive into two other blocks: a description of the security components that the system requires to achieve those objectives, and the identification of the threats that the system will be exposed to, in the specific environment where it will be used.

Both the description of the security components and the identification of the threats are carried out at a high level and with no formal description. The goal is to be able to *identify* the threats and the security elements to fulfill the security requirements of the systems within a specific environment. Anyhow, this step may require the participation of a security expert to help the CPSs designer. A formal description is provided in the following steps of the framework.

## SHIELD Ontology

The SHIELD framework is built on a set of *distributed ontologies*, which help us to characterize the system and the security parameters. Although security semantics are known, they often address one specific system. SHIELD defines ontologies for (1) the system description, (2) the security functionality of the system components, and (3) the description of SPD. Such a formal description of the SHIELD components has the aim to

1. Identify the SPD attributes (functionalities) and their mutual relations
2. Identify the functional and technological dependencies between SPD functionalities and system components

These two descriptions are necessary because attributes and relations prepare the system for an easier "quantification" of SPD metrics (see the next step), while SPD functionalities dependencies prepare the system for the "implementation" of the composition decisions.

This information is represented by means of "ontology" or, more in general, a "semantic description" because it is simply a matter of "knowledge representation": this is information to be stored in a structured way, and the ontology allows it, better than traditional databases, to have great expressiveness with reduced size (in terms of bytes).

The SHIELD ontology, as depicted in Figure 4.3, comprises two sections: one about the system and one about the threat. This distinction follows the Common Criteria guidelines, according to which, in order to assess the

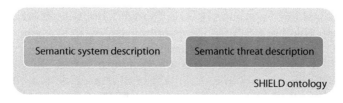

**FIGURE 4.3**
SHIELD ontology.

security level of a system, it is necessary to start from the menaces that affect that system.

In particular, the SHIELD logical procedure for ontology structuring follows the logical process depicted in Figure 4.4:

1. The system is decomposed into atomic elements, named SPD functionalities.
2. These functionalities, individually or composed with the other, realize an SPD means, which is a means to prevent a menace.
3. This means is mapped over the SPD threats to countermeasure or mitigate them.
4. According to the mitigated threats, it is possible to quantify the impact on the overall SPD level, thanks to the SPD metrics that assign a value to the threats.

From an operational point of view, this step is translated into the following guidelines:

- Each manufacturer producing a SHIELD-compliant device provides it with a file containing a semantic description of the related SPD functionalities.
- Each SHIELD device is able to provide this semantic description to the SHIELD security agents by means of service discovery or specifically tailored signaling protocols.

In this chapter, the SHIELD security agent is described as one of the major enabling technologies.

## Metric-Based Assessment

Once provided a formal ontological description of the SHIELD components with their SPD functionalities, their mutual relations, and the functional and technological dependencies between the SPD functionalities and system components, we can proceed with the quantification of the SPD metrics (Figure 4.5).

One particular system might be composed of multiple subsystems (e.g., an aircraft is considered a system [RTCA/DO-178B] but composed of multiple subsystems). Therefore, SPD concepts are measured in systems of systems as a whole: this means that metrics measure each subsystem (component) independently and are composed until the entire system measurement is gathered. This requires both a definition of SPD metrics that allows us to collect quantifiable measurements and a composition model for computing the all-inclusive system.

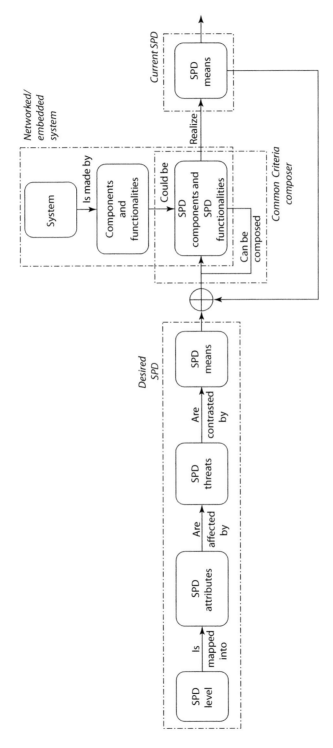

**FIGURE 4.4**
SHIELD ontology logical process.

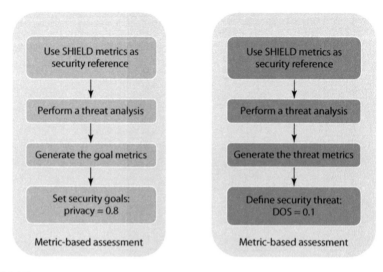

**FIGURE 4.5**
Metric-based assessment.

SHIELD defines scalable SPD metrics that allow us to determine the SPD level of any individual, as well as the overall system, in a holistic mode. SHIELD provides a formal method to compose SPD metrics in both design and runtime scopes, being able to provide an input toward the overlay layer, which is also capable of composing other SPD functionalities in a semantic way.

Metrics-based assessment is the enforcement and validation of the measurement itself. The main problem for the measurement is to establish the correct value (setting the goal) for the metric. This value is often provided by manufacturers and integrators according to standards and regulations, or is based on empirical and experimental stress tests. An associated problem is that the threat SPD levels or the goal SPD level can change at runtime. This entails that the expected behavior of the system shall be corrected through metrics and a monitoring mechanism. SHIELD deals with this problem in an interdomain, heterogeneous, and distributed environment.

Nowadays, there is no instrument or tool able to measure systems as SHIELD does. This mechanism will impact the industry, setting a new reference in two ways: generating a new formal and quantifiable model for measuring systems of systems and industrializing a tool for implementing this model.

Details on some of the metrics developed to support the SHIELD framework are provided in Chapter 5.

## Composability

Once metrics are derived for the components abstracted so far, all the puzzle pieces are available and it is possible to combine them in order to achieve the configuration that satisfies the goal security need (in terms of SPD level) (Figure 4.6).

The SHIELD architecture provides a modular, scalable framework for the composition of SPD components, but it does not force a unique composition algorithm, leaving the possibility of further modifications and updates.

However, three basic classes of "composition drivers" have been chosen:

- Composition driven by *security and safety standards* (like Common Criteria): Following this approach, system components are put together according to a vulnerability assessment, in order to mitigate the identified menaces. To this extent, once the SPD value associated with a specific status of the system is known, when some events occur that modify this status, it is possible to take action aimed at maintaining the desired SPD level. This can be done by enabling or disabling interfaces, or security functionalities, or components of the system and considering their impact in mitigating the identified threats, and thus affecting the SPD level.

- Composition driven by *policies*: According to this logic, the enabling or disabling of the SHIELD components is driven by a set of preloaded, or dynamically modified, policies. They specify, for certain circumstances of specific components, the action that has to be enforced in the system.

- Composition driven by *context-aware control algorithms*: According to this logic, SPD functionalities are modeled by means of proper mathematical models, and the decision about the configuration is made as a result of optimization or control algorithms.

SHIELD permits us to use a single composition driver, or to run more than one sequentially.

From an operational point of view, this step requires the presence of a software module named "security agent" that is able to

FIGURE 4.6
Composability.

**FIGURE 4.7**
Testing and verification.

- Retrieve information from the semantic, policies, and services databases
- Use this information to elaborate a decision
- Translate this decision into a control command to be enforced in the system

Since the composition is oriented to satisfy the SPD level, it is strictly connected to the metrics definition.

## Verification

SHIELD has created a set of requirements and metrics and designed a reference system architecture. The components that constitute the SHIELD platform are subject to a formal validation procedure, which allows us to prove their correctness. The testing and verification plan is formalized upon a specific methodology, taking into account all necessary validation parameters (Figure 4.7).

The components of a SHIELD system can be highly heterogeneous. Some of them are software components, some are hardware components, and others are algorithms or models. For this reason, a common methodology for validation and verification activities is not provided, giving the choice of the most suitable means of verification to the experts of the different layers. Figure 4.8 shows the SHIELD testing and verification procedure.

## Control Theoretic Approach to Composable Security

The key point of the SHIELD approach is the "embedded intelligence" by which decisions are made to compose systems functionalities; this intelligence is given by control algorithms derived from a formulation of the composability problem as a closed-loop problem.

The problem of composability of security (or SPD) functionalities in the CPSs domain is a field with big potential in terms of industrial exploitations, but at the same time, it is still unexplored in the literature, mainly due to the lack of a mathematical theory, or a methodology, able to tame the domain heterogeneity. This book aims at being a tentative first in this direction.

The generic problem of composability has indeed been addressed in the last decade by several scientific groups, but limited to the computer science

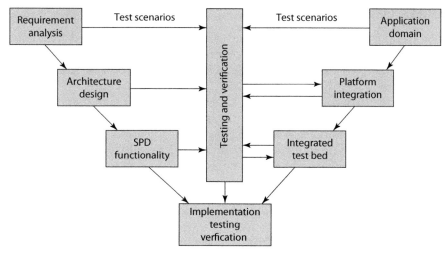

**FIGURE 4.8**
Platform verification procedure.

domain, with the objective of designing a new class of software systems able to provide any kind of functionality simply being reconfigured in real time, without the need to reassess the code ex novo, or modify it. This is the case, for example, with the European project CONNECT (2009–2013), in which an architecture for composability of networked software systems was developed [1,2], together with a formal language and an algebra to model the system components [3].

Another domain in which the composability problem has been stressed is systems engineering: in this field, the purpose is to define formal or abstract representation of system components in order to reduce the design effort and increase the reusability of models and solutions [4], for either virtual prototyping or simulation phases. A well-known technological solution of this problem is the Unified Modeling Language (UML), derived to create a convergence language in systems engineering, together with plenty of formal approaches for the design activities (e.g., see [5,6]).

The security domain is not new to composability approaches, but the biggest limitation is that these approaches only address *uniform properties* (security of interacting protocols, security of composed software systems, etc.), and this cannot be adopted in the CPSs environment, where heterogeneity is the driving requirement.

The essence of composability is

1. The possibility of ensuring that the individual properties are not lost when put together
2. The rising of enriched properties, bigger than the simple sum of the original ones

## SHIELD Closed-Loop Formulation

The composability problem defined so far can be formulated in a good way by using a control theoretic representation (Figure 4.9). This representation is even closer to the effective implementation of the SHIELD framework.

The difficulty of such a problem is that SPD is concepts that are not commonly measured (1) jointly and (2) in a quantitative way. In fact, for each of them specific rules and standards exist, only for some specific domains (ICT, railways, etc.), and provide methods and metrics for measurements, but they are mainly qualitative and highly focused.

Moreover, the control of such properties is more related to procedural aspects or policy implementation, rather than to control commands and mathematical solutions; however, a formal approach is needed to provide a reference methodology to compose and control SPD, hopefully in a machine-understandable way (in order to be implemented in automatic software engines).

To overcome these issues, a decoupling solution has been proposed (Figure 4.9), with the objective of splitting the problem into

- A logic-driven part, related to the measurement and composition of SPD parameters according to standards, rules, and policies (lower loop)
- A quantitative part in charge of performing further optimization on the quantitative parameters describing the system (upper loop)

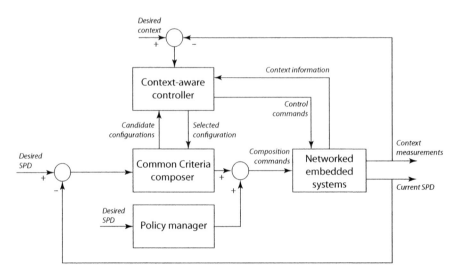

**FIGURE 4.9**
Composability: Closed-loop view.

The logic-driven part is referred to as the *Common Criteria composer*, with reference to the ISO-IEC 15408 Common Criteria standard [7] that drives the composition, while the qualitative part is referred to as the *context-aware controller*, to underline the additional optimization performed by the controller on the basis of context information.

Going into further details, the *reference signal* of this closed-loop formulation is the SPD desired level, obtained and quantified according to the SHIELD metrics (mainly based on the Common Criteria standard, where specific procedures to quantify system vulnerability are defined) and to the SHIELD composability rules for complex systems (like the medieval castle approach derived in [8]).

This signal is used by the logic-driven composer, to elaborate a decision in two different ways:

- By means of the Common Criteria approach, enriched with
- A context-aware controller (i.e., hybrid automata based)

A second (optional) reference signal is used by these latter controllers for additional optimizations.

In addition, *policies* can also control the SPD properties, but since they do it in a pseudoheuristic way (meaning that the SPD countermeasures are predefined and mapped), their behavior is quite similar to that of a disturbance independent of the system dynamic.

All is depicted in Figure 4.9.

The interaction between the two loops is simple:

*Step 1*: The Common Criteria approach, by adequately measuring the SPD level for each component and applying specific rules, is able to derive a *set of possible (and suitable) configurations* that satisfy the desired SPD level.

*Step 2*: The context-aware controller receives this set of configurations as input, as well as other system parameters (context information), and performs further analysis (e.g., optimization on non-SPD relevant parameters) and *chooses the configuration* (among the suitable ones) that best satisfies the additional targets.

*Step 3*: The resulting configuration is then (eventually) arbitrarily modified by the application of policies that, acting as a disturbance, make some decision driven by predefined rules.

*Step 4*: If the resulting configuration still satisfies the SPD needs, the solution (in terms of the configuration) is then implemented in the system by enforcing the activation or deactivation of SPD functionalities and components. Otherwise, the process is reiterated from the beginning on a new set of SPD functionalities.

The enabling technology that allows the information exchange between the controller and system, in a joint qualitative and quantitative way, is the ontology, which contains all the relevant information to extrapolate metric values and describe system components, functionalities, and logical relations compliant with the defined steps.

This last point represents, in particular, one of the major achievements of this work, since for the first time a semantic representation has proved to be the only solution to address a control problem (for this particular scenario, due to the heterogeneity and the double qualitative and quantitative description of the system).

### Logic-Driven Control Loop

The logic-driven part of the SHIELD approach is based on the logical chain described in the SHIELD ontology (see introduction). This chain is indeed the translation of the Common Criteria methodology (asset threats countermeasures chain) into a set of logical steps to be carried out to couple menaces with countermeasures and to derive a system configuration that is compliant with the SPD needs. This logic-driven control loop is depicted in Figure 4.10.

First, the reference and the output signal, which has to be measured, are translated into the concept of *means*; that is, the desired level of SPD or the current level of SPD is represented by the set of countermeasure deployed, or to be deployed.

The evolution of the system is modeled through the atomic components by which it is made: in fact, from an SPD perspective, the system evolves by activating or deactivating SPD functionalities, so its state is univocally identified by these elements.

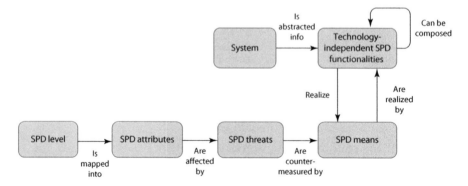

**FIGURE 4.10**
SHIELD logic-driven control loop.

Last, but not least, the Common Criteria composer has been represented by the simple relation *"realize"*: from this perspective, the matching between functionalities and the means that they realize allows us to pass from the user's need to system implementation.

## Quantitative Control Loop: Context-Aware Controller

The most innovative results of the composability problem are given by the quantitative control loop developed to support the SHIELD framework. With reference to Figure 4.9, the quantitative controller is referred to as context-aware controller.

In order to derive this controller, three points must be addressed:

1. Providing a (formal) model of the individual components
2. Providing an algebra or some operators finalized to the assessment of the composition of atomic elements
3. Providing control algorithms, or rules, to act on component parameters

Despite this, the context-aware controller presented here brings innovation by proposing

- The development of a formal model for the SHIELD node, based on hybrid automata
- The identification of a concurrent or parallel composition as basic algebra for composition
- The identification of rules to act on the modeled parameters

With respect to the integrated approach presented earlier, for this chapter, it is assumed that the Common Criteria composer has performed its logical inferences and is able to identify *one or more possible configurations* of the system components. These configurations are given as input to the context-aware controller to perform ancillary optimization, together with some parameters directly measured on the system.

Regarding the closed-loop formulation derived in Chapter 2, the blocks can be rearranged a bit to better depict this duality (Figure 4.11).

So the reference signal for the context-aware controller is not the desired SPD level (i.e., not directly measurable in a quantitative way) but a generic desired context, which can be translated into quantifiable and measurable variables (bandwidth, energy consumption, etc.).

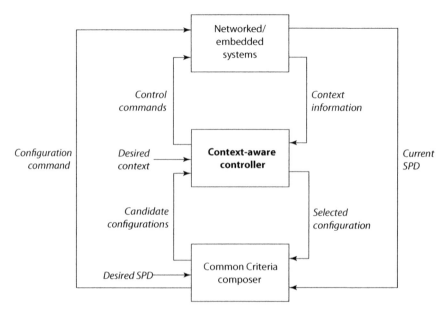

**FIGURE 4.11**
Context-aware controller.

## Hybrid Automata Model of the Embedded Systems Domain

A CPS is, commonly, a computing platform used to perform some dedicated tasks in a specific environment. Thanks to the miniaturization of components and the increase in the computational capabilities, modern CPSs look more like mini-computers than simple nodes. Despite the size and purpose, the architectural description of CPSs and embedded systems in the literature (e.g., see [9]) is fixed and is based on the consolidated von Neumann scheme, including

- Elements in charge of *input/output (I/O) communication*
- Components devoted to the short-term and long-term *information storing*
- One or more *processing units*
- A *power supply* unit

This element will be the building block of our mathematical model. After having provided a description of this node, we will also model the network layer, connecting several CPSs exchanging information.

Modeling, especially for control purposes, is usually a balance between complexity and expressiveness. In particular, from a control theoretic perspective, the objective of the modeling activity is to formalize the system behavior in order to clearly identify

- Variables that can be used to measure and inspect the evolution of the system
- Variables that can be used to manipulate the evolution of the system
- Dynamics and relations that describe these evolutions

In particular, for the SHIELD purposes, it is not necessary to perform an exhaustive or fine-tuned control of each and every single variable of a CPS, but to represent the high-level or functional behavior of the component with a simplified (but meaningful) dynamic. For that reason, a simplification is done in the elicitation of the mathematical model, mainly based on the same valuable approach adopted in other domains with similar objectives, and in particular in the manufacturing domain by [10], who addressed the modeling of a production line; in this reference paper, which inspired our model, a hierarchical procedure is followed to decompose the system into two hierarchical layer responsible for

- *Macroevents and macrostates*: Modeled with the different states of an automaton
- *Microevents and continuous dynamic*: Modeled with the time-discrete and time-continuous equation of a hybrid system

This allows us to overcome the intrinsic complexity of modeling a heterogeneous networked system.

### Model for the Microscopic Behavior of a Generic Embedded System

At first, a finite time frame is identified: $\Delta_k = [t_k, t_{k+1})$, $k = 0, 1, 2, \ldots$, representing the time period between the beginning of a *macroevent* associated with the embedded system behavior and the beginning of the subsequent macroevent (they usually correspond to the activation or deactivation of a functionality). This time interval is called macroperiod.

During a macroperiod, the atomic elements that compose a CPS evolve, following certain dynamics, described in the following for the subsystems identified before. These dynamics have the same structure for all nodes $N_i$ that compose the system, with differences only in coefficients and parameters (to discriminate different technologies).

#### Communication Buffer and I/O Communication

The node buffer represents the element in charge of storing the information received from the outside world or from another node, which is waiting to be processed by the CPS before being forwarded to the final recipient.

Let us introduce the I/O variables $r_{in,i}(k)$, to represent the input rate feeding the $i$-th CPS, and $r_{out,i}(k)$, to represent the output rate leaving the $i$-th CPS. Both these rates are considered constant during the $k$-th time interval $\Delta_k$.

The buffer dynamic, that is, the queue length $q_i(t)$ of the $i$-th CPS, is given by the difference between the input rate and output rate, computed during the time interval $\Delta_k$ and described, as usual, according to the first-order fluid approximation ([11]), a linear approximation adopted for the description of discrete flow networks (like a network of embedded systems exchanging packets). The dynamic is given in Equation 4.1:

$$q_i(t) = q_i(t_k) + \left[ r_{in,i}(k) - r_{out,i}(k) \right](t - t_k) \tag{4.1}$$

where $t \in [t_k, t_{k+1})$, $q_i(t_k)$ is the initial state and the following physical constraint (4.2) of buffer overflow holds:

$$0 \le q_i(t) \le q_{MAX,i} \tag{4.2}$$

*Processing Unit*

The processing unit represents the computational capability of the node, that is, the possibility of performing operations (mainly for security purposes, e.g., cyphering) on the managed data. For this reason, the dynamic of the CPU is directly in relation to (1) the amount of data entering and leaving the node and (2) the type of operations performed on these data.

Considering $c_i(t)$, the value of the CPU utilization during the time interval $\Delta k$, its evolution is described in Equation 4.3:

$$c_i(t) = \gamma c_i(k) + \alpha r_{in,i}(k) + \beta r_{out,i}(k) \tag{4.3}$$

where $t \in [t_k, t_{k+1})$ and $\alpha, \beta \ge 1$. In particular,

- $\gamma c_i(k)$ is an arbitrary offset for CPU utilization, representing ancillary operations different from the ones performed on the data to be served (e.g., authentication or logging)
- $-\alpha$ and $\beta$ are parameters modeling the entity of the operations performed on the data and scaling adequately the I/O rate managed by the node

An example will simplify understanding. Let us assume that the $i$-th CPS is a sensor node collecting measurements and sending them encrypted to another node; then $\alpha = 1$ (i.e., no actions are performed on the input data during the sensing process) and $\beta = 2$ because the procedure of cyphering data before sending them to another destination requires a bigger effort by the CPU (in this case, a double effort).

On the other hand, considering the $i$-th node acting as the receiver of these data, then $\alpha = 2$ in order to unencrypt the information and $\beta = 1$ because the information is directly forwarded, for example, to the application layer.

It is evident that different values of α and β lead to different utilizations of the CPU; values less than 1 are not allowed for these parameters, since data must at least be read or written and the effort is directly equal to their entity. Finally, two remarks are needed.

*Remark 1*: Since the I/O rate is considered constant during the time interval $\Delta_k$, $c_i(t)$ is constant during the time interval as well, so no real dynamic is represented and this piece of information could have been modeled simply as a constraint on the weighted sum of I/O rates. However, this formulation is ready for further modification where, for example, α and β or the computational offset $\gamma c_i(k)$ is driven by more complex dynamics.

*Remark 2*: We assume that the CPU is enabled with multitasking and consequently is able to perform operations at the same time on data collected by the environment and on data sent to the environment.

Of course, even for the CPU the following physical constraint (4.4) of CPU saturation holds:

$$0 \leq c_i(t) \leq c_{MAX,i} \tag{4.4}$$

*Node Power Supply*

The power supply is the node component in charge of providing the energy necessary to perform all the operations. In particular, it is well known in the literature that the power consumption of an electronic device is given by (1) data transmission and (2) data processing, with a factor 10 of difference (transmitting is much more expensive than processing).

The power supply is considered to have an initial value and to decrease continuously during the operations, typically being a battery. In the case of fixed power supply (e.g., connection with power network), the discharge rate is assumed to equal zero (easier to model than an infinite initial state).

Considering the energy $x(t)$ of the $i$-th node, Equation 4.5 holds.

$$x_i(t) = x_i(t_k) - \mu \left[ \delta_1 c_{in,i}(k) + \delta_2 \left( r_{in,i}(k) + r_{out,i}(k) \right) \right] (t - t_k) \tag{4.5}$$

where in the square brackets the individual contribution from the data elaboration (first term) and from the data transmission (second terms) is evident. We assume that $\delta_2 > \delta_1 > 0$.

Since the CPU level is a function of the I/O rate, Equation 4.5 can be rewritten as a function of only the I/O rate, obtaining Equation 4.5′:

$$x_i(t) = x_i(t_k) - \mu \left[ \delta_1 \gamma c_i(k) + (\alpha \delta_1 + \delta_2) r_{in,i}(k) + (\beta \delta_1 + \delta_2) r_{out,i}(k) \right] (t - t_k) \tag{4.5'}$$

In both cases, $t \in [t_k, t_{k+1})$, $x(t_k)$ is the initial charge level, and the following physical constraint (4.6) of energy consumption holds:

$$x_i(t) \geq 0 \tag{4.6}$$

Moreover, the parameter $\mu$ indicates a finite power source (i.e., a battery) or an infinite power supply (i.e., power network):

$$\mu = \begin{cases} 1 & \text{if the power supply is a finite source} \\ 0 & \text{if the power supply is an infinite source} \end{cases} \tag{4.7}$$

The resulting description of the evolution of an embedded system during a specific state is summed up in Equation 4.8:

$$\left\{ \begin{array}{c} q_i(t) = q_i(t_k) + \left[ r_{in,i}(k) - r_{out,i}(k) \right](t - t_k) \\ c_i(t) = \gamma c_i(k) + \alpha r_{in,i}(k) + \beta r_{out,i}(k) \\ x_i(t) = x_i(t_k) - \mu \left[ \delta_1 \gamma c_i(k) + (\alpha \delta_1 + \delta_2) r_{in,i}(k) + (\beta \delta_1 + \delta_2) r_{out,i}(k) \right](t - t_k) \\ 0 \leq q_i(t) \leq q_{MAX,i} \\ 0 \leq c_i(t) \leq c_{MAX,i} \\ x_i(t) \geq 0 \\ 0 \leq r_{in,i} \leq r_{in,MAX,i} \\ 0 \leq r_{out,i} \leq r_{out,MAX,i} \\ \alpha, \beta \geq 1 \\ \delta_2 > \delta_1 > 0 \\ \mu = \{0,1\} \\ t \in [t_k, t_{k+1}) \end{array} \right. \tag{4.8}$$

The model of the possible states is presented in the following section.

### Model for the Macroscopic Behavior of a Generic CPS

The macroscopic behavior for the CPS can be assimilated to the evolution through a finite (reduced) set of admissible *macrostates*, reflecting the potential operating conditions of the system. Such an evolution is modeled by means of an *automaton*, a formal representation of a machine that can assume different states. In the case of the embedded system considered for the SHIELD purposes, five statuses have been identified ($OU_i$, $OL_i$, $OS_i$, $FL_i$, and $FF_i$), described below:

- *Node operational underutilized ($OU_i$)*: The node resources are somehow being wasted because either the buffer or the CPU is empty.

- *Node operational linear* (*OL$_i$*): The node is working in a linear zone, where all its resources are well used.
- *Node operational in saturation* (*OS$_i$*): One or more node resources is being overutilized, causing data loss (e.g., buffer overflow or CPU saturation).
- *Node near failure for low resources* (*FL$_i$*): The power supply level is decreasing and the node is going to be switched off.
- *n = Node fully failure* (*FF$_i$*): The power level has reached its zero level. The node is no longer able to operate.

*Remark 3*: The identifier F reflects a failure or near-failure condition, while the identifier O reflects an operating condition.

Several macroevents allow us to move to or from these macrostates. These events are $\varepsilon_i = \{bf_i, bu_i, bls_i, blu_i, le_i\}$:

- *bf$_i$* (*buffer full*): The buffer reaches its maximum capacity.
- *bu$_i$* (*buffer underutilization*): The buffer is empty.
- *bls$_i$* (*buffer linear from saturation*): A change in the I/O rate occurs, so the buffer evolution changes from its current conditions.
- *blu$_i$* (*buffer linear from underutilization*): A change in the I/O rate occurs, so the buffer evolution changes from its current conditions.
- *le$_i$* (*low energy*): The power level passes a predefined threshold and resources utilization is reduced in order to save energy.
- *ze$_i$* (*zero energy*): The power supply is over and the node switches off.

The resulting automaton is depicted in Figure 4.12.

From a formal point of view, this automaton is described by the tuple ($Q$, $\varepsilon$, $\delta$, $q_0$), where

- $Q = \{OU, OL, OS, Fl, FF\}$ is the set of possible states

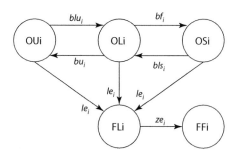

**FIGURE 4.12**
Automaton for the SHIELD CPS.

- $\varepsilon = \{bf, bu, bls, blu, le\}$ is the set of macroevents that drive a status change
- $\delta:Q x \varepsilon \rightarrow Q$ is the transition function that allows, given an initial state and a macroevent, us to univocally move to another state
- $q_0$ is the initial state

### Hybrid Automata Associated with a Generic CPS

By putting together the continuous dynamics of Equation 4.8 with the discrete states of Figure 4.12, it is possible to derive the complete hybrid automata associated with the generic CPS.

A hybrid automaton is a structure $H = (Q, \varepsilon, act, x_{inv}, u, E)$ where

- $Q$ is the set of possible states.
- $\varepsilon$ is the set of macroevents that drive a status change.
- $act:Q \rightarrow Diff\_eq$ specifies the dynamics at each location $act = f(x, u)$.
- $x_{inv}:Q \rightarrow x_{inv}$ specifies the set of all admissible state vectors at each location.
- $u_{inv}:Q \rightarrow U_{inv}$ specifies the set of all admissible input vectors at each location.
- $E:Q \times \varepsilon \times Guard \times Jump \times Q$ is the set of edges. An edge $e = (q, \varepsilon, g, j, q')$ is an edge from location $q$ to $q$ labeled with symbol $\varepsilon$, with guard $g$ and jump relation $j$.

For each state, some of these values change, because the set of admissible state and input vectors, or the dynamic equation and the guard and jump relations, depends on the state itself.

The different states will be presented one by one, by considering that the state variables are the ones describing the status of the system—buffer queue length, CPU, and energy $(q, c, x)$—while the input variables are the ones on which it is possible to act: the rate in input and output to and from the system $(r_{in}, r_{out})$.

### State Operational Linear (OL)

The OL state characterizes the system evolving in linear conditions, when neither underutilization nor overutilization (saturation) of resources is experienced. The state and the allowed transitions are depicted in Figure 4.13.

The dynamic equation $act(OL)$ that describes the evolution of the system is obtained by differentiating Equation 4.8. In particular, Equation 4.9 holds:

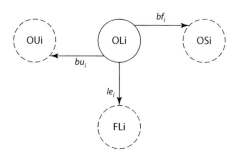

**FIGURE 4.13**
Location OL and allowed transitions.

$$act\left(OL\right) = \begin{cases} \dot{q}_i = r_{in,i} - r_{out,i} \\ \dot{c}_i = 0 \\ \dot{x}_i = ac_i + br_{in,i} + dr_{out,i} \end{cases} \tag{4.9}$$

with $a = -\mu\delta_1\gamma$, $b = -\mu(\alpha\delta_1 + \delta_2)$, and $d = -\mu(\beta\delta_1 + \delta_2)$.

The set of admissible values for the state variables is the following:

$$X_{inv}\left(OL\right) = \begin{cases} 0 \le q_i \le q_{MAX,i} \\ \gamma c_i \le c_i \le c_{MAX,i} \\ x_i \ge x_{Th} \end{cases} \tag{4.10}$$

where the condition on $c_i$ is almost useless, since it is constant during the whole state and will never trigger any guard. However, this constraint is kept in order to allow us to specify a range of variations for CPU utilization.

*Remark 4*: Since the CPU level is given by the weighted sum of I/O rates plus an offset (Equation 4.3), the relation of the maximum and minimum CPU level is translated into a condition of the maximum and minimum sum of the I/O rate (i.e., admissible values of the input variable). This information is indeed included in the set $U_{inv}$.

The set of admissible values for the input variables is then as follows:

$$U_{inv}\left(OL\right) = \begin{cases} 0 \le r_{in,i} \le r_{in,MAX,i} \\ 0 \le r_{out,i} \le r_{out,MAX,i} \\ 0 \le \alpha r_{in,i} + \beta r_{out,i} \le c_{MAX,i} - \gamma c_i \end{cases} \tag{4.11}$$

Please note that no mutual restrictions are set for the I/O rates (e.g., $r_{in} > r_{out}$ or vice versa) in order to allow a state change in both directions (saturation and underutilization).

There are three edges to move to three adjacent states. The guard conditions, jump relations, and associated macroevents are listed below:

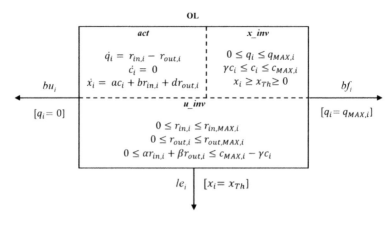

**FIGURE 4.14**
Hybrid automata for an embedded system: OL location.

- Guard condition $[x_i = x_{Th}]$ indicates that the battery has reached a warning level and triggers a transition from $le_i$ to the $FL_i$ state, where adequate constraints will take care of the energetic needs (e.g., by limiting the amount of data to be processed).
- Guard condition $[q_i = 0]$ indicates that the buffer is empty (the CPU will never be *zero*) and triggers a transition from $bu_i$ to the $OU_i$ state.
- Guard condition $[q_i = q_{MAX,i}]$ indicates that the buffer is full or the CPU is saturated and triggers a transition from $bf_i$ to the $OS_i$ state.

The resulting state is depicted in Figure 4.14.

*State Operational Underutilized (OU)*

The OU state characterizes the system evolving in a resource-wasting condition, when a lack of data to be processed is experienced. The state and the allowed transitions are depicted in Figure 4.15.

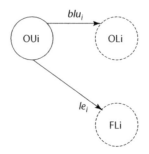

**FIGURE 4.15**
Location OU and allowed transitions.

The dynamic equation $act(OU)$ that describes the evolution of the system is obtained in the same way as in the linear case. In particular, Equation 4.12 holds:

$$act(OU) = \begin{cases} \dot{q}_i = r_{in,i} - r_{out,i} \\ \dot{c}_i = 0 \\ \dot{x}_i = ac_i + br_{in,i} + dr_{out,i} \end{cases} \tag{4.12}$$

with $a = -\mu\delta_1\gamma$, $b = -\mu(\alpha\delta_1 + \delta_2)$, and $d = -\mu(\beta\delta_1 + \delta_2)$.

The set of admissible values for the state variables is as follows:

$$X_{inv}(OU) = \begin{cases} 0 \le q_i \le q_{Th,i} \\ \gamma c_i \le c_i \le c_{MAX,i} \\ x_i \ge x_{Th} \end{cases} \tag{4.13}$$

*Remark 5*: Theoretically, the $X_{inv}(OU)$ for the queue or the CPU should be limited to the *zero* value. However, this would result in a Zeno dynamic, since this value is instantaneously reached and the transition immediately activated in both directions. Consequently, a hysteresis mechanism has to be introduced, and this can be done by adding a threshold in one of the two relations that connects two states.

The difference from the linear case is that the node tries to store as many packets as possible in order to have enough data to process by reducing the amount of data sent outside, and this will be reflected in the admissible input conditions.

The set of admissible values for the input variables is as follows:

$$U_{inv}(OU) = \begin{cases} 0 \le r_{in,i} \le r_{in,MAX,i} \\ 0 \le r_{out,i} \le r_{out,MAX,i} \\ 0 \le \alpha r_{in,i} + \beta r_{out,i} \le c_{MAX,i} - \gamma c_i \\ r_{out,i} \le r_{in,i} \end{cases} \tag{4.14}$$

by which it is imposed (last constraint) that the I/O balance allows the buffer to be filled with data.

There are two edges to move to two adjacent states. The guard conditions, jump relations, and associated macroevents are listed:

- Guard condition $[x_i = x_{Th}]$ indicates that the battery has reached a warning level and triggers a transition from $le_i$ to the $FL_i$ state, where adequate constraints will take care of the energetic needs (e.g., by limiting the amount of data to be processed).

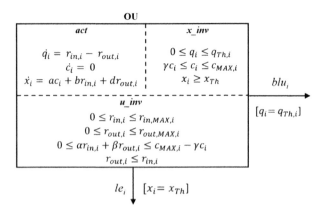

**FIGURE 4.16**
Hybrid automata for a CPS: OU location.

- Guard condition $[q_i = q_{Th,i}]$ indicates that the buffer has reached a minimum acceptable value (hysteresis) and triggers a transition from **blu**$_i$ back to the **OL** state.

The resulting state is depicted in Figure 4.16.

*State Operational Saturation (OS)*

The OS state characterizes the system evolving in a resource overutilization condition, when either a buffer saturation or a CPU saturation is experienced. The state and the allowed transitions are depicted in Figure 4.17.

The dynamic equation *act(OS)* that describes the evolution of the system is obtained in the same way as in the linear case. In particular, Equation 4.15 holds:

$$act(OS) = \begin{cases} \dot{q}_i = r_{in,i} - r_{out,i} \\ \dot{c}_i = 0 \\ \dot{x}_i = ac_i + br_{in,i} + dr_{out,i} \end{cases} \tag{4.15}$$

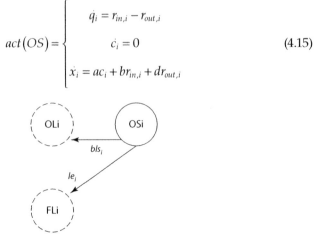

**FIGURE 4.17**
Location OS and allowed transitions.

with $a = -\mu\delta_1\gamma$, $b = -\mu(\alpha\delta_1 + \delta_2)$, and $d = -\mu(\beta\delta_1 + \delta_2)$.

The difference from the linear case is that the node has to free its resources and, consequently, send outside more data than those coming in.

The set of admissible values for the state variables is the following:

$$X_{inv}(OS) = \begin{cases} q_{Bound,i} \leq q_i \leq q_{MAX,i} \\ \gamma c_i \leq c_i \leq c_{MAX,i} \\ x_i \geq x_{Th} \end{cases} \tag{4.16}$$

In particular, the admissible values for queue length and CPU level are changed, because until a minimum lower bound threshold is reached, the system does not move from this state (i.e., is not allowed to receive further).

> *Remark 6*: Even in this case, theoretically, the $X_{inv}(OU)$ for the queue or the CPU should be limited to the $MAX$ value. However, this would result in a Zeno dynamic, since this value is instantaneously reached and the transition immediately activated in both directions. Consequently, a hysteresis mechanism has to be introduced, and this can be done by adding a threshold in one of the two relations that connects two states.

The set of admissible values for the input variables is the following:

$$U_{inv}(OS) = \begin{cases} 0 \leq r_{in,i} \leq r_{in,MAX,i} \\ 0 \leq r_{out,i} \leq r_{out,MAX,i} \\ 0 \leq \alpha r_{in,i} + \beta r_{out,i} \leq c_{MAX,i} - \gamma c_i \\ r_{in,i} \leq r_{out,i} \end{cases} \tag{4.17}$$

by which it is imposed (last constraint) that the I/O balance allows the buffer to be emptied.

There are two edges to move to two adjacent states. The guard conditions, jump relations, and associated macroevents are listed:

- Guard condition $[x_i = x_{Th}]$ indicates that the battery has reached a warning level and triggers a transition from $le_i$ to the $FL_i$ state, where adequate constraints will take care of the energetic needs (e.g., by limiting the amount of data to be processed).
- Guard condition $[q_i = q_{Bound,i}]$ indicates that the buffer has sufficiently reduced the queue length (hysteresis) and triggers a transition from $bls_i$ back to the $OL$ state.

The resulting state is depicted in Figure 4.18.

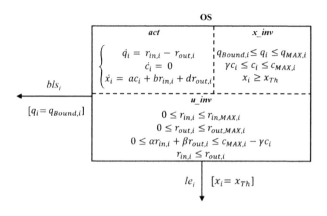

**FIGURE 4.18**
Hybrid automata for an embedded system: OS location.

*State Near Failure for Low Resources (FL)*

The FS state characterizes the system evolving in a low-resource status, with low available energy. The state and the allowed transitions are depicted in Figure 4.19.

The dynamic equation $act(FL)$ that describes the evolution of the system is obtained in the same way as in the linear case. In particular, Equation 4.18 holds:

$$act(FL) = \begin{cases} \dot{q}_i = r_{in,i} - r_{out,i} \\ \dot{c}_i = 0 \\ \dot{x}_i = ac_i + br_{in,i} + dr_{out,i} \end{cases} \tag{4.18}$$

with $a = -\mu \delta_1 \gamma$, $b = -\mu(\alpha \delta_1 + \delta_2)$, and $d = -\mu(\beta \delta_1 + \delta_2)$.

The set of admissible values for the state variables is indeed the following:

$$X_{inv}(FL) = \begin{cases} 0 \leq q_i \leq q_{MAX,i} \\ \gamma c_i \leq c_i \leq \varphi_1 c_{MAX,i} \\ 0 \leq x_i \leq x_{Th} \end{cases} \tag{4.19}$$

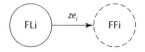

**FIGURE 4.19**
Location FL and allowed transitions.

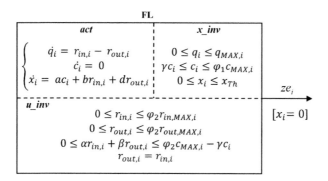

**FIGURE 4.20**
Hybrid automata for an embedded system: FL location.

In particular, the admissible values for the CPU level are scaled by a factor of $0 \geq \varphi_1 \leq 1$, which forces the evolution to a low profile. The intrinsic objective is to maximize the availability of remaining resources by consuming less power, and this is achieved by limiting the I/O rate, that is, the input variables.

The difference from the linear case, in fact, is that the node has to work with limited resources, so the dynamic remains the same, with different conditions: it is imposed to have an equilibrium between the input and output rates, so that the buffer level remains constant and neither an underutilization nor an overutilization is experienced.

At the same time, the set of admissible values for the input variables should be forced to lower values as well, by using a scaling factor of $0 \geq \varphi_2 \leq 1$.

$$U_{inv}(FL) = \begin{cases} 0 \leq r_{in,i} \leq \varphi_2 r_{in,MAX,i} \\ 0 \leq r_{out,i} \leq \varphi_2 r_{out,MAX,i} \\ 0 \leq \alpha r_{in,i} + \beta r_{out,i} \leq \varphi_1 c_{MAX,i} - \gamma c_i \\ r_{out,i} = r_{in,i} \end{cases} \qquad (4.20)$$

with the further condition that $\varphi_1 c_{MAX,i} - \gamma c_i > 0$.

There is only one edge to move to the adjacent states. The guard conditions, jump relations, and associated macroevents are listed:

- Guard condition $[x_i = 0]$ indicates that the battery is over and no other dynamic is possible. The jump conditions impose $x_i, q_i, c_i, r_{in,i}, r_{out,i} = 0$, and the triggered transition is *ze*.

The resulting state is depicted in Figure 4.20.

*State Fully Failure (FF)*

All the dynamics in this state are equal to zero, since the system is switched off and no evolution occurs.

*Complete Hybrid Automata Model for a CPS*

By putting together all the states, the preliminary model proposed by this work is derived: in Figure 4.21, the representation of the generic *i*-th CPS is proposed.

All the parameters are model dependent and contribute to differentiate the nodes.

Even if expressive enough, this model needs a couple of refinements in order to capture all the information necessary to perform context-aware composition in an SPD perspective.

## Model Refinement to Address Dependability

A first refinement could be done to this preliminary model in order to represent, in a formal way, the dependability associated with the CPS.

The dependability, in our case, is the possibility of performing a task in a continuative way: in the case of failure, the node stops working. Reasonably, the probability that a failure occurs should be related to the amount of tasks, or data, processed by the node during its operating life.

From a mathematical point of view, we need to introduce a variable that is able to keep a memory of the amount of data processed by the system during the whole evolution, and to adequately trigger a transition toward the fully failure state, to represent that the node is broken (the behavior of a broken node is the same as that of a node with no energy).

The guard condition that enables the transition will be driven by the *maximum lifetime* parameter of the specific embedded system and will originate the macroevent *f* failure that triggers a transition to the **FF** state.

The new macroscopic layer with the new transitions associated with the dependability is depicted in Figure 4.22.

The new set of macroevents is $\varepsilon_i = \{bf_i, bu_i, bls_i, blu_i, le_i, ze_i, f_i\}$.

First, Equation 4.21 is added to the microscopic layer:

$$d_i(t) = d_i(t_k) + \left[ r_{in,i}(k) + r_{out,i}(k) \right](t - t_k) \tag{4.21}$$

where the variable $d_i(t)$ represents the total amount of data (either input and output) processed by the *i*-th embedded system during the time interval $\Delta k$.

By differentiating it for inclusion in the $act(Q)$ list, Equation 4.22 holds:

$$\dot{d}_i = r_{in,i} + r_{out,i} \tag{4.22}$$

and $X_{inv}(Q)$ is enriched with the constraint (4.23):

$$0 \le d_i \le d_{MAX,i} \tag{4.23}$$

what is associated with the guard condition $[d_i = 0]$, which triggers a transition from $f_i$ to **FF**.

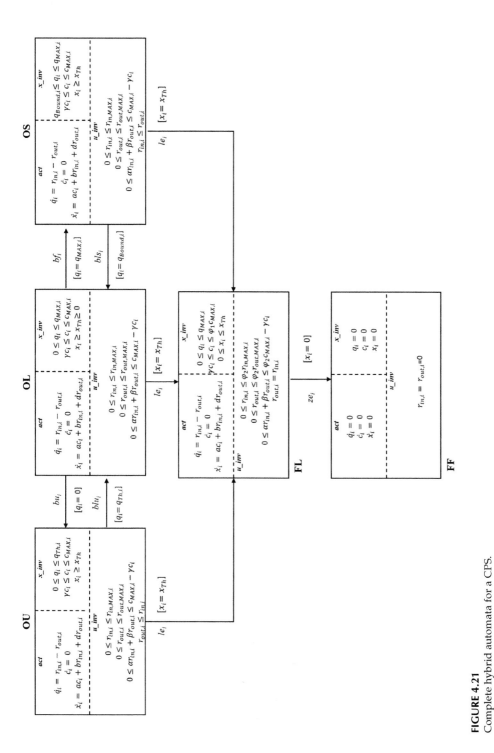

**FIGURE 4.21**
Complete hybrid automata for a CPS.

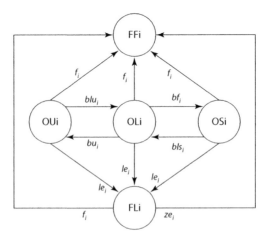

**FIGURE 4.22**
Hybrid automata with dependability transitions.

These modifications have to be propagated to all the states, with the obvious exception of the FF state, thus giving the system an enriched behavior.

In fact, during the evolution, in every state the variable $d_i$ is continuously increased by the amount of processed data, and when it reaches a reference value, the node suddenly breaks and enters into the fully failure state.

This variable will be useful for control purposes (as will be described in the next paragraphs), since it could allow us to discriminate the evolution of two embedded systems with the same parameters and dynamics, but with different dependability thresholds.

Before providing an overall picture of the modified system, a second refinement must be introduced.

### Model Refinement to Address Measurable Context Information

The second modification that can be done to this model is the introduction of a set of variables (to be chosen everytime, depending on the specific needs) to collect context information during system evolution. There are two ways to achieve that:

1. By introducing a *cumulative variable* that sums time after time the values of a reference variable
2. By introducing a *time counter* that increases, as time increases, during a particular condition

Both of these variables can be reinitialized or not, depending on their purposes.

*Variable to Measure the Processed Output Data before Congestion*
One may be interested in measuring the amount of data produced by the system until a buffer saturation occurs, with a consequent data loss. This

can be considered, for example, as a context measure of the reliability of the system.

So, given the output rate $r_{out,i}$, the cumulative variable $v_i$ is introduced in the microscopic layer, as formalized in Equation 4.24:

$$v_i(t) = v_i(t_k) + r_{out,i}(k)(t - t_k) \tag{4.24}$$

and differentiated for inclusion in the $act(Q)$ list (Equation 4.25):

$$\dot{v}_i = r_{out,i} \tag{4.25}$$

Then the $X_{inv}(Q)$ is enriched accordingly with the constraint

$$v_i \geq 0 \tag{4.26}$$

In this case, the counter should be reset once the saturation occurs, in order to measure the amount of data successfully transmitted without losses, so the jump condition $v_{j:} = 0$ is included in the transition $bf_{j}$.

The variable is not upper-bounded, so theoretically it could grow to infinite during the evolution of the system. However, the presence of the failure event will ensure that, definitely, the system will go into the zero state, so a finite value is reached.

*Variable to Measure the Time Duration of an Event*

Much more easy, one could be willing to measure directly the duration of an event. For example, in this model we could add a measure of the time spent in linear operating condition. This is achieved in two simple steps.

The first step is, as usual, the updating of the microscopic-layer equations with the introduction of Equation 4.27, which describes the evolution of the counter:

$$s_i(t) = s_i(t_k) + (t - t_k) \tag{4.27}$$

This equation, once differentiated, can be included in the $act(Q)$ list:

$$\dot{s}_i = 1 \tag{4.28}$$

Note that this equation is present only in the state involved in the counting (i.e., OL). In the others, Equation 4.29 holds:

$$\dot{s}_i = 0 \tag{4.29}$$

Then the $X_{inv}(Q)$ is enriched accordingly with one of the constraints:

$$\begin{cases} s_i = 0 \text{ to keep the counter switched off} \\ s_i \geq 0 \text{ to keep the counter alive} \end{cases} \tag{4.30}$$

The second step is the introduction of the guard and jump conditions. The guards are inherited from the existing system, while the associated (new) jump conditions is $s_i := 0$, allowing us to reset the counter.

*Enriched Hybrid Automata for CPSs*

The resulting enriched model for the CPS is then described by Equation 4.31, which considers a system for which is possible to measure the buffer queue length, the CPU utilization, the energy consumption, the amount of processed data, the dependability level (data processed before a fault occurs), and the amount of time spent in operating conditions.

$$
\begin{cases}
q_i(t) = q_i(t_k) + \left[ r_{in,i}(k) - r_{out,i}(k) \right](t - t_k) \\
c_i(t) = \gamma c_i(k) + \alpha r_{in,i}(k) + \beta r_{out,i}(k) \\
x_i(t) = x_i(t_k) - \mu \left[ \delta_1 \gamma c_i(k) + (\alpha \delta_1 + \delta_2) r_{in,i}(k) + (\beta \delta_1 + \delta_2) r_{out,i}(k) \right](t - t_k) \\
d_i(t) = d_i(t_k) + \left[ r_{in,i}(k) + r_{out,i}(k) \right](t - t_k) \\
v_i(t) = v_i(t_k) + r_{out,i}(k)(t - t_k) \\
s_i(t) = s_i(t_k) + (t - t_k) \\
0 \le q_i(t) \le q_{MAX,i} \\
0 \le c_i(t) \le c_{MAX,i} \\
x_i(t) \ge 0 \\
0 \le d_i(t) \le d_{MAX,i} \\
s_i(t) \ge 0 \\
v_i(t) \ge 0 \\
0 \le r_{in,i} \le r_{in,MAX,i} \\
0 \le r_{out,i} \le r_{out,MAX,i} \\
\alpha, \beta \ge 1 \\
\delta_2 > \delta_1 > 0 \\
\mu = \{0,1\} \\
t \in [t_k, t_{k+1})
\end{cases}
\tag{4.31}
$$

The input variables are the input and output rates, which allow us to decide how much data must be forwarded to the node, or must be produced by the node. Several parameters allow us to tailor the model by defining the contribution of the data processing on the CPU load, the dependability value, the maximum charge level of the battery, and the buffer size.

The resulting automaton that considers context information is built by the following states.

*Enriched State Operational Linear (OL)*

The difference from the simplified model is that the volume of processed data drives the fault, the amount of processed data between two saturation events is monitored, and the time spent in nominal conditions is monitored as well.

$$act(OL) = \begin{cases} \dot{q}_i = r_{in,i} - r_{out,i} \\ \dot{c}_i = 0 \\ \dot{x}_i = ac_i + br_{in,i} + dr_{out,i} \\ \dot{d}_i = d_{in,i} + d_{out,i} \\ \dot{v}_i = v_{out,i} \\ \dot{s}_i = 1 \end{cases} \tag{4.32}$$

with $a = -\mu\delta_1\gamma$, $b = -\mu(\alpha\delta_1 + \delta_2)$, and $d = -\mu(\beta\delta_1 + \delta_2)$.

$$X_{inv}(OL) = \begin{cases} 0 \le q_i \le q_{MAX,i} \\ \gamma c_i \le c_i \le c_{MAX,i} \\ d_i \le d_{MAX,i} \\ v_i \ge 0 \\ s_i \ge 0 \\ x_i \ge x_{Th} \end{cases} \tag{4.33}$$

$$U_{inv}(OL) = \begin{cases} 0 \le r_{in,i} \le r_{in,MAX,i} \\ 0 \le r_{out,i} \le r_{out,MAX,i} \\ 0 \le \alpha r_{in,i} + \beta r_{out,i} \le c_{MAX,i} - \gamma c_i \end{cases} \tag{4.34}$$

The guard and jump conditions are as follows:

- Guard condition $[d_i = d_{MAX,i}]$ triggers a transition from $f_i$ to $FF_i$.
- Guard condition $[x_i = x_{Th}]$ triggers a transition from $le_i$ to $FL_i$.
- Guard condition $[q_i = 0]$ triggers a transition from $bu_i$ to $OU_i$.

- Guard condition $[q_i = q_{MAX,i}]$ triggers a transition from $bf_i$ to $OS_i$ with jump condition $v_i = 0$.

*Enriched State Operational Underutilized (OU)*
The same difference as for the OL state applies.

$$act(OU) = \begin{cases} \dot{q}_i = r_{in,i} - r_{out,i} \\ \dot{c}_i = 0 \\ \dot{x}_i = ac_i + br_{in,i} + dr_{out,i} \\ \dot{d}_i = d_{in,i} + d_{out,i} \\ \dot{v}_i = v_{out,i} \\ \dot{s}_i = 0 \end{cases} \tag{4.35}$$

with $a = -\mu\delta_1\gamma$, $b = -\mu(\alpha\delta_1 + \delta_2)$, and $d = -\mu(\beta\delta_1 + \delta_2)$.

$$X_{inv}(OU) = \begin{cases} 0 \le q_i \le q_{Th,i} \\ \gamma c_i \le c_i \le c_{MAX,i} \\ d_i \le d_{MAX,i} \\ v_i \ge 0 \\ s_i \ge 0 \\ x_i \ge x_{Th} \end{cases} \tag{4.36}$$

$$U_{inv}(OU) = \begin{cases} 0 \le r_{in,i} \le r_{in,MAX,i} \\ 0 \le r_{out,i} \le r_{out,MAX,i} \\ 0 \le \alpha r_{in,i} + \beta r_{out,i} \le c_{MAX,i} - \gamma c_i \\ r_{out,i} \le r_{in,i} \end{cases} \tag{4.37}$$

The guard and jump conditions are as follows:

- Guard condition $[d_i = d_{MAX,i}]$ triggers a transition from $f_i$ to $FF_i$.
- Guard condition $[x_i = x_{Th}]$ triggers a transition from $le_i$ to $FL_i$.
- Guard condition $[q_i = q_{Th,i}]$ triggers a transition from $bu_i$ to $OL_i$.

*Enriched State Operational Saturation (OS)*
The same difference as for the OL state applies.

$$act(OS) = \begin{cases} \dot{q}_i = r_{in,i} - r_{out,i} \\ \dot{c}_i = 0 \\ \dot{x}_i = ac_i + br_{in,i} + dr_{out,i} \\ \dot{d}_i = d_{in,i} + d_{out,i} \\ \dot{v}_i = v_{out,i} \\ \dot{s}_i = 0 \end{cases}$$

(4.38)

with $a = -\mu\delta_1\gamma$, $b = -\mu(\alpha\delta_1 + \delta_2)$, and $d = -\mu(\beta\delta_1 + \delta_2)$.

$$X_{inv}(OS) = \begin{cases} q_{Bound,i} \leq q_i \leq q_{MAX,i} \\ \gamma c_i \leq c_i \leq c_{MAX,i} \\ d_i \leq d_{MAX,i} \\ v_i \geq 0 \\ s_i \geq 0 \\ x_i \geq x_{Th} \end{cases}$$

(4.39)

$$U_{inv}(OS) = \begin{cases} 0 \leq r_{in,i} \leq r_{in,MAX,i} \\ 0 \leq r_{out,i} \leq r_{out,MAX,i} \\ 0 \leq \alpha r_{in,i} + \beta r_{out,i} \leq c_{MAX,i} - \gamma c_i \\ r_{in,i} \leq r_{out,i} \end{cases}$$

(4.40)

The guard and jump conditions are as follows:

- Guard condition $[d_i = d_{MAX,i}]$ triggers a transition from $f_i$ to $FF_i$.
- Guard condition $[x_i = x_{Th}]$ triggers a transition from $le_i$ to $FL_i$.
- Guard condition $[q_i = q_{Bound,i}]$ triggers a transition from $bls_i$ to $OL_i$.

*Enriched State Near Failure for Low Resources (FL)*
The same difference as for the OL state applies:

$$
act(FL) = \begin{cases}
\dot{q}_i = r_{in,i} - r_{out,i} \\
\dot{c}_i = 0 \\
\dot{x}_i = ac_i + br_{in,i} + dr_{out,i} \\
\dot{d}_i = d_{in,i} + d_{out,i} \\
\dot{v}_i = v_{out,i} \\
\dot{s}_i = 0
\end{cases}
\tag{4.41}
$$

with $a = -\mu\delta_1\gamma$, $b = -\mu(\alpha\delta_1 + \delta_2)$, and $d = -\mu(\beta\delta_1 + \delta_2)$.

$$
X_{inv}(FL) = \begin{cases}
0 \le q_i \le q_{MAX,i} \\
\gamma c_i \le c_i \le \varphi_1 c_{MAX,i} \\
d_i \le d_{MAX,i} \\
v_i \ge 0 \\
s_i \ge 0 \\
0 \le x_i \le x_{Th}
\end{cases}
\tag{4.42}
$$

$$
U_{inv}(FL) = \begin{cases}
0 \le r_{in,i} \le \varphi_2 r_{in,MAX,i} \\
0 \le r_{out,i} \le \varphi_2 r_{out,MAX,i} \\
0 \le \alpha r_{in,i} + \beta r_{out,i} \le \varphi_1 c_{MAX,i} - \gamma c_i \\
r_{out,i} = r_{in,i}
\end{cases}
\tag{4.43}
$$

with the further condition that $\varphi_1 c_{Max,i} - \gamma c_i > 0$.
   The guard and jump conditions are as follows:

- Guard condition $[d_i = d_{MAX,i}]$ or $[x_i = 0]$ triggers a transition from $f_i$ or $ze_i$ to **FF**$_i$ with jump conditions $x_i, q_i, c_i, r_{in,i}$ and $r_{out,i} = 0$.

*Enriched State Fully Failure (FF) and Final Model*
No modification to the FF state is done. The final model is depicted in Figure 4.23.

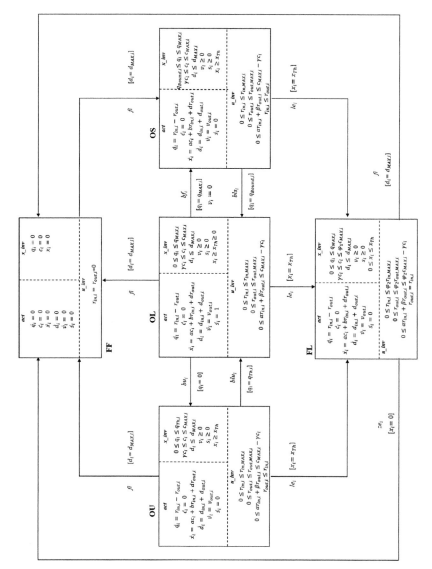

**FIGURE 4.23**
Enriched hybrid automata for CPSs.

**Hybrid Automata Control and Composition: Preliminary Considerations**

The purpose of this chapter was to propose a mathematical model of a CPS oriented to context-aware optimization and security-, dependability-, or privacy-related aspects.

In this respect, the model presented in the previous section represents the basic building block that can be used to model any generic system composed of several nodes. This can be done in two ways:

1. By means of parallel composition
2. By means of concurrent composition

The difference between these compositions is mainly related to the interdependencies and scalability of the resulting problem.

### *Parallel Composition versus Concurrent Composition*

A system described by a hybrid automaton evolves among several states by means of guards and jump conditions. The problem of composing two automata is that, in case one variable is shared between the automata, an unexpected transition in *System 1* could be triggered by *System 2*, or alternatively, jump conditions in one system could unexpectedly modify the values of the variable in the other system (e.g., a reset).

Therefore, the main difference between the two approaches is in the variables involved: if the intersection between them is *empty*, then the systems can evolve in *parallel*, each one with its own dynamics, and their I/O can go to other systems independently. Moreover, this type of composition does not involve any state modification, because the resulting number of states is equal to the sum of the states $Q$ of the original systems.

On the contrary, in the case of concurrent composition, an explicit and significant interaction between the two systems' variables (and their evolution) is expected. In this respect, jump conditions and guards must be checked carefully, in order to ensure their coherence and consistency. Furthermore, the state space resulting from the composition is given by $Q_1 \times Q_2$, and this results in an explosion of state number and, consequently, increased complexity in the resulting model.

### *How to Control the Resulting System from a Context-Aware Perspective*

Assuming that the model presented by this work is used to create a complex representation of a networked embedded system for a context-aware control, the best way to approach this problem could be the use of optimization strategies. This way to proceed is well consolidated in the manufacturing domain, from which the rationale of this model has been derived [11].

In particular, given the hybrid automata model, with its input variables $r_{in}$ and $r_{out}$, one could decide the average rate of data managed by the node, which maximizes or minimizes specific conditions (i.e., the throughput, the saturation occurrences, the unavailability time, and so on).

So, the generic linear programming problem could be defined for the system (Equation 4.44):

$$\underset{u}{\text{Maximize}}\, J\left(x,u\right)$$

$$\text{s.t.} \tag{4.44}$$

$$u \in U_{inv}\left(Q\right)$$

where either the objective function or the admissible set depends on the current location $Q$. For this reason, no conditions are imposed on the evolution of $x$; otherwise, the solution would prevent any state jump.

Examples of an objective function could be

- $J(x, u) = V$ to maximize the volume of processed data
- $J(x, u) = -S$ to minimize the time spent in a saturation condition
- $J(x, u) = S$ to maximize the time spent in an operational linear condition

This shows the importance of the state to the overall problem, because the same function could be maximized or minimized depending on the specific context condition.

The objective functions are then used to find a solution for any context-aware control problem.

Recalling the closed-loop formulation of our problem (Figure 4.24), the objectives of the context-aware controller are to (1) select among several candidate configurations and (2) select control command for direct enforcement in the system.

This is done in a three-step procedure:

*Step 1*: The first point is achieved by creating one model of the system (using the hybrid automata) for each candidate configuration and computing the same objective function over all of them. Then, by comparing results on the basis of the desired context (e.g., bandwidth, throughput, or duration), the choice of the best solution is trivial (Figure 4.25).

*Step 2*: At this point, the same solution used to evaluate the candidate configuration is translated into control commands (in this case, the input commands are the level of input and output rates to be managed by the node), and these command are sent back to the system for enforcement.

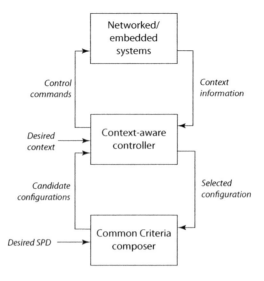

**FIGURE 4.24**
Context-aware controller.

> *Step 3*: In case a new node is dynamically discovered, or the user's
> needs (objective function) change, the process is reiterated, by build-
> ing another set of feasible configurations, or modifying the adequate
> parameters in the current one and evaluating the objective functions
> again.

A simple example is provided in the following section.

### Example of Context-Aware Composition

Just to give an example of the potential of such an approach, a simple scenario
is depicted, where there are two nodes available, performing cyphered data
exchange: both of them satisfy the SPD requirements, so the node to be used
will be chosen on the basis of context information. In particular, the objective is to maximize the throughput or lifetime of the system (i.e., the energy
duration).

Both nodes have the same buffer parameters, but the difference is that
Node 2 has reduced $r_{in}$, $r_{out}$ capabilities and an initial energy value of 90%,
to simulate a less performing node. For the sake of simplicity, only discrete
values for $r_{in}$ and $r_{out}$ are allowed, with a 10-step increment for $r_{in}$ and a 5-step
increment of $r_{out}$. In Table 4.2, the main parameters are reported.

The model has been built with the HYSDEL® Hybrid System ToolBox for
MATLAB®, developed by A. Bemporad. On both models, the same objective
function aimed at maximizing the throughput (i.e., the sum of the I/O rate)

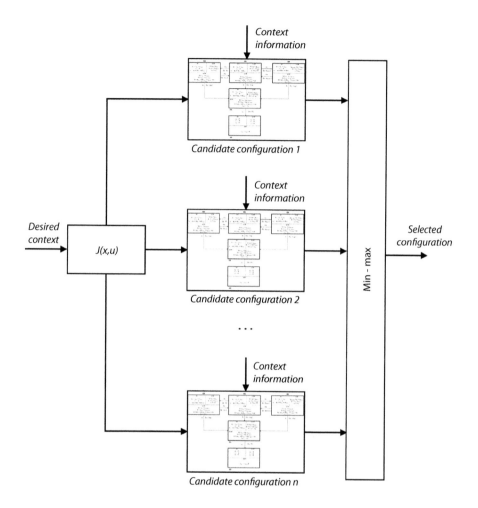

**FIGURE 4.25**
Context-aware controller: Choice of the best candidate solution.

**TABLE 4.2**

Node Parameters

| Parameter | Node 1 | Node 2 |
|---|---|---|
| $q_{MAX}$ (pkt) | 65 | 65 |
| $rin_{MAX}$ (pkt/s) | 40 | 30 |
| $rout_{MAX}$ (pkt/2) | 40 | 20 |
| $energy_{INIT}$ (%) | 100 | 90 |
| $rin_{INIT}$ (pkt/s) | 40 | 20 |
| $rout_{INIT}$ (pkt/2) | 30 | 20 |

has been applied, resulting in a control policy that defines the values for $r_{in}$ and $r_{out}$ for both nodes.

The evolution of Node 1 is depicted in Figure 4.26: at the beginning of the evolution, the node accumulates packets due to the imposed initial conditions; then a saturation occurs, so the output rate is increased and the input rate is reduced in order to empty the buffer while ensuring a high overall rate. These values are almost frozen until the node runs in low energy: at this point, the I/O rate is set to the fixed reference value foreseen for the "energy-saving" condition, and the node remains in this state until the complete expiration of resources.

In Figure 4.27, the evolution of Node 2 is reported: at the beginning of the evolution, the node accumulates packets for the imposed initial conditions, and consequently, saturation occurs. Then the value for the I/O rate is computed to empty the buffer: since $r_{out}$ has a small upper bound, it remains constant, while the input rate is reduced (in order to keep both values at a high level for throughput maximization). Then the node

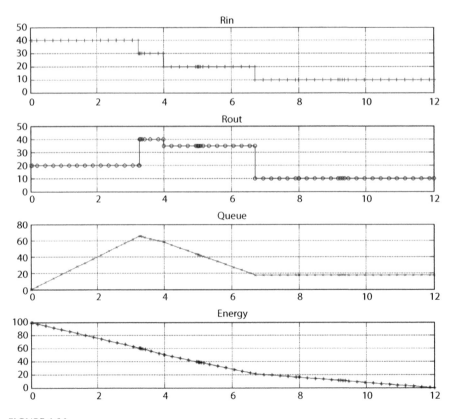

**FIGURE 4.26**
Evolution of Node 1.

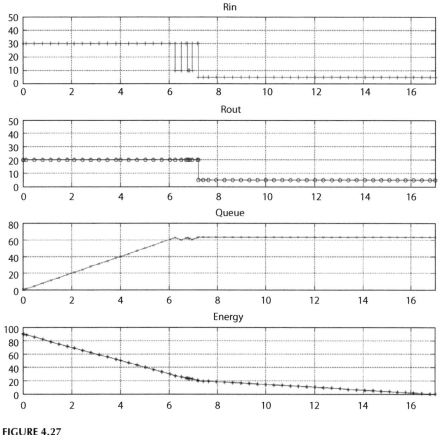

**FIGURE 4.27**
Evolution of Node 2.

remains in near-saturation conditions until the energy-saving condition is verified: as for Node 1, the I/O rates are set to the fixed reference values for this condition and the evolution continues until complete expiration of the resource.

So given a desired context to *maximize throughput*, Node 1 should be selected, due to the higher I/O rate. However, the high rate implies a higher energy consumption, and the node operational life is reduced.

On the contrary, given a desired context that aims to *preserve the lifetime* of the system, Node 2 should be selected, since the reduced rates imply less power consumption (while preserving the maximum allowable throughput anyway).

This is a very simple example of a reduction scenario, but it is clear enough to give an idea of the potential of this approach: by carefully tailoring the objective functions, all the desired behaviors can be achieved.

## Enabling Technologies for Composability

### Security Agent

The security agent is the entity, within the SHIELD architecture, that could be considered the "embedded intelligence." Its role is somehow to monitor the actual status of the devices, in a few words, what is happening, and to decide counteractions in case something goes not exactly as expected. More precisely, the main objective of the security agent is to try to maintain a certain level of SPD and/or to enforce in the overall system a set of actions to permit the system to reach a desired level. Its architecture is depicted in Figure 4.28.

The *monitoring engine* is in charge of interfacing the overlay layer with the middleware layer, retrieving sensed metadata from heterogeneous SHIELD devices belonging to the same subsystem, aggregating and filtering the provided metadata, and providing the subsystem situation status to the context engine.

The *context engine* is in charge of keeping the situation status updated, as well as storing and maintaining any additional information meaningful to keeping track of the situation context of the controlled SHIELD subsystem. The situation context contains both status information and configuration information (e.g., rules, policies, and constraints) that are used by the decision maker engine.

The *decision maker engine* uses the valuable, rich input provided by the context engine to apply a set of adaptive (closed-loop or rule-based) and technology-independent algorithms. The latter, by using (as input) the above-mentioned situation context and by adopting appropriate advanced methodologies able to profitably exploit such input, produce (as output) *decisions* aimed at guaranteeing, whenever possible, target SPD levels over the controlled SHIELD subsystem.

The decisions mentioned above are translated by the *enforcement engine* into a set of proper *enforcement rules* actuated by the SHIELD middleware layer all over the SHIELD subsystem controlled by the considered security agent.

The so-called *"security agent intelligence"* coordinates all the actions of the security agent itself. The roles and the functioning of the remaining components of the security agent remain unchanged.

A further focus is needed on the decision maker engine. It is composed of various modules, as indicated in Figure 4.29.

As it is possible to see in the figure, the decision maker engine is composed of five different modules.

The *decision core* is actually the module deciding what to do, based on several pieces of information:

- The *metrics* provided by the various SHIELD entities and devices
- The *system models* used to calculate the value of the SPD level starting from the provided metrics

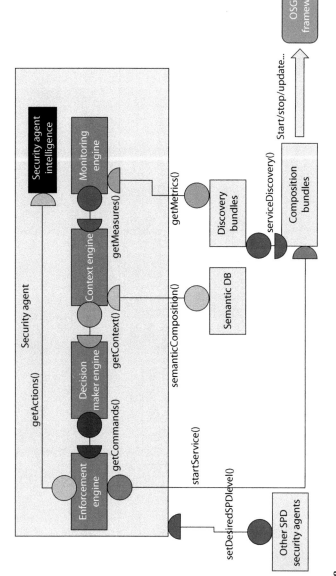

**FIGURE 4.28**

SHIELD security agent architecture. DB, database; OSGi, Open Source Gateway initiative.

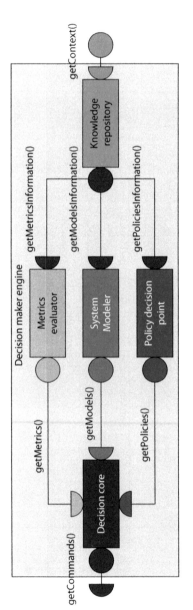

**FIGURE 4.29**
Decision maker engine.

- The *policies*, which are some specific rules the decision core should take care of

The decision made by the decision core is used as input by the enforcement engine.

The *metrics evaluator* retrieves *metric information* (measures, raw data, etc.) from the knowledge repository and, based on that, calculates the associated metrics.

The *system modeler* retrieves *model information* (rules used to compose the metrics) from the knowledge repository and, based on that, calculates the model to be associated with the metrics of the current system.

The *policy decision point* retrieves *policy information* (basic policies) from the knowledge repository and, based on that, calculates the overall policies to be used within the current system.

The *knowledge repository* contains all the information for the above-mentioned modules, retrieving *context information* from the context engine.

The main interfaces of the previously defined components and the relative main implementation classes are as follows:

- *ISecurityAgent*: The interface defining the main methods of a security agent, in particular *setDesiderMinimumSPDLevel*.
- *IEnforcementEngine*: The interface defining the main methods of the enforcement engine, in particular *getActions*.
- *IDecisionMakerEngine*: The interface defining the main methods of the decision maker engine, in particular *getCommands*.
- *IMetricsEvaluator*: The interface defining the main methods of the metrics evaluator, in particular *getMetrics*.
- *IPolicyDecisionPoint*: The interface defining the main methods of the policy decision point, in particular *getPolicies*.
- *ISystemModeler*: The interface defining the main methods of the system modeler, in particular *getModels*.
- *IKnowledgeRepository*: The interface defining the main methods of the policy decision point.
- *IContextEngine*: The interface defining the main methods of the policy decision point, in particular *getContext*.
- *IMonitoringEngine*: The interface defining the main methods of the policy decision point, in particular *getMeasures*.

### Implementation Remarks

- *The usage of drools as a business intelligence engine*: Drools is the freeware counterpart of the well-known business intelligence engine, providing more or less the same flexibility and power. The usage of an approach based on business intelligence permitted us to "fix" the code of the security agent and to encode the rules it has to use to enforce the proper actions (rules that are necessarily scenario dependent) in dedicated "configuration" files. This approach, even if has implied a slightly higher development effort, in the long run has saved a lot of resources. The same code has in fact been used for all the SHIELD demonstration scenarios; only the business logic had to be changed, described with proper drools files.

- *Service location protocol*: It has been decided to use the SLP proto-col for enforcing actions, rather than for registering or discovering available SHIELD devices.

### Drools Example

Here follows an example file for the drools configuration files. At the time of writing, the definitive drools files that are going to be used within the actual SHIELD scenarios are still being finalized (and for this reason, it was not possible to depict them here).

```
package droolstest
import java.util.LinkedList;
import java.util.Hashtable;
import java.util.List;
import java.util.Vector;
import org.labreti.securityagent.commontypes.ADevice;
import org.labreti.securityagent.commontypes.DeviceModel;
import org.labreti.securityagent.commontypes.IDeviceModel;
import org.labreti.securityagent.commontypes.Metrics;
import org.labreti.securityagent.commontypes.ScenarioModel;
import org.labreti.securityagent.commontypes.ICommand;
import org.labreti.securityagent.commontypes.CCommand;
import org.labreti.scenariocontroller.ISPDLogger;
import org.labreti.scenariocontroller.ILogger;
global LinkedList metricsList;
global LinkedList modelsList;
global LinkedList commandsList;
global Integer minimumSPDLevel;
global ISPDLogger spdLogger;
global ILogger logger;
rule "UAV Scenario"

    dialect "java"

    when
    eval(true);
then
    int s = modelsList.size();
    ScenarioModel sm = new ScenarioModel(modelsList);
    double spdMax=0;
```

```
    int treshold = minimumSPDLevel.intValue();
    int pos = -1;
    int onMission = -1;
 int statusOnMission = -1; //-1 = invalid; 3=approaching; 4 =
patrolling; 5=returning; 6= landed
    double curSPDLevel = sm.getSPDLevel();
    System.out.println("DRLOOLS>: Current scenario SPD Level:
"+curSPDLevel);
    logger.evolutionln("Security Agent: Current scenario SPD
Level: "+curSPDLevel);
    spdLogger.setNextSPDLevel(curSPDLevel);
    if (curSPDLevel>=treshold){ //Case 1
        System.out.println("DROOLS:> Case 1 (DO NOTHING)!");
        logger.evolutionln("Security Agent: Case 1 (DO
NOTHING)!");
    }else { //Case 2:
        for (int i = 0 ; i<s;i++){
            double spd;
            IDeviceModel d = (IDeviceModel) modelsList.
get(i);
            spd=d.getSPDLevel();
            ADevice dev = d.getDevice();
            System.out.println("DROOLS:>Current Device:
"+dev.name);
            String stat = dev.status.toLowerCase();
            System.out.println("DROOLS:>Current Device's
status: "+stat);
            System.out.println("DROOLS:>Current Device's SPD
Level: "+spd);
            if (stat.equals("UNAVAILABLE")){
               System.out.println("DROOLS:> UNAVAILABLE
UAV!");
                }else if (stat.equals("ready")){
                    if (spd>spdMax) {
                      spdMax=spd;
                      pos = i;
                      }
                  }else {
                      onMission = i;
                      if (stat.equals("approaching"))
statusOnMission = 3;
                      if (stat.equals("patrolling"))
statusOnMission = 4;
                      if (stat.equals("returning"))
statusOnMission = 5;
                      if (stat.equals("landed"))
statusOnMission = 6;
                  }
              }
            if (onMission==-1){ //none on mission
```

```
        if (pos==-1) {//CASE 2.c
        System.out.println("DROOLS:> Case 2.c -> RAISE
ALARM!!!");
logger.evolutionln("Security Agent: Case 2.c -> RAISE
ALARM!!!");
        }
        else { //CASE 2.a
        System.out.println("DROOLS:>Case 2.a!!!");
        String nm=((IDeviceModel) modelsList.get(pos)).
getDevice().name;
    logger.evolutionln("Security Agent: Case 2.a -> START_
MISSION \"Patrolling Mission\""+nm);
        System.out.println("Starting "+nm);
    ICommand cc = new CCommand(nm,"START_MISSION","Patrolling
Mission");
        //System.out.println(cc);
        commandsList.add(cc);
        }
    }
        else {
        switch(statusOnMission){
        case 3: //CASE 2.b.i
            //DO NOTHING
            System.out.println("DROOLS:>Case 2.b.i!!!");
            logger.evolutionln("Security Agent: Case
2.b.i");
            break;
        case 5: //CASE 2.b.iii
            //DO NOTHING
            System.out.println("DROOLS:>Case 2.b.iii!!!");
            logger.evolutionln("Security Agent: Case
2.b.iii");
            break;
        case 6: //CASE 2.b.iv
            System.out.println("DROOLS:>Case 2.b.iv!!!");
            logger.evolutionln("Security Agent: Case
2.b.iv");
    ((IDeviceModel) modelsList.get(onMission)).getDevice().
status ="UNAVAILABLE";
            break;
            case 4: //CASE 2.b.ii
            System.out.println("DROOLS:>Case 2.b.ii!!!");
            //STOP patrolling UAV
String nm=((IDeviceModel) modelsList.get(onMission)).
getDevice().name;
            System.out.println("Stopping "+nm);
ICommand cc = new CCommand(nm,"ABORT_MISSION","Patrolling
Mission");
logger.evolutionln("Security Agent: Case 2.b.ii -> ABORT_
MISSION \"Patrolling Mission\""+nm);
```

```
                    //System.out.println(cc);
                    commandsList.add(cc);
                    //START another UAV, if possible
                    if (pos>-1){
                        System.out.println("DROOLS:>Case
2.b.ii.I!!!");
nm=((IDeviceModel) modelsList.get(pos)).getDevice().name;
                        System.out.println("Starting "+nm);
cc = new CCommand(nm,"START_MISSION", "Patrolling Mission");
logger.evolutionln("Security Agent: Case 2.b.ii.I -> START_
MISSION \"Patrolling Mission\""+nm);
                        //System.out.println(cc);
                        commandsList.add(cc);
                    }else {
                        System.out.println("DROOLS:>Case 2.b.ii.II
-> RAISE ALARM!!!");
logger.evolutionln("Security Agent: Case 2.b.ii.II  -> RAISE
ALARM!!!");
                    }
                    break;
                }
            }
        }
End
```

## References

1. Bertolino A., Towards ensuring eternal connectability. Presented at Proceedings of the 6th International Conference on Software and Data Technologies (ICSOFT 2011), Seville, Spain, July 2011.
2. Grace P., Georgantas N., Bennaceur A., Blair G., Chauvel F., Issarny V., Paolucci M., Saadi R., Souville B., Sykes D., The CONNECT architecture. In *11th International School on Formal Methods for the Design of Computer, Communication and Software Systems: Connectors for Eternal Networked Software Systems*, Springer, Berlin, 2011, 27–52.
3. Autili M., Chilton C., Inverardi P., Kwiatkowska M., Tivoli M., Towards a connector algebra. In *Proceedings of 4th International Symposium on Leveraging Applications of Formal Methods, Verification and Validation (ISoLA 2010)*, vol. 6416 of LNCS, Springer, Berlin, 2010, 278–292.
4. Oster C., Wade J., Ecosystem requirements for composability and reuse: An investigation into ecosystem factors that support adoption of composable practices for engineering design, *Syst. Eng.* 16, 439–452, 2013.
5. De Carli A., Andreozzi S., Fiaschetti A., Documentation of requirements in complex systems, *Rivista Automazione e Strumentazione*, December 2007.

6. De Carli A., Andreozzi S., Fiaschetti A., Requirements documentation of a controlled complex motion system. Presented at Proceedings of the IEEE International Workshop on Advanced Motion Control (ACM '08), Trento, Italy, March 26–28, 2008.

7. ISO (International Organization for Standardization), ISO-IEC 15408: Common criteria for information technology security evaluation, v3.1, ISO, Geneva, July 2009.

8. Morgagni A. et al., pSHIELD SPD metrics, pSHIELD project deliverable, September 2011.

9. Noergaard T., *Embedded Systems Architecture: A Comprehensive Guide for Engineers and Programmers*, 2nd ed., Newnes, Oxford, 2013.

10. Balduzzi F., Giua A., Seatzu C., Modelling automated manufacturing systems with hybrid automata. In *Workshop on Formal Methods and Manufacturing*, Zaragoza, Spain, September 6, 1999, pp. 33–48.

11. Chen H., Mandelbaum A., Discrete flow networks: Bottleneck analysis and fluid approximations, *Math. Oper. Res.* 16, 408–446, 1991.

# 5

# *Security, Privacy, and Dependability Metrics*

**Andrea Morgagni, Andrea Fiaschetti, Josef Noll,
Ignacio Arenaza-Nuño, and Javier Del Ser**

## CONTENTS

## Introduction

It is well known that one can control only what he can measure; hence, measuring security, privacy, and dependability (SPD) is a key enabler of the SHIELD methodology.

The focus in the SHIELD methodology was on domain-specific adaptability and prototypical demonstrations, but the overall concept and architecture is still based on *modularity* and *scalability*, meaning that the SHIELD behavior should not be linked to any specific implementation. This is particularly true for metrics evaluation: SPD composability is supposed to work properly with any possible metrics computation algebra or composition algorithm (see Chapter 4), thus ensuring future platform *expandability*. Within the SHIELD methodology several metrics computation approaches have been investigated, with two major methodologies developed, particularly suitable for (1) automated control rooms, for example, smart grids, and (2) the assessment of systems, answering the need for an automated (machine-supported) approach. These methodologies have been identified as the *medieval castle* approach and the *multimetrics* approach, and they are presented in this chapter, with links to application scenarios provided.

## Medieval Castle Approach

The measurement of SPD functions, both qualitatively and quantitatively, has been a long-standing challenge to the research community and is of practical importance to the information technology (IT) industry today. The IT industry has responded to demands for improvement in SPD by increasing the effort for creating what in general can be defined as "more dependable" products and services. How the industry can determine whether this effort is paying off and how consumers can determine whether the industry's effort has made a difference is a question that does not yet have a clear answer. That is why the determination of security privacy and dependability metrics is not a trivial task. There may be different quantifiable representations of SPD metrics [1].

This section describes the method that takes inspiration from the Common Criteria (CC) standard [2–4] to ensure a consistently measured SPD metric for embedded systems (ESs) that is expressed as a cardinal number.

The CC permits comparability between the results of independent security evaluations. The CC allows this by providing a common set of requirements for the security functionality of IT products and for assurance measures applied to these IT products during a security evaluation. These IT products may be implemented in hardware, firmware, or software. The evaluation process establishes a level of confidence that the security functionality of these IT products and the assurance measures applied to these IT products meet these requirements. The evaluation results may help consumers to determine whether these IT products fulfill their security needs.

The CC addresses the protection of assets from unauthorized disclosure, modification, or loss of use. The categories of protection relating to these three types of failure of security are commonly called confidentiality, integrity, and availability. The CC may also be applicable to aspects of IT security outside of these three. The CC is applicable to risks arising from human activities (malicious or otherwise) and to risks arising from non-human activities. Apart from IT security, the CC may be applied in other areas of IT, but makes no claim of applicability in these areas.

In the SHIELD methodology, the CC standard was not completely applied as defined in ISO 15408, but the CC philosophy was adapted to the particular demands of the SHIELD technical requirements with the purpose to obtain:

- An SPD metric for each SHIELD component implementing an SPD function (as described in CC vulnerability assessment). In this way, an SPD metric following the approach of a rigorous international standard was derived.

- The formal description of SHIELD SPD functions according to CC security functional requirements, which are a standard way of expressing the SPD functional requirements.

The definition of this method is based on a SHIELD defined taxonomy and framework for integrating dependability and security. Only apparently, it could seem that privacy was forgotten, but in this work, privacy was intended as a reason for security rather than a kind of security. For example, a system that stores personal data needs to protect the data to prevent harm, embarrassment, inconvenience, or unfairness to any person about whom data are maintained, and for that reason, the system may need to provide a data confidentiality service.

An important issue faced in this metric definition is the quantification of the SPD measure (indicated even as SPD level) of a complex system by first quantifying the SPD level of its components.

Composability in SPD is another long-standing problem. We start from the point that the SHIELD ES is designed to be predictably composable, such that the properties of the resulting composite system are easily determined, and the overall SPD level is calculated according to a given method. This method starts with an intuitive graphical representation of the system (medieval castle), and then it is converted into an algebraic expression using several abstract operators. This composability modeling method has an immediate and complete semantic ontological description represented in specific documents.

This approach is proposed for the estimation of SPD functions indifferently if the SPD function is implemented in the node, network, or middleware SHIELD layer, with the purpose of making easier the monitoring of the current SPD levels of the various layers and of the overall system, as well as the assessment of the various SPD levels.

## Integrating Dependability and Security

With rapidly developed network technologies and computing technologies, network-centric computing has become the core information platform in our private, social, and professional lives. The increased complexity of these platforms and their easy access have made them more vulnerable to failures and attacks, which in turn has become a major concern for society.

Traditionally, there are two different communities separately working on the issues of dependability and security. The community of dependability is more concerned with nonmalicious faults. The security community is more concerned with malicious attacks or faults.

Dependability was first introduced as a general concept covering the attributes of reliability, availability, safety, integrity, maintainability, and so on. With ever-increasing malicious catastrophic Internet attacks, Internet providers have realized the need to incorporate security issues. Effort has been made to provide basic concepts and taxonomy to reflect this convergence.

Measures for security and dependability have also been discussed. A framework based on the work by Hu et al. [5] was proposed in this metric approach. It integrates dependability and security (privacy will be considered part of security, as mentioned in the introduction). This approach does not intend to cover every detail of dependability and security. It places major relevant concepts and attributes in a unified feedback control system framework and illustrates the interaction via well-established control system domain knowledge.

### Dependability and Security Attributes Definition

To address SPD in the context of ESs and cyber-physical systems (CPSs), it is essential to define the assets that these kinds of systems aim to protect.

The SHIELD assets can be categorized into two principal groups: logical and physical assets. Inside these categories, it is possible to define information, services, and software as logical assets and human beings, while hardware or particular physical objects are physical assets.

These assets are characterized by their dependability and security attributes.

The original definition of dependability refers to the ability to deliver a service that can be justifiably trusted. The alternative definition is the ability to avoid service failures that are more frequent and severe than is acceptable. The concept of trust can be defined as accepted dependence, and dependability encompasses the following attributes:

- *Availability*: Readiness for correct service. The correct service is defined as what is delivered when the service implements a system function.
- *Reliability*: Continuity of correct service.
- *Safety*: Absence of catastrophic consequences on the users and environment.
- *Integrity*: Absence of improper system alterations.

- *Maintainability*: Ability to undergo modifications and repairs.

Security has not been introduced as a single attribute of dependability. This is in agreement with the usual definitions of security, which is viewed as a composite notion of the three following attributes:

- *Confidentiality*: The prevention of the unauthorized disclosure of information
- *Integrity*: The prevention of the unauthorized amendment or deletion of information
- *Availability*: The prevention of the unauthorized withholding of information

Avizienis et al. [6] merged the attributes of dependability and security together (see Figure 5.1). Similarly, the above attributes can be reframed as follows:

- *Availability*: Readiness for correct service. The correct service is defined as delivered system behavior that is within the error tolerance boundary.
- *Reliability*: Continuity of correct service. This is the same as the conventional definition.
- *Safety*: Absence of catastrophic consequences on the users and the environment. This is the same as the conventional definition.
- *Integrity*: Absence of malicious external disturbance that makes a system output off its desired service.
- *Maintainability*: Ability to undergo modifications and repairs. This is the same as the conventional definition.

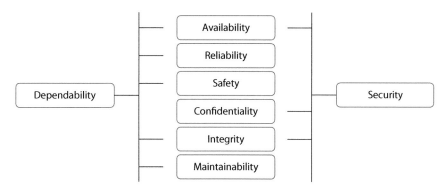

**FIGURE 5.1**
Attributes of security and dependability.

- *Confidentiality*: Property that data or information is not made available to unauthorized persons or processes. In the proposed framework, it refers to the property that unauthorized persons or processes will not get system output or will be blocked by the filter.

The importance of the SPD attributes of such assets is usually expressed in terms of the consequential damage resulting from the manifestation of threats (Figure 5.2).

In this taxonomy, a fault is defined as a cause of an error and a failure is linked to the error that is outside of the error tolerance boundary and is caused by a fault.

### Integration of Dependability and Security Concepts

Integrating dependability and security is very challenging, and ongoing research effort is needed. Avizienis et al. [6] and Jonsson [7] proposed a system view to integrate dependability and security. They tried to use the system function and behavior to form a framework. A schema of the taxonomy of dependable and secure computing is proposed. Although the taxonomy and framework are discussed in the context of a system interacting with its environment, there seems to be a lack of a cohesive and generic integration [6]. Jonsson [7] made an attempt to provide a more generic integration framework using an input/output system model. In this scheme, faults are introduced as inputs to the system and delivery of service and denial of service are considered system outputs. However, the interactions among other system components are still unclear. In fact, both system models proposed by Avizienis et al. [6] and Jonsson [7] are of open-loop systems, which are unable to provide a comprehensive description of the interaction relationship. For example, they cannot describe the relationship between fault detection and fault elimination using the proposed system.

Also, security attributes have rarely been addressed. In the next section, a new framework to address these issues is proposed.

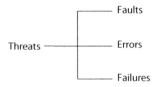

**FIGURE 5.2**
Threats taxonomy.

*Proposed Framework Feedback Control System Model*

In this section, a feedback control system, shown in Figure 5.3, is proposed as a framework that can generically integrate dependability and security. The following are conventional notations of feedback control systems, and they are tailored whenever needed for the SHIELD framework.

*Key Notations of the Feedback Control System Model* The meanings of the elements of the feedback control system are as follows:

- *System*: A composite entity constructed from functional components. The interaction of these components may exhibit new features or functions that none of the composite components possess individually.
- *Plant*: A system that is represented as a function P(.) that takes its input as a functional argument, that is, P(s(t)), and transfers it into an output accordingly. Note that the predefined function P(.) can be modified via the plant adjustment channel.
- *Regulator*: A control function whose role is to regulate the input of the plant, under the influence of the environment, such as instructions of the user, observed error, and input disturbance, to achieve the desired objective.
- *Control system*: A system that uses a regulator to control its behavior.
- *Feedback*: Use of the information observed from a system's behavior to readjust or regulate the corrective action or control so that the system can achieve the desired objectives.
- *Feedback control system*: A control system that deploys a feedback mechanism. This is also called a closed-loop control system.

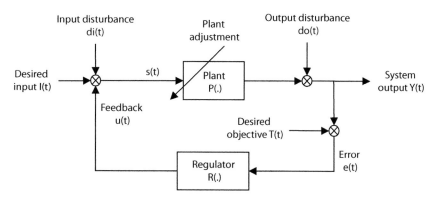

**FIGURE 5.3**
Feedback control system.

- *Open-loop control system*: A control system without a feedback mechanism.
- *Desired objectives*: Desired objectives normally specified by the user.
- *Disturbance*: Anything that tends to push the system behavior off the track is considered a disturbance. A disturbance can occur within the system or from the external environment.
- *System output*: System behavior perceived by the user.

*Definitions of Basic Concepts of Dependability and Security*   In order to understand the relationship of various components and functions of the feedback control system shown in Figure 5.3, the following relationship descriptions are provided according to the conventional feedback control system theory. We use parameter $\varepsilon$ as a threshold of error. Then we will have the following equations:

$$s(t) = I(t) + di(t)\varepsilon + u(t) \tag{5.1}$$

$$u(t) = R(e(t)) \tag{5.2}$$

$$e(t) = Y(t) - T(t) \tag{5.3}$$

$$Y(t) = do(t) + P(s(t)) \tag{5.4}$$

We use the following variable expressions to represent the desired or normal values: $s_0(t)$, $u_0(t)$, $e_0(t)$, $Y_0(t)$, and $P_0(.)$. Under the desired condition or normal condition, all disturbances do not appear and we have the following relationships:

$$s_0(t) = I(t) \tag{5.5}$$

$$u_0(t) = R(0) = 0 \tag{5.6}$$

$$E_0(t) = Y_0(t) - T(t) = 0 \tag{5.7}$$

$$Y_0(t) = P(s_0(t)) = P(I(t)) \tag{5.8}$$

The above description is for a single-input, single-output (SISO) control system. In general, systems have multiple functional components where each component itself forms a system and interconnects the others. Distributed control theory is also able to describe such multiple-input, multiple-output (MIMO) relationships. However, it will make little difference for our conceptual discussions in this book whether it is a SISO system or a MIMO system. This is because multiple inputs, multiple outputs, and disturbances can be represented by individual vectors, and Equations 5.1 through 5.8 still hold, where symbols then become vectors. The error vector $\varepsilon_0(t)$ can be quantitatively measured by its norm $||\varepsilon_0(t)||_2$. By default, we assume a SISO system environment unless stated otherwise.

Now we can provide the following definitions using this framework:

- *Error*: A deviation of the system behavior or output from the desired trajectory represented by e(t).

- *Error tolerance boundary*: A range within which the error is considered acceptable by the system or user. This boundary is normally specified by the user. To be more precise, this refers to e(t) < ε, where ε is a threshold parameter value, which has been specified by the user, and is used to denote the error tolerance boundary. For example, given that a normal service response time is 1 min, an error tolerance boundary of 0.5 will mean a maximum tolerance response time of 1.5 min.

- *Desired service*: The delivered system behavior is at or close to the desired trajectory; that is, we have the feedback control system error e(t) → 0.

- *Correct service*: The delivered system behavior is within the error tolerance boundary, that is, e(t) < ε.

- *Service failure or failure*: An event that occurs when the system output deviates from the desired service and is beyond the error tolerance boundary, that is, e(t) > ε.

- *Fault*: Normally, the hypothesized cause of an error is called a fault [6]. It can be internal or external to the system. Observing that many errors do not reach the system external state and cause a failure, Avizienis et al. [6] have defined active faults that lead to error and dormant faults that are not manifested externally. Under our proposed framework, dormant faults are do(t), di(t), which do not propagate to e in Equations 5.1 through 5.4.

The conventional attributes of dependability and security description can be integrated using the following approach:

- *Availability*: Readiness for correct service. The correct service is defined as a delivered system behavior that is within the error tolerance boundary. This can be interpreted as e(t) < ε if a required service is provided to the input of the feedback control system shown in Figure 5.3.

- *Reliability*: Even under the disturbances shown in Figure 5.3, we have e(t) < ε; that is, the correct service is still maintained.

- *Safety*: Absence of catastrophic consequences on the users and the environment. This is the same as the conventional definition.
- *Integrity*: Absence of malicious external disturbance, which makes the system output off its desired service. This can be interpreted as the absence of malicious plant adjustment, which modifies the function of the plant, leading to $e(t) > \varepsilon$.
- *Maintainability*: Ability to undergo modifications and repairs. This is the same as the conventional definition.
- *Confidentiality*: Property that data or information is not made available to unauthorized persons or processes. In the proposed framework, it refers to the property that unauthorized persons or processes will not be able to observe the values or contents of the sensitive variables, such as $s_0(t)$ or $Y_0(t)$.

### Fault Classification

We have provided a set of elementary fault classes with minimum overlapping. An intuitive choice is to start two classes, namely, human-made faults (HMFs) and non-human-made faults (NHMFs).

#### Human-Made Faults

HMFs result from human actions. They include the absence of actions when actions should be performed (i.e., omission faults). Performing wrong actions leads to commission faults.

HMFs are categorized into two basic classes: faults with unauthorized access (FUAs) and not faults with unauthorized access (NFUAs).

*Faults with Unauthorized Access*    The class of FUAs attempts to cover traditional security issues caused by malicious attempt faults. A malicious attempt fault has the objective of damaging a system. A fault is produced when this attempt is combined with other system faults. We have investigated FUAs from the perspective of availability, reliability, integrity, confidentiality, and safety:

- *FUAs and confidentiality*: Confidentiality refers to the property that information or data are not available to unauthorized persons or processes, or that unauthorized access to a system output will be blocked by the system filter. Confidentiality faults are mainly caused by access control problems originating in cryptographic faults, security policy faults, hardware faults, and software faults. Cryptographic faults can originate from encryption algorithm faults, decryption algorithm faults, and key distribution methods. Security policy faults are normally management problems and can appear in different forms (e.g., as contradicting security policy statements).

- *FUAs and integrity*: Integrity refers to the absence or improper altera-tion of information. An integrity problem can arise if, for instance, internal data are tampered with and the produced output relies on the correctness of the data.

- *FUAs and availability and reliability*: In general, availability refers to a system readiness to provide correct service, and reliability refers to continuity of correct service, but according to interpre-tation proposed before, these attributes have been considered as one because both guarantee the correct service with an error $e(t) < \varepsilon$. A typical cause of such faults is some sort of denial of ser-vice (DoS) attack that can, for example, use some type of flooding (SYN, ICMP, or UDP) to prevent a system from producing correct output. The perpetrator in this case has gained access to a sys-tem, albeit very limited, and this access is sufficient to introduce a fault.

- *FUAs and safety*: Safety refers to the absence of catastrophic conse-quences on a system user's end environment. A safety problem can arise if, for instance, an unauthorized system access can cause the possibility of human lives being endangered.

*Not Faults with Unauthorized Access*   There are HMFs that do not belong to FUAs. Most of such faults are introduced by error, such as configuration problems, incompetence issues, and accidents.

### Non-Human-Made Faults

NHMFs refer to faults caused by natural phenomena without human par-ticipation. These are physical faults caused by a system's internal natural processes (e.g., physical deterioration of cables or circuitry) or by external natural processes. They can also be caused by natural phenomena. For example, in communication systems, a radio transmission message can be destroyed by an outer space radiation burst, which results in system faults, but has nothing to do with system hardware or software faults (Figure 5.4).

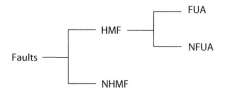

**FIGURE 5.4**
Faults classification.

*Tree Representation of Faults*

From the above discussions, we propose the elementary fault classes shown in Figure 5.5.

From these elementary fault classes, we have constructed a tree representation of faults, as shown in Figure 5.6.

Figure 5.7 shows different types of availability faults. The Boolean operation block performs either OR or AND operations or both on the inputs. We provide several examples to explain the above structure. We consider the case when the Boolean operation box is performing OR operations. F1.1 (a malicious attempt fault with intent of availability and reliability damage) combined with software faults will cause an availability or reliability fault. F1.1, in combination with hardware faults, can also cause an availability fault. F7 (natural faults) can cause an availability fault. F1.1 and F8 (networking protocol) can cause a denial of service fault.

The interpretation of S2 (Figure 5.8) is similar to that of S1. The combination of F1.2 and F2 can alter the function of the software and generate an integrity fault. Combining F1.2 and F4 can generate a person-in-the-middle attack and so on.

Figure 5.9 shows types of confidentiality faults.

The interpretation of S3 is very similar to those of S1 and S2. A combination of F1.3 and F2 can generate a spying type of virus that steals users' logins and passwords. It is easy to deduce other combinations.

Figure 5.10 shows types of maintainability faults.

The interpretation of S4 is very easy because faults combinations are not present. So a maintainability fault can be generated by a single kind of fault belonging to F2, F3, F5, F6, or F7, or a combination of them.

Figure 5.11 shows types of safety faults.

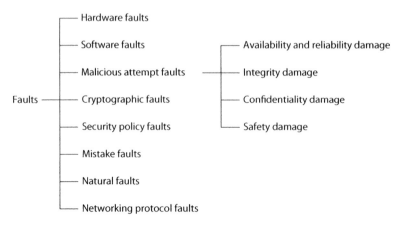

**FIGURE 5.5**
Elementary fault classes.

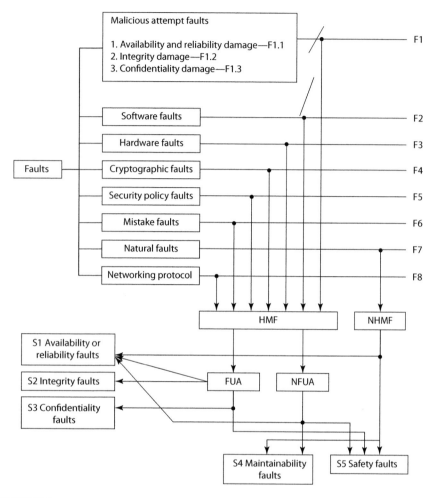

**FIGURE 5.6**
Tree representation of faults.

The interpretation of S5 is very similar to those of S1, S2, and S3. A combination of F1.4 and F2 can generate an unpredictable system state causing a catastrophic consequence on users and the environment. A similar deduction can be done for the other combinations.

## SPD Metric Definition for Basic Components

Each SPD function description can be extracted by a catalog (based on CC Part 2 for FUA mitigation and CC Part 3 for NFUA and NHMF mitigation) where SPD functions are grouped in classes, families, and components (Figure 5.12—light gray evidenced part).

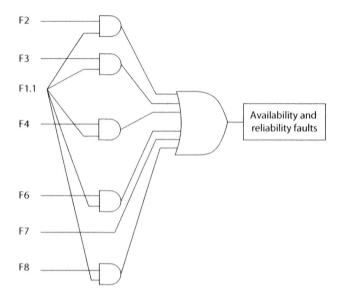

**FIGURE 5.7**
Availability and reliability faults.

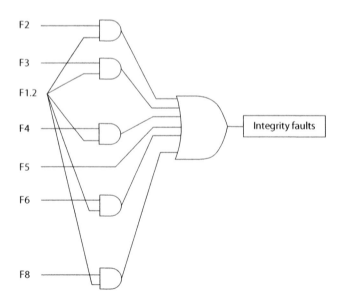

**FIGURE 5.8**
Integrity faults.

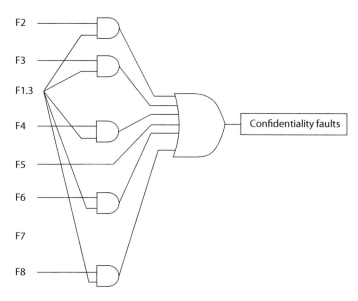

**FIGURE 5.9**
Confidentiality faults.

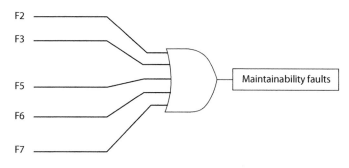

**FIGURE 5.10**
Maintainability faults.

The two parallel processes (left, FUAs; right, NFUAs and NHMFs) identified in Figure 5.12 (gray evidenced part), which lead to the SPD function measure, are both based on CC.

In particular, the first one is based on a vulnerability assessment and the second one on the evaluation of defined life cycle support elements (LCSEs) (these elements are generally defined in documents, e.g., guidance, manuals, and development environments).

The vulnerability assessment of SPD functions is conducted with the purpose of estimating their robustness, and determining the existence and exploitability of flaws or weaknesses in their operational environment. This determination is based on a vulnerability analysis and supported by testing.

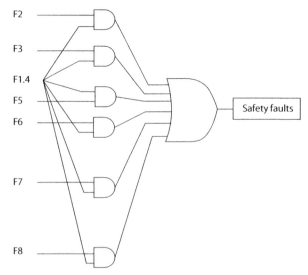

**FIGURE 5.11**

Safety faults.

The vulnerability analysis consists of the identification of potential vulnerabilities, identified as the follows:

- Encountered in the public domain or in another way
- Found through an analysis of SPD functions taken into account
- Bypassed through
  - Tampering
  - Direct attacks
  - Monitoring

Penetration tests are performed to determine whether identified potential vulnerabilities are exploitable in the operational environment of the SHIELD. They are conducted assuming an attack potential.

The following factors should be considered during the calculation of the minimum attack potential required to exploit a vulnerability:

1. Time taken to identify and exploit (elapsed time)
2. Specialist technical expertise required (specialist expertise)
3. Knowledge of the SPD functionality design and operation (knowledge of the functionality)
4. Window of opportunity
5. IT hardware or software or other equipment required for exploitation (equipment)

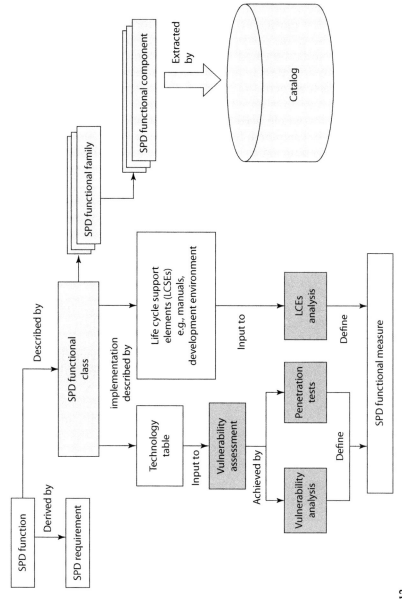

**FIGURE 5.12**

SPD function measure definition scheme.

Table 5.1 identifies the factors listed above and associates numeric values with the total value of each factor.

The value of the minimum attack potential required to exploit a vulnerability is the measure of SPD function tested. In the next section, this measure is indicated by the letter d. This value indicates how much is valid in the implementation of that particular SPD function for incrementing SHIELD fault resistance. These values are summarized in Table 5.2.

This process has to be repeated for each SPD function implemented by SHIELD components and, if there are different implementations of the same SPD function, for each of them.

After calculating measures for all SPD functions, it is possible to define a single measure for the whole SHIELD system considering the composition approach described in Chapter 4. Its value is indicated as d.

LCSE analysis is conducted to prevent the misuse of the SHIELD SPD functions evaluating life cycle documents.

Misuse may arise from

- Incomplete guidance documentation
- Unreasonable documentation
- Unintended misconfiguration of the SHIELD
- Forced exception behavior of the SHIELD

The measure of life cycle documents is estimated through the CC standard, giving a numeric value for each passed activity and then summing it. This value is indicated as $d_{LC}$.

By CC experiences, the value assigned for a single component is included in the [0; 1] interval; in fact, these functionalities are only supporting, and so it cannot increment the SPD assurance of the SHIELD system.

We assume that all these activities are carried out by independent and unbiased IT security specialists (evaluators) in order to obtain an evaluated value of the SPD function measure (SPD level), which has the qualities of repeatability, reproducibility, impartiality, and objectivity.

In the context of IT security evaluation, as in the fields of science and testing, repeatability, reproducibility, impartiality, and objectivity are considered to be important principles.

An evaluated SPD level is repeatable if the repeated evaluation yields the same result as the first evaluation.

An evaluated SPD level is reproducible if the evaluation of the same implemented SPD function in the same operational environment performed by a different evaluator yields the same results as the evaluation performed by the first evaluator.

An evaluated SPD level is performed impartially if the evaluation is not biased toward any particular result.

**TABLE 5.1**

Factor/Value for Calculation of the Minimum Attack Potential

| Factor | Value |
| --- | :---: |
| **Elapsed Time** | |
| ≤1 day | 0 |
| ≤1 week | 1 |
| ≤1 month | 4 |
| ≤2 months | 7 |
| ≤3 months | 10 |
| ≤4 months | 13 |
| ≤5 months | 15 |
| ≤6 months | 17 |
| >6 months | 19 |
| **Expertise** | |
| Layman | 0 |
| Proficient | 3[a] |
| Expert | 6 |
| Multiple experts | 8 |
| **Knowledge of Functionality** | |
| Public | 0 |
| Restricted | 3 |
| Sensitive | 7 |
| Critical | 11 |
| **Window of Opportunity** | |
| Unnecessary/unlimited access | 0 |
| Easy | 1 |
| Moderate | 4 |
| Difficult | 10 |
| Unfeasible | 25[b] |
| **Equipment** | |
| Standard | 0 |
| Specialized | 4[c] |
| Bespoke | 7 |
| Multiple bespoke | 9 |

[a] When several proficient persons are required to complete the attack path, the resulting level of expertise still remains "proficient" (which leads to a 3 rating).

[b] Indicates that the attack path is considered not exploitable due to other measures in the intended operational environment of the SHIELD.

[c] If clearly different test benches consisting of specialized equipment are required for distinct steps of an attack, this should be rated as bespoke.

**TABLE 5.2**

Rating of Vulnerabilities and SHIELD Resistance

| Value | Attack Potential Required to Exploit Scenario | SHIELD Resistant to Attackers with Attack Potential of |
|---|---|---|
| 0–9 | Basic | No rating |
| 10–13 | Enhanced basic | Basic |
| 14–19 | Moderate | Enhanced basic |
| 20–24 | High | Moderate |
| ≥25 | Beyond high | High |

An evaluated SPD level is performed objectively if the result is based on actual facts uncolored by the evaluators' feelings or opinions.

## Quantifying the SPD Measure of Composed SPD Functions

Several measures for the SPD of systems have been presented in the literature, including an adversary work factor, adversary financial expenses, an adversary time, probability-like measures, or simply defining a finite number of categories for the SPD of systems.

In this section, two types of measurements are described:

- A deterministic approach to give a single measure of the SPD assurance level of a composed ES, starting from an intuitive graphical representation of the system itself [8]
- A system for security assurance assessment

### *SPD for Medieval Castle*

Starting from the consideration that, in the Middle Ages, SPD was obtained by building castles, this metric approach shows how the SPD of a castle with up to two doors can be computed under the assumption that the SPD of each door are known.

The castle at point (a) illustrated in Figure 5.13 is the simplest castle we can think of. It has a wall (depicted as a square), a treasure room (depicted by a circle) that is the main target of an attacker, and a door d (the only access for the attackers to reach the treasure room). In this case, the SPD of the castle is equal to the SPD of the door.

The castle at point (b) has two entrances, $d_1$ and $d_2$, where $d_1$ is weaker than $d_2$. In this case, the attackers can assault the castle at two points simultaneously. Thus, the castle's SPD measure will be weaker than or equal to the SPD measure of $d_1$ (we assume that $d_1$ and $d_2$ are totally independent, so the castle's SPD measure will be equal to the SPD measure of $d_1$).

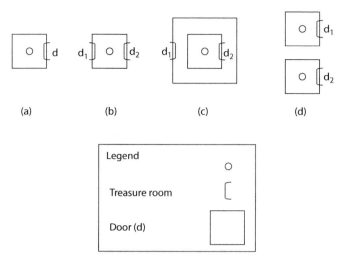

**FIGURE 5.13**
Medieval castle examples.

In case (c), the SPD of the castle may be stronger than the security of a castle with only one door (an attacker must break into two doors to get into the treasure room).

In the last example (castle [d]), the attackers have two possible castles to attack. Considering that,

1. An attack is successful if the attackers get into one of the two treasure rooms
2. The distance of the castles is too large to allow for a simultaneous attack of both castles

The SPD of both castles is in the interval $[d_1; d_2]$.

### SPD of Systems with Two SPD Functions

Starting from the previous section, there is the need to consider further information to compute the SPD level of a (medieval or modern) system:

- An SPD measure (leveling) M
- The SPD measure of its SPD functions (or doors*): $d_1, d_2, \varepsilon, d_n$
- A function d: $M^n \rightarrow M$, which maps the SPD measure of the doors to the SPD level of the overall system

---

\* A castle's door can be seen without distinction as an interface of a function implementing SPD.

To deal with the complexity of today's systems, we assume that the function d can be depicted as a term using different operators. A modeling method for secure systems is thus constructed by first defining an SPD measure and then defining a set of operators to combine the SPD function measures.

Figure 5.14 shows three operations for the unbounded continuous case with interval $[0; +\infty)$.

Considering $d_{min}$ the smaller operand and $d_{max}$ the larger operand, we can set the following operations:

- MIN operations, if $d = d_{min}$
- MEAN operations, if $d_{min} \leq d \leq d_{max}$
- OR operations, if $d_{max} \leq d \leq +\infty$

where both operands are unbounded deterministic variables and values are computed by the measure of SPD metrics of basic components, as previously summarized.

*MIN Operation*

A MIN operation should be used for a system that resembles castle (b) in Figure 5.13. In this case, the defenders have to defend both doors. Thus, the system is as weak as $d_{min}$. In fact, a potential attacker can attack both doors (SPD functions) at the same time, without additional efforts or costs, and the weaker door is the first to be broken. So

$$d_{MIN} = MIN\,(d_{min},\, d_{max})$$

When a potential attacker can attack n-doors (SPD functions) with the previous hypothesis, the formula becomes

$$d_{MIN} = MIN\,(d_1,\, d_2, ...., d_n)$$

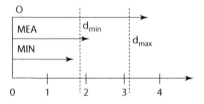

**FIGURE 5.14**
SPD set of operations.

*OR Operation*

For systems corresponding to the situation named castle (c) in Figure 5.13, OR operations can be used.

To defend the castle, either $d_{min}$ or $d_{max}$ has to be defended. A corresponding system is at least as strong as $d_{max}$.

This kind of system models the concept of "defense in depth," where multiple defense mechanisms are used in layers across the system to protect the assets. We use multiple defenses so that if one defensive measure fails, there are more behind it to continue to protect them.

With the assumption that the doors have to be attacked sequentially, the security of the castle can be computed by

$$OR\left(d_{min}, d_{max}\right) = d_{OR} = d_{min} + d_{max}$$

In the presence of an n-doors sequence, the following formula can be considered:

$$OR\left(d_1, d_2, ...., d_n\right) = d_{OR} = d_1 + d_2 + \cdots + d_n$$

If doors are protected with the same lock, lockpicking the second lock might be much easier after successfully opening the first door, and so on. This case can model the redundancy of SPD functions; we can indicate this with $OR_n$, where n is the number of SPD functions carrying out the redundancy.

As a first approximation, we propose calculating the SPD measure for an $OR_n$ system by

$$OR_n\left(d\right) = d_{OR_n} = d + \sum_{i=2}^{n} \frac{d}{i}$$

To model this situation, we can indicate castle (c) (Figure 5.13) as depicted in Figure 5.15 just to underline that the system is protected by the redundancy of the same SPD function.

*Mean and Power Mean Operation*

As introduced in castle (d) in the figure, an attacker may first choose one of two doors and, in a second step, attack the system as if it had only one door. In this situation, the system is stronger than any system with two doors concurrently attackable, but weaker than any system where the attackers must break into two doors. Thus, its SPD lies in the interval $[d_{min}; d_{max}]$.

Therefore, these kinds of systems can be modeled using a mathematical mean operation, which satisfies the same property $(d_{min} \le d \le d_{max})$.

If the attackers choose a door (in a random way) with equal probability 0.5 for each door, the SPD of the system is

$$\mathrm{MEAN}\left(d_{min}, d_{max}\right) = d_{MEAN} = \frac{d_{min} + d_{max}}{2}$$

which is the arithmetic mean of $d_{min}$ and $d_{max}$.

Many situations could be considered, for example,

1. The attackers might have some knowledge of which doors are more vulnerable and prefer the doors with lower security.
2. The defenders will have some information on the attackers' preferences and be able to strengthen the doors that are most likely to be attacked.

Both situations can be taken into account for using the general power mean $M_p$, defined as

$$d_{MEAN} = M_p = \left(\frac{d_{min}^P + d_{max}^P}{2}\right)^{\frac{1}{P}}$$

where the parameter p indicates the amount of knowledge the attackers and defenders have.

In particular,

- If p equals 1, the power mean equals the arithmetic mean; that is, neither the attackers nor the defenders have an influence on which door is chosen.
- If p becomes greater, the knowledge of the defenders increases, and thus the probability that the attackers choose the strong door. This situation can model the honeypot solution to distract attackers from attacking a more valuable system. For example, if p is 2, $M_p$ becomes the root of squared means, which can be computed by

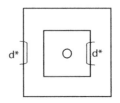

**FIGURE 5.15**
Redundancy scheme.

$$d_{MEAN} = M_2 = RSM = \sqrt{\frac{d_{min}^2 + d_{max}^2}{2}}$$

- If $p \to +\infty$, the attackers always choose the strong door, and thus

$$d_{MEAN} = d_{max}$$

- If $p$ is smaller than 1, the attackers know the castle well and the weak door is more likely to be under attack. For example, if $p \to 0$, $M_p$ is called the geometric mean G, defined as

$$d_{MEAN} = M_0 = G = \sqrt{d_{min} \cdot d_{max}}$$

- If $p \to -\infty$, the attackers will choose the weakest door, and thus

$$d_{MEAN} = d_{min}$$

However, choosing the right $p$ weights is very difficult in practice. Where it is possible, we can estimate the value of $p$ from the vulnerability analysis. In the presence of n elements for castle (d) in Figure 5.13, where we can consider $d_1, d_2, \ldots, d_n$ doors, the formulas for different cases become the following:

- The attackers randomly choose a door with equal probability $1/n$ for each door; the security of the system is

$$MEAN(d_1, \ldots, d_n) = d_{MEAN} = \frac{d_1 + \cdots + d_n}{n}$$

which is the arithmetic mean of $d_1, \ldots, d_n$.
In a more general case,

- The attackers might have some knowledge about more vulnerable doors and prefer the doors with lower security.
- The defenders will have some information on the attackers' preferences and will be able to strengthen the doors that are most likely to be attacked. In this case, it is possible that the attackers choose the strong door.

Both scenarios can be taken into account for using the general power mean $M_p$, defined as

$$d_{MEAN} = M_p(d_1, \ldots, d_n) = \sqrt[p]{\frac{1}{n} \sum_{i=1}^{n} d_n^p}$$

The parameter p determines the amount of knowledge the attackers and defenders have.

In the extreme cases,

- $p \to +\infty$: The attackers always choose the stronger door and thus

$$d_{MEAN} = d_{MAX} = MAX\left(d_1,\ d_2,\ldots,\ d_n\right)$$

- $p \to -\infty$: The attackers choose the weakest door and thus

$$d_{MEAN} = d_{MIN} = MIN\left(d_1,\ d_2,\ldots,\ d_n\right)$$

### SPD Measure of Systems with n-SPD Functions

In this example, a castle with eight doors and two treasure rooms (Figure 5.16a) is considered, and the attackers are successful if they are able to get into one of the two treasure rooms.

A semantic representation of the system is shown in Figure 5.16b. This representation can be easily implemented by the proposed ontology for the SPD function modeling.

Figure 5.16b was created by starting at the doors, which now form the leaves of a tree. Then, the nodes were repeatedly connected in pairs by an OR, MIN, or MEAN gate, respectively. For system evaluation, we can replace the abstract doors by the set of SPD function interfaces that expose an attack surface.

A mathematical expression for the SPD measure of this system can be defined as follows:

$$d = OR\left(MIN\left(OR\left(MIN\left(d_7,d_8\right),d_6\right),MIN\left(d_1,d_2\right)\right),MEAN\left(d_3,MIN\left(d_4,d_5\right)\right)\right)$$

In this case, we can replace OR with a + operator so this function becomes

$$d = MIN\left(MIN\left(d_1,d_2\right),\left(d_6+MIN\left(d_7,d_8\right)\right)\right)+MEAN\left(d_3,MIN\left(d_4,d_5\right)\right)$$

### Corrective Value Introduced by SPD Life Cycle Support Element

In order to take into account the life cycle element (operational manuals, installation guides, etc.), the d value of the SPD measure obtained by the method described previously must be multiplied with $d_{LC}$. The total value is

$$d_{TOT} = d \times d_{LC}$$

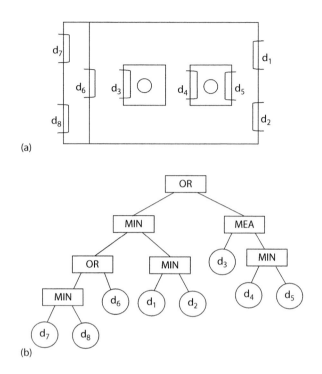

(a)

(b)

**FIGURE 5.16**
n-SPD scheme.

Because 0 and 1 comprise $d_{LC}$, we are sure that well-done life cycle elements cannot increase the total SPD metric value, but ones that are misunderstood can decrease it.

### Example of an Application Scenario

In this section, we show a practical example of the application of the SPD functions composition approach. We consider a reduced control system installed on a train, as depicted in Figure 5.17.

It is composed of a control unit connected by means of a cyphered connection to a sensor in a redundancy configuration and the related configuration manuals. The assets to protect are data that are sent by the sensor to the central unit and that are recorded inside the central unit itself. In this system, we considered the following SPD functions:

1. Antitampering redundancy (sensor)
2. Configuration manuals (system)
3. Cypher (data transfer)
4. Identification and authentication (central unit)
5. Access control (central unit)

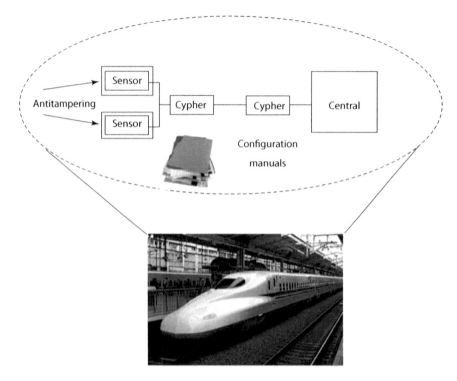

**FIGURE 5.17**
Application scenario.

According to the described approach, this system can be modeled as shown in Figure 5.18.

The doors in the figure refer to

- $d_1^* = $ SPD: Measure of the sensor antitampering strength in a redundant configuration
- $d_2 = $ SPD: Measure of the cypher strength
- $d_3 = $ SPD: Measure of the access control strength
- $d_4 = $ SPD: Measure of the identification and authentication strength

The correspondent system tree representation of the application scenario is shown in Figure 5.19.

The mathematical expression for the SPD measure of this application scenario system can be defined as follows:

$$d_{TOT} = \text{MEAN}\left(\text{MIN}\left(\text{OR}_2\left(d_1^*\right), d_2\right), \text{MIN}\left(\text{OR}\left(d_3, d_4\right), d_2\right)\right) \times d_{LC}$$

where $d_{LC} = $ SPD is a measure of the life cycle documentation.

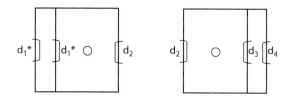

**FIGURE 5.18**
Medieval castle model of the application scenario.

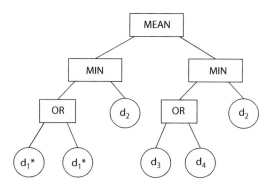

**FIGURE 5.19**
Application scenario system tree representation.

## Multimetrics Approach

This section presents the multimetrics approach, addressing "measurable security" in industrial sectors previously working with threat-based assessments of security.

A threat-based assessment of security is based on the identification of threats, and applying security functionality to mitigate these threats. Such methodologies, focusing on attacker capabilities or different attack surfaces [9], are also entitled attack-centric approaches.

The multimetrics approach is more applicable to the security and privacy assessment of systems, assuming a security class for a given application. In short, multimetrics uses metrics as means of measuring the security or privacy of a system. The multimetrics focuses on an analysis of a system, its components, and different system configurations (system-centric approach), compared with methodologies that take into account the attacker capabilities or different attack surfaces (attack-centric approach).

The SHIELD multimetrics envisages the translation of technological parameters into security parameters, as demonstrated in Figure 5.20. A component has various configuration parameters (here, Conf A ... C), for example, different encryption mechanisms or different types of communication.

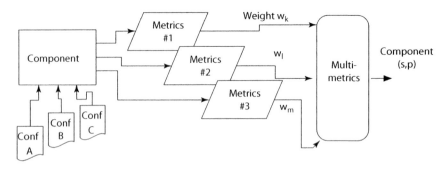

**FIGURE 5.20**
Principle operation of the multimetrics method.

Each security- or privacy-related aspect of the configuration will be assessed through a metrics, for example, an encryption metrics. The importance of that metrics for the security of the system is specified through a weight $w_k$. The multimetrics analysis then summarizes the various metrics, weighted through their specific weights, and creates a value for the component security. Composing a system of subsystems with components and applying the multimetrics subsequentially then allows for an SPD analysis of a system of systems.

The core multimetrics methodology was developed through the SHIELD initiative, and applied by Garitano et al. for a privacy elaboration of a smart vehicle [10] (see Chapter 9, "Privacy in Smart Vehicles: A Case Study") and by Noll et al. for a smart grid communication [11] (see Chapter 9, "Measurable Security in Smart Grids"). Both application domains were used as examples to demonstrate the applicability of the approach.

## Multimetrics Assessment

The first step in the process is the SPD metrics assessment. The objective is to achieve an overall system SPD level, $SPD_{System}$. $SPD_{System}$ is a triplet composed of individual SPD levels $(s, p, d)$. Each of the levels is represented by a range between 0 and 100; the higher the number, the higher the SPD level. However, in order to end up with $SPD_{System}$, during the whole process, the criticality is evaluated.

The criticality of a system, as indicated in Figure 5.21, is a mapping of system parameters to the operational importance of the system. Take the example of a process control, where the response time is the *operational parameter*. We have identified the following operational statuses: ideal, good, acceptable, critical, and failure. If the process needs a response time of less than 100 ms to be operational, each response coming later than 100 ms is seen as a failure. We might define a response time of 50 ms as critical, knowing that a retransmittal of a lost message will probably exceed the operational limit of 100 ms. Accordingly, response times for ideal, good, and acceptable

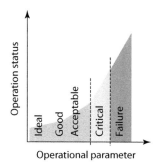

**FIGURE 5.21**
Operational status of a system in terms of criticality.

are defined. The goal of criticality is now to map the operational parameter "response time" to a criticality for SPD.

The criticality is a triplet $(Cs, Cp, Cd)$, defined as the complement of SPD, and expressed as $(Cs, Cp, Cd) = (100, 100, 100) - (s, p, d)$. Metrics are the objects or entities used to measure the criticality of components. The criticality of a component for a given configuration is evaluated through one or more metrics. The proposed metric definition methodology is composed of four phases: parameter identification, parameter weighting, component–metric integration, and metric running, as shown in Figure 5.22.

The parameter identification phase involves the component analysis, which will be measured by the metric itself. Based on the component purpose and its characteristics, the identified parameters will be classified and transferred to the next phase. The next step, parameter weighting, evaluates the repercussion of the possible values of each parameter on the component SPD level. This step needs to be done by an expert in the field, since it requires a good knowledge not only about the system itself, but also of the SPD domains. The impact of each parameter value for a given SPD aspect is defined by its weight. Notice that the weight will be different for each SPD aspect.

The next phase, the component–metric integration, consists of identifying the possible values that a parameter could have based on all the configurations in which a system will run.

After all the parameters, the weight of their values within a component, and the possible configuration values have been identified, the last step is to run the metric. In this phase, parameter values specified by each of the

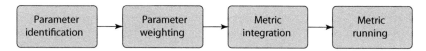

**FIGURE 5.22**
Multimetric definition and run phases.

configuration files are evaluated through the respective metric and provide the criticality level of the component. This step is repeated for every SPD aspect.

## Identified Challenges

In order to achieve industrial uptake, several challenges need to be addressed. The main challenge is to define metrics converting a security parameter into an opera.

Other challenges, for example, the adaptation of security classes (or grades) in the various domains, will be a topic of harmonization and standardization, given that metrics and assessment methodology are available as frameworks.

The example of an encryption metrics is used to describe the challenges of linking technology parameters to security levels, as indicated in Figure 5.23 for the type of encryption to a security level.

The challenges with the metrics are mainly linked to a minimum (min) and maximum (max) security value, and to the values representing various grades of encryption. As an example, does "no encryption" relate to a security of 0, 10, or 20? And what would an increase from, for example, Advanced Encryption Standard (AES) with 128 bits to AES with 192 bits mean for security—increasing from 90 to 100, for example? These types of questions need to be further elaborated and standardized.

All the elements involved in the system evaluation have a weight. The weight describes the importance of a configuration for a subsystem, for example, authentication or remote access.

To start with, each component, metric, or subsystem has a different weight, used for the SPD$_{System}$ evaluation. In addition to SPD levels, the weight is described by a value in a range between 0 and 100; the higher the number, the higher the importance of an element. The usage of the same range for SPD and element importance helps us to assign the corresponding value.

Having adopted the multimetrics approach to several domains, we experienced that users tend to select values in the middle of the scale, rather than

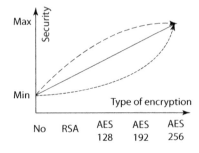

**FIGURE 5.23**
Symbolic representation of mapping between type of encryption and component security.

using the full scale. Thus, in order to increase the weight effect in multi-metrics, the initial weight value is escalated quadratically. Our sensitivity analysis showed that a linear weighting results in an averaging effect for the SPD value of subsystems and systems, rather than pinpointing criticalities.

### Application Examples

The examples presented in this book take into consideration both approaches, a linear weighting for the smart vehicle use case (see Chapter 9, "Privacy in Smart Vehicles: A Case Study") and a quadratic weighting for the smart grid communication example (see Chapter 9, "Measurable Security in Smart Grids").

## References

1. Jaquitti A. *Security Metrics: Replacing Fear, Uncertainty, and Doubt*. Boston: Addison-Wesley, 2007.
2. ISO (International Organization for Standardization). Common criteria for information technology security evaluation—Part 2: Security functional components. Version 3.1, revision 3, final—CCMB-2009-07-002. Geneva: ISO, 2009.
3. ISO (International Organization for Standardization). Common criteria for information technology security evaluation—Part 3: Security assurance components. Version 3.1, revision 3, final—CCMB-2009-07-003. Geneva: ISO, 2009.
4. ISO (International Organization for Standardization). Common methodology for information technology security evaluation—Evaluation methodology. Version 3.1, revision 4. Geneva: ISO, 2012.
5. Hu J, et al. Seamless integration of dependability and security concepts in SOA: A feedback control system based framework and taxonomy. *Journal of Network and Computer Applications*, 34(4), 1150–1159 (2011).
6. Avizienis A, Laprie JC, Randell B, and Landwehr C. Basic concepts and taxonomy of dependable and secure computing. *IEEE Transactions on Dependable and Secure Computing*, 1(1), 11–33 (2004).
7. Jonsson E. An integrated framework for security and dependability. In *Proceedings of the 1998 Workshop on New Security Paradigms*, Charlottesville, VA, September 22–26, 1998.
8. Walter M and Trinitis C. Quantifying the security of composed systems. In *Proceedings of PPAM-05*, Poznan, Poland, September 2005.
9. Manadhata PK and Wing JM. An attack surface metric. *IEEE Transactions on Software Engineering*, 37(3), 371–386 (2010).
10. Garitano I, Fayyad S, and Noll J. Multi-metrics approach for security, privacy and dependability in embedded systems. *Wireless Personal Communications*, 81, 1359–1376 (2015).
11. Noll J, Garitano I, Fayyad S, Åsberg E, and Abie H. Measurable security, privacy and dependability in smart grids. *Journal of Cyber Security*, 3(4), 371–398 (2015).

# Section II

# SHIELD Application Scenarios, New Domains, and Perspectives

# 6

## Airborne Domain

Cecilia Coveri, Massimo Traversone, Marina
Silvia Guzzetti, and Paolo Azzoni

### CONTENTS

### Introduction

In recent years, the massive reduction in production costs, together with the exponential increase of computational capabilities, has contributed to the diffusion of embedded technologies in every part of our life: from automotive to healthcare, and from entertainment to energy. From this perspective, one of the most important objectives is the development of new technologies and new strategies to address security, privacy, and dependability (SPD) in the context of embedded systems, producing relevant consequences in all the systems used for safety, reliability, and security applications. This approach is based on modularity and expandability and can be adopted to bring built-in SPD solutions in all the strategic markets of embedded systems, including transportation, communication, healthcare, energy, and manufacturing.

The role of embedded systems in aerospace industry has become crucial to increasing operational reliability; reducing the cost of avionics components and related maintenance; reducing weight, and thus augmenting fuel efficiency; and reducing volume and power consumption [1].

Research projects [2] propose new solutions to strengthen the consistency of the avionics platforms that can be designed, programmed, and configured with software tools following open standards. The evolution from federated architectures to modern distributed integrated modular avionics (IMA) can be characterized by three steps:

1. Making avionics architectures modular
2. Making avionics architectures integrated
3. Making avionics architectures distributed

To follow these indications, an avionics system must include SPD solutions, expected to produce a relevant impact on all the systems used for safety, reliability, and security applications. The trend from modern civilian aircraft is to support the aircraft application with an IMA platform, allowing several hosted applications to share physical and logical resources.

In this context, the adoption of the SHIELD methodology in a dedicated avionics surveillance system, composed of a set of heterogeneous subsystems, including the IMA platform as the core component, represents an adequate answer to the emerging trends and clearly highlights the added value of the SHIELD methodology in an avionics system of systems (SOS).

---

## Dependable Avionics Scenario: A Case Study

This case study has been selected to demonstrate some specific SHIELD functionalities in the context of a complex system, focusing specifically on composability and interoperability. The dependable avionics scenario can be considered an SOS: a set of heterogeneous subsystems, logically or physically connected, that cooperate for the execution of one or more tasks. A typical example of application involving an avionics SOS is the "surveillance system" adopted for search and rescue or security monitoring. In this context, the term *surveillance* refers to a set of techniques, devices, and methodologies used to detect, investigate, recognize, and prevent specific behavior, events, or situations related to dependability, safety, and security. In this perspective, the scenario of the dependable avionics system for surveillance can be considered a reference vertical application to highlight the potentialities and added value of the SHIELD methodology.

Typically, an avionics-based system for surveillance is composed of a large number of heterogeneous subsystems (radar, sonar, etc.), including aircraft of many different categories and unmanned aerial vehicles (UAVs). Each operational subsystem must be able to communicate with its own network, operating according to a certain protocol and providing specific functionalities. For the security of the surveillance operations, it is vital that the communication between the different levels of the system satisfies an adequate level of security and dependability.

The dependable avionics scenario is based on a set of UAVs connected to a ground control station that are responsible for monitoring a large restricted area to avoid the access of unauthorized people. From the SPD perspective, the components of the surveillance systems are orchestrated by the SHIELD middleware, which ensures the proper level of SPD in all the system statuses and in the presence of threats or faults (Figure 6.1).

To test and evaluate the SHIELD capabilities, during normal operational status, hardware faults or threats are injected into the hardware platform or simulated. This kind of events can be managed individually with avionics subsystems that natively support the SHIELD methodology, which allows us to identify the event, isolate it, and react with a specific countermeasure. From an SOS perspective, threats and/or faults can be identified, isolated, and managed with the composability functionalities of the SHIELD methodology. To test and evaluate the two levels of complexity in terms of event management, two different threats and faults categories are considered:

1. Threats and/or faults that affect single avionics subsystems of the surveillance system and that can be managed directly at the subsystem level (e.g., a failure of the positioning subsystem).

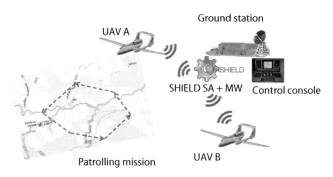

**FIGURE 6.1**
Avionics-based surveillance scenario.

2. Threats and/or faults that affect multiple subsystems of the surveillance system and that can be managed only through the SHIELD orchestration (e.g., extending the surveillance scenario to the use of two UAVs, a ground control and surveillance operators).

## System for Dependable Avionics Surveillance

The dependable avionics surveillance system has a complex and heterogeneous architecture composed of UAVs, embedded avionics, and ground appliances. In order to manage this complexity, the test and evaluation of the surveillance system have been performed on a prototype that is composed of the real avionics subsystems and also includes simulated components (e.g., the flight operations are simulated). Albeit a simplification, the composability and dependability features of the surveillance systems have been fully tested and validated (Figure 6.2).

The subsystems of the dependable avionics surveillance system are:

- *OMNIA*: This subsystem is based on an IMA platform, which collects, analyzes, and distributes the data acquired from the aircraft sensors. For the SHIELD avionics dependable prototype, the OMNIA subsystem has been equipped with two or more computer unit/remote interface unit (CU/RIU), in order to simulate the onboard avionics on a UAV. This configuration can be changed (composed) to include more "nodes" if necessary. OMNIA provides a fault recovery engine

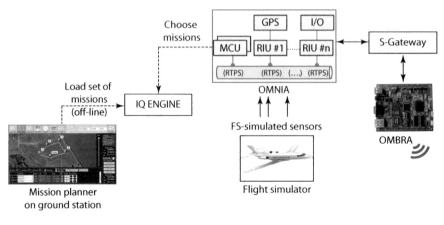

**FIGURE 6.2**
Dependable avionics surveillance system architecture.

that is in charge to dynamically replace faulty components with spare ones, if available.

- *IQ engine*: This subsystem is a cognitive pilot that relies on a knowledge-based system (configured with a set of missions by means of a software called Mission Planner) to perform the autopilot activity on the UAVs, simulated with a flight simulator.
- *S-Gateway*: The gateway subsystem is a hardware adaptation layer that translates the system statuses and faults into the SHIELD "vocabulary."
- The SPD-driven smart transmission layer is a set of network services that utilize the SDR and cognitive radio technology in order to provide secure and dependable communication in critical channel conditions.
- *SHIELD middleware*: This subsystem is a software layer that implements the SHIELD methodology and the SPD management mechanisms. In particular, this layer is in charge of performing the discovery and composability activities. The SHIELD middleware is ground based.
- To fulfill the scenario requirements, other commercial off-the-shelf (COTS) components, such as the flight simulator (X-plane, by Laminar Research), have been adopted.

The individual subsystems of this architecture are assembled to mimic a real surveillance system, adopting the following solutions:

- A reduced version of the UAV1 is represented by an OMNIA system, which acts as a mission management system, connected to IQ_Engine, which acts as an autopilot, and to the S-Gateway, which represents the interface toward the SHIELD ecosystem. The communication to the ground station is managed by the SPD-driven smart transmission layer component.
- A second UAV2, identical to UAV1.
- The ground station is implemented by a PC hosting all SHIELD middleware services, equipped with the S-Gateway and the same communication link used by the UAVs.

The use of these prototypes makes the dependable avionics surveillance scenario natively SHIELD compliant, highlighting the enhancements of SPD solutions obtained, with the application of the SHIELD methodology:

- Security is ensured by continuous, interference-free communication between the UAV and the ground station. This aspect is covered with the adoption of appropriate communication mechanisms in the scenario.

- *Privacy*: Different access levels to the ground station for the mission operators are guaranteed.
- *Dependability*: The data acquired by onboard sensors are protected against corruption (integrity, safety, availability, and reliability). The system operations are supported by automated system recovery.

## Hardware Components

### OMNIA Hardware Components

The OMNIA prototype is composed of a network of aircraft and mission management computers (AMMC) and a related RIU connected between them with a high-speed deterministic serial line; each "unit" is connected to the aircraft sensors.

The system or network, by means of an additional middleware layer built around the concept of publisher–subscriber architecture (tailored in accordance to the certification avionics constraint: Avionics Data Distribution Protocol [ADDP]), virtualizes the connection of a related sensor with all the "computer units" that are present in the system, also allowing the fault tolerance functionalities. More specifically,

- The main unit AMMC is used as an IMA central unit.
- The other units (referenced as RIU-IMA) are mainly used as sensor interfaces.
- The IMA central unit and RIU-IMA are both computer units.
- All the computer units are connected via Ethernet (rate constraint or best-effort methodology); each computer unit can implement the interface with the avionics sensor.

The ADDP functionality has been integrated, as a library, into the equipment software (EQSW) environment. This library, according to the IMA concept, is segregated in a partition in order to increase the reliability and flexibility of the system. The ADDP represents the communication pattern used by the IMA and the RIU to exchange data. In particular, the publish–subscribe architecture is designed to simplify one-to-many data-distribution requirements. In this model, an application "publishes" data and "subscribes" to data. Publishers and subscribers are decoupled from each other. Real-time applications require more functionalities than those provided by the traditional publish–subscribe semantics. The ADDP adds publication and subscription timing parameters and properties so that the application developer can control different types of data flows, and therefore the application's performance and reliability goals can be achieved.

The ADDP developed has been implemented on the top of the User Datagram Protocol/Internet Protocol (UDP/IP) and tested on the Ethernet; the determinism of the Ethernet is provided by the ARINC664P7 network.

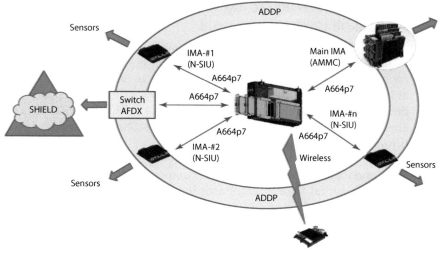

**FIGURE 6.3**
OMNIA architecture.

Figure 6.3 illustrates a logical view of OMNIA demonstrator architecture. Exchanged data between all the IMA computer units could be

- Discrete signals
- Bus1553 data words
- ARINC429 data words
- More in general, all the input/output (I/O) signals

In particular, data are managed by the RIU and sent and received by the IMA central unit.

As already mentioned, the solution is characterized by two pieces of AMMC reduced equipment connected to each other via ARINC664P7, where:

- The IMA central unit consists mainly of a rack with:
  - A processor module based on the PPC microprocessor (APM460), Ethernet 10/100/1000 BASE-T link, and RS232 serial I/F (monitor–debug channel).
  - An ARINC664P7 end system interface (mezzanine PCI card on CPU).
- The RIU consists of a rack with:
  - A processor module based on the PPC microprocessor (APM460), Ethernet 10/100/1000 BASE-T link, and RS232 serial I/F (monitor–debug channel).
  - An ARINC664P7 mezzanine card.

- Two I/O boards (RS422, ARINC429, discrete, and analog).
- A Mil-Std-1553B mezzanine card.
- The high-speed deterministic Ethernet switch is the ARINC664P7-Switch 3U VPX Rugged that enables critical network-centric applications in harsh environments and provides 16 Ethernet ports. The avionics full-duplex switched Ethernet interface provides a reliable, deterministic connection to other subsystems. ARINC664P7 will be an ELECTRICAL (OPTICAL optional) connection with a baseline throughput of 100 Mbps.

Being a modular architecture, every component has been developed according to the actual avionics standards in terms of processing cycles, data buses, signal types, and memory use.

### S-Gateway Hardware Components

The S-Gateway prototype represents an implementation of one of the central components defined in the SHIELD framework. It has been specifically designed to operate in the avionics scenario and plays the pivotal task of interconnecting SHIELD components with legacy embedded systems, with either proprietary physical (pp-ESD) or logical (pl-ESD) communication capabilities.

Regarding the specific task of the avionics scenario, the S-Gateway prototype is directly interfaced with two types of prototypes, the OMNIA prototype and the SPD-driven smart transmission layer software-defined radio (SDR).

In such a context, the S-Gateway provides both physical and logical adaptation layers, allowing intra-gateway and extra-gateway communication. Therefore, the S-Gateway operates as a proxy-like system for the connected nodes, by translating the SHIELD middleware message into a legacy message and vice versa, and adapting the proprietary interfaces, so that the legacy component can interact with the SHIELD network.

Besides these proxy features, the S-Gateway prototype performs a set of additional tasks to ensure real-time and safe behaviors for internal data management and data exchange.

The board implemented for this purpose is the ZedBoard, a low-cost development board based on the Xilinx Zynq™-7000 All Programmable SoC (AP SoC). This board provides all the features to make a bare metal, Linux, or other OS/RTOS-based design. Additionally, several expansion connectors expose the processing system and programmable logic I/Os for easy user access.

The great strength of such an architecture derives from the efficient coupling, on the same chip, of the hard-coded ARM® A9 MPCore processing system and the high flexibility of the 7-series programmable logic, interfaced through the high-speed AMBA AXI4 interface protocol.

This architectural choice is justified by the high customizability of the Zynq SoC and from the availability of a series of prebuilt interfaces (SPI, I2C, CAN, Eth, etc.), which can be used to interface the gateway with a large variety of devices. Furthermore, this choice has been strengthened by the availability of a powerful development ecosystem that greatly simplifies the development process and reduces the time to market.

### SPD-Driven Smart Transmission Layer Hardware Components

The SPD-driven smart transmission layer encompasses a set of services deployed at the network level designed for SHIELD SDR-capable power nodes. The goal of the prototype is to ensure smart and secure data transmission in critical channel conditions, exploiting the reconfigurability properties, and the learning and self-adaptive capabilities of the SDR. To achieve this objective, the SPD-driven smart transmission layer consists of a number of secure wideband multirole–single-channel handheld radios (SWAVE HHs), each interconnected with the OMBRA v2 multiprocessor embedded platform (see Chapter 3).

In the context of the SHIELD dependable avionics surveillance scenario, the following functionalities are available:

- *Self-awareness*: The network learns the current topology, the number and identity of the participants, and their position, and reacts to variations of their interconnection.
- *Jamming detection and counteraction*: The nodes recognize the presence of hostile signals and inform the rest of the network nodes; the nodes cooperate in order to come with the optimal strategy for avoiding disruption of the network.

SWAVE HH (hereafter referred to as HH) is a fully operational SDR terminal capable of hosting a multitude of wideband and narrowband waveforms. The maximum transmission power of HH is 5 W, with the harmonics suppression at the transmit side more than −50 dBc. The superheterodyne receiver has specified an image rejection of better than −58 dBc. The receiver is fully digital; in very high frequency (VHF), 12-bit 250 MHz analog-to-digital converters (ADCs) perform the conversion directly at radio frequency (RF), while in ultrahigh frequency (UHF), analog-to-digital conversion is performed at intermediate frequency (IF).

The radio is powered by Li-ion rechargeable batteries (battery life is 8 h at the maximum transmission power for a standard 8:1:1 duty cycle); however, it may also be externally powered through a 12.6 V direct current (dc) source. Hypertach expansion at the bottom of HH provides several interfaces, namely, 10/100 Ethernet, USB 2.0, RS-485 serial, dc power interface (maximum 12.7 V), and push to talk (PTT). The radio provides operability in both the VHF (30–88 MHz) and UHF (225–512 MHz) bands. The software architecture of the radio is compliant with the software communication architecture (SCA) 2.2.2 standard.

The considered power node—OMBRA v2 platform—is composed of a small form factor system-on-module (SOM) with high computational power (developed by Selex ES) and the corresponding carrier board. It is based on an ARM Cortex A8 processor running at 1 GHz, encompassed with powerful programmable Xilinx Spartan 6 FPGA and Texas Instruments TMS320C64 + DSP. It can be embodied with up to 1 GB of LPDDR RAM, has support for microSD cards up to 32 GB, and provides interfaces for different RF front ends. The node is dc-powered, and has Windows CE and Linux distribution running on it (Figure 6.4).

Connection to HH is achieved through the Ethernet, as well as a serial port. The Ethernet is used for remote control of the HH, using the Simple Network Management Protocol (SNMP). For the serial connection, due to different serial interfaces—RS-232 and RS-485—a RS-232-to-RS-485 converter is needed. Serial connection is used for transferring the spectrum snapshots from HH to power node.

**FIGURE 6.4**
SPD-driven smart transmission layer: Handheld and SHIELD power node.

## Software Components

### OMNIA Software Components

The OMNIA software main blocks are separated into the following different layers (Figure 6.5):

- SHIELD application layer
- Middleware layer
- Module support layer

The *SHIELD application layer* provides all the OMNIA functionalities of communication with the SHIELD prototypes, collecting the UAV flight and system information, computing the UAV and OMNIA health status, and managing and ensuring the equipment redundancy.

The SHIELD IMA software application holds the following software components:

- A component that implements communication with the SHIELD autopilot prototype.
- A component that implements communication with the S-Gateway prototype.
- A component in charge of visualizing flight simulator data coming from RIU.
- A component in charge of managing the errors occurring in the acquisition of I/O data and redundancy management.
- An interface with a real-time publisher–subscriber (RTPS) application programming interface (API) to subscribe data.
- A UDP/IP library interface to accomplish the communication via the Ethernet.

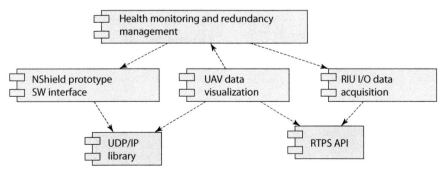

**FIGURE 6.5**
IMA software component diagram.

The SHIELD RIU software application is composed of the following software components:

- A component in charge of acquiring avionics data from the flight simulator.
- A component in charge of detecting and managing the errors occurring in the acquisition of DASIO I/O data (health monitoring).
- An interface with RTPS API to publish data.
- A UDP/IP library interface to accomplish communication via the Ethernet.
- An EQSW library interface to acquire data from RIU I/O sensors (Figure 6.6).

The *middleware layer* consists of EQSW (API and virtual device drivers) and RTPS software. The middleware is responsible for three core programming functionalities:

1. Maintain the database that maps publishers to subscribers, resulting in logical data channels for each publication between publishers and subscribers.
2. Serialize (also called marshal) and de-serialize (or de-marshal) the data on its way to and from the network to reconcile publisher and subscriber platform differences.
3. Deliver the data when it is published.

The RTPS middleware is built to run on the top of an unreliable transport, such as UDP/IP, but it must be designed to be platform independent. In OMNIA, the network communication is a 100 Mbps, optically connected, dual-redundant ARINC664-compliant UDP/IP packet-switched network, and in this ARINC664P7 context, all normal transmissions are always carried across both networks. The ARINC664 transmissions require the definition of

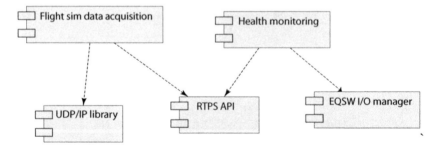

**FIGURE 6.6**
RIU software component diagram.

two links in order to create a transmit–receive path. Each link in ARINC664 is termed a virtual link. A virtual link may be used for multiple transmission types between any two fixed line replaceable units (LRUs), by addressing the data to different UDP ports, or by application separation of the data.

The *module support layer* includes all the target-dependent software modules, that is, the operating system (RTOS Integrity 178B), the board support package, and the resident software layers.

In local I/O communications, all activity boards can be implemented using a different topology as an access shared memory or even via PCI Express.

In order to implement SPD features, the OMNIA system is able to provide a relevant check relating to either the data integrity exchanged or the integrity of the OMNIA system (fail, reconfiguration, etc.). Health monitoring and fault management within the OMNIA platform are performed at the node level by means of continuous built-in tests.

The dependability of the system communication is guaranteed by the ARINC664P7 network, which provides a dual link redundancy and quality of service. Besides, thanks to its topology and structure, ARINC664P7 significantly reduces wires, improving the system reliability.

The composability functionality is provided thanks to the integration in the IMA architecture of a new module developed in accordance with the SHIELD methodology.

### S-Gateway Software Components

The S-Gateway prototype connects devices that can be dynamically discovered by the SHIELD middleware. To accomplish this functionality, the S-Gateway must

1. Continuously monitor the device's status.
2. Publish and make exploitable to the SHIELD middleware the services exposed by linked devices, so that their capabilities can be used to preserve the system SPD status.

The S-Gateway periodically checks the connected devices in order to obtain their status and communicate it to the SHIELD middleware, by translating it into the SPD metrics format and by using the methods exposed by the SHIELD middleware to register them in the ground control station, where the SHIELD security agent analyzes them to perform the SPD control activity on the system.

Since the SHIELD security agent should promptly react to restore the SPD level in the case of critical events, the system status is transmitted with different rates based on the system awareness. This "SPD-driven" periodical monitoring functionality is performed by sending interrogation messages with a configurable rate.

The variations of the SPD level are automatically interpreted by the internal *Controller* core, which regulates the communication timings by configuring the *Freshness* and *Policies save* cores, which produce the periodical interrupt events that are promptly served by the processor that executes the bare metal application.

Any potential issue that occurs during the resolution of these interrupts (such as a mismatch between a request and the corresponding response), together with errors on cyclic redundancy code (CRC) check or on the decrypted data coherence check, is identified by the *Fault_Detector* core, which informs the system *Controller* in order to obtain a counteraction to limit the bad consequences due to these faults (e.g., impose less stringent timing constraints or induce the retransmission of corrupted data).

Besides the previously described proxy-like behavior, through which the S-Gateway converts the device-dependent messages into SHIELD messages and vice versa, an "aggregation activity" is performed, which consists of the elaboration of all the SPD data relating each system component to provide a unique set of SPD values that represent the current state of the whole system.

This composition activity is necessary to satisfy the security agent requirement to analyze only a single set of SPD values that are representative of the system's actual status and depend on the status of each one of its components.

To accomplish this task in the context of the dependable avionics surveillance scenario,

- The S-Gateway in the ground control station must specifically be configured to know how to compose the metrics of all the system components.
- The S-Gateway on board the UAV must provide it with the SPD data of all the devices that compose the cluster on the UAV, both periodically and on specific request coming from the SHIELD middleware.

In Figure 6.7, a schematic representation of the S-Gateway periodical monitoring activity is described.

The interaction with the SHIELD middleware happens through the service location protocol, a service discovery protocol that allows us to find services offered by a device in a local network, without prior configuration.

This protocol is used to announce the services that the system exposes for the scenario. Each service must have a URL that is used to locate the service. Additionally, it may have an unlimited number of name–value pairs, called attributes.

For this reason, the result coming from the composition activity of component SPD parameters, which describe the services that the system exposes, has to be registered in the repository in execution in the ground control station.

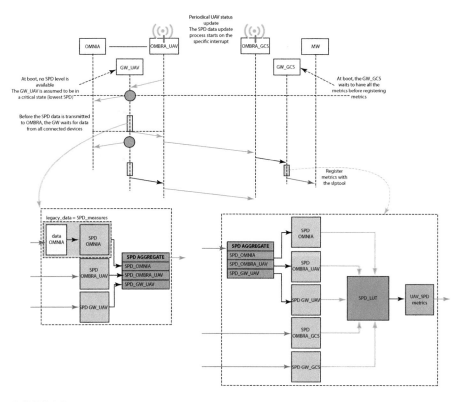

**FIGURE 6.7**
S-Gateway periodical monitoring activity.

Then the SHIELD security agent will periodically perform a service discovery on this repository to find the services offered by the various scenario actors (e.g., the two UAVs of the surveillance scenario), and once these services have been analyzed, some decisions can be made to act on the scenario status.

Similar to what happens for the other scenario components, these decisions represent the services exposed by the SHIELD security agent, so they must be registered in the SLP repository in order to be made discoverable by all the scenario's actors.

In the surveillance scenario, the discovery activity in the SLP repository is performed by the S-Gateway in the ground control station, which can interpret the commands provided by the SHIELD security agent and then deliver them to the correct UAV cluster (Figure 6.8).

The S-Gateway prototype leverages a highly customizable behavior, not modifiable at runtime, to preserve security aspects during its functioning, by avoiding malicious attempts to modify the system behavior with modification of the system configuration.

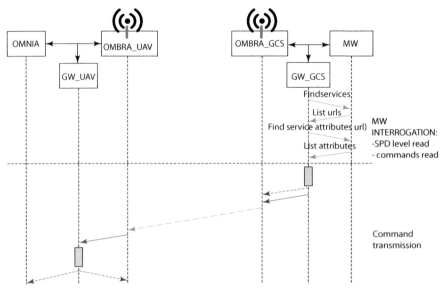

**FIGURE 6.8**
S-Gateway discovery activity.

### SPD-Driven Smart Transmission Layer Software Components

The handheld radio terminal is based on SCA 2.2.2—the most widespread software architecture for SDRs.

The interfacing of the SPD-driven smart transmission layer and the SHIELD architecture is provided by a client–server connection that is established through the SHIELD data interchange (S-DI). The S-DI is used for communications between the SPD-driven smart transmission layer and the S-Gateway. S-DI is a full-duplex protocol designed to handle asynchronous reads and writes from both parties. S-DI does not specify any means of control flow, ACKs or NACKs, or commands and responses. These features must be implemented at the application layer. The nS-DI protocol is based on the Simple Object Access Protocol (SOAP). SOAP is a protocol specification for exchanging structured information in the implementation of web services in computer networks. It relies on the XML information set for its message format, and usually relies on other application layer protocols, most notably the Hypertext Transfer Protocol (HTTP) or the Simple Mail Transfer Protocol (SMTP), for message negotiation and transmission. Within this scenario, S-DI uses the SOAP-over-UDP standard covering the publication of SOAP messages over the UDP transport protocol, providing for one-way and request–response message patterns.

In order to implement SPD features, the SPD-driven smart transmission layer performs periodic checks of the relevant node-related and network-related parameters. Each combination of parameters represents a unique SPD level of the prototype. For certain SPD levels (certain combinations of parameters), adequate policies are defined at the upper SHIELD layers (middleware and overlay), which in turn trigger the reconfiguration of the parameters in order to restore the satisfying SPD level.

For demonstration purposes, relevant parameters that should be periodically checked are:

- Packet delivery rate (over or under a predefined threshold).
- Status of cryptography (on or off).

Hence, four combinations of SPD levels of the SPD-driven smart transmission layer prototype are defined, each of them relying on a policy that may be triggered in case the SPD level drops under the minimum acceptable level.

### IQ_Engine Cognitive Pilot Software Components

IQ_Engine has the potential to fill more roles than a pure autopilot, hence the "cognitive pilot" definition. It differs from most other autopilot systems, since it is based on a knowledge database rather than proportional-integral-differential (PID) regulators and mathematical algorithms. This allows a more flexible and efficient exploitation of the technology, while still being compliant with strict avionics standards. The knowledge database is tailored for its purpose, and is extremely efficient.

IQ_Engine also fits easily in the IMA architecture, since it relies on only analog, discrete, or digital data as input, and vice versa as output. IQ_Engine cooperates with the surrounding subsystems, as illustrated in Figure 6.9. Of these IQ_Engine subsystems, only those included by the dashed line are avionics systems. The other subsystems are off-line and ground based.

The Mission Planner system is used during preflight to draw and upload the mission to the UAVs, while IQ_Edit is also used during preflight to define the knowledge database. The X-plane simulator has been adopted for visualization of the flights of both UAV1 and UAV2.

The software structure of IQ_Engine is based on the components shown in Figure 6.10.

The agents are all started as the engine is started, and each of them is responsible for its own part of the main mission. The agents and, accordingly, IQ_Engine communicate with the external world through their own dataset. The dataset is kept up to date by input routines, and sends messages and commands through specific output routines.

The I/O routines are also responsible for conversions, filtering, and other adaption, as required. Some special inputs are represented by the commands received from the SHIELD ground station, which arrive through OMNIA, and may cause the flight to start, stop, or initiate status reports.

OMNIA also has direct links to the flight simulator for the collection of flight information by itself. These data are also used to influence the SPD level of the flight operation.

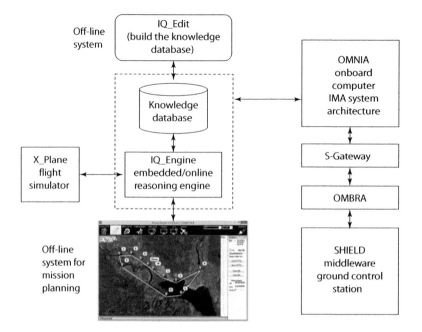

**FIGURE 6.9**
IQ_Engine and surrounding systems.

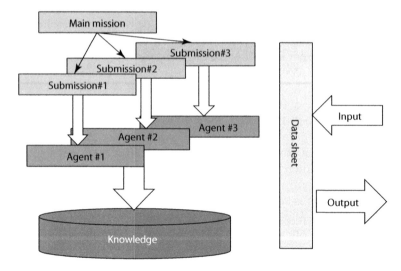

**FIGURE 6.10**
IQ_Engine components.

### Dependable Avionics Scenario: Demonstrator Integrated Architecture

The dependable avionics surveillance prototype is an integrated solution that combines the subsystems described in the previous sections, orchestrated and managed from the SPD perspective by the SHIELD framework.

The integration between the subsystems has been performed according to the avionics surveillance system architecture, as illustrated in Figure 6.11.

The integration of IQ_Engine, OMNIA, the S-Gateway, and the SPD-driven smart transmission represents the UAV that is in charge of the surveillance mission.

Health monitoring and fault management within the OMNIA platform are performed at the node level by means of continuous built-in tests. The integrity of the sensors' data is handled at the OMNIA middleware level.

Within the avionics scenario, the UAV is capable of:

- Simulating the main flight functionality.
- Collecting the UAV states that are changing during the mission and sending them to the middleware in the ground station.
- Actuating commands generated by the middleware in the ground station.

The smart SPD-driven transmission, directly connected to the S-Gateway, ensures safe and robust communication from the UAV to the ground control station.

The integration of the UAV prototypes with the middleware services is granted by the S-Gateway prototype, which is designed to execute the following operations:

- Acquire the measurement from the nodes and calculate the related SPD metrics.
- Register the services provided by the nodes to the middleware, using an appropriate protocol (SLP for the scenario).
- Discover the commands available on the network.
- Translate the request of service into a low-level command and forward it to the node available in its cluster.

Finally, Figure 6.12 illustrates the architecture in terms of SHIELD technology layers (node, network, and middleware or overlay), highlighting the allocation of the prototypes involved in the dependable avionics surveillance system.

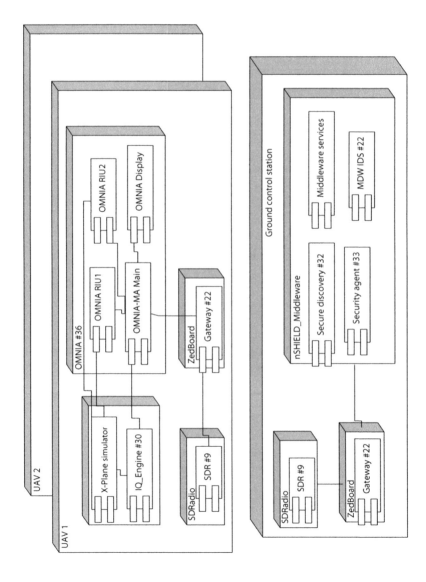

**FIGURE 6.11**

Dependable avionics surveillance integrated architecture.

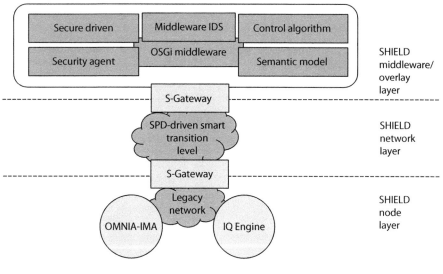

**FIGURE 6.12**
Mapping of the surveillance subsystems on the SHIELD technology layers.

## SPD Metrics Calculation

The SPD levels associated with the states of the dependable avionics surveillance system have been calculated adopting the SHIELD methodology. Although these values are calculated automatically by the SHIELD middleware at runtime, this section illustrates the process adopted when the rules of composition are applied at the SOS level.

### Subsystem SPD Levels

The process starts with the calculation of the SPD levels for the avionics subsystems. Considering the complexity of the avionics system, which comprises a total of 11 components, some of the subsystems have been merged into a single component to simplify the calculation of the SPD levels. This decision is also justified by their SPD behavior: some SPD functionalities are not generated by the single subsystems but only from their cooperation. Following this approach, the architecture of the dependable avionics surveillance system can be partitioned into five logical macrosubsystems, which have been considered for the SPD-level calculation (Table 6.1).

**TABLE 6.1**

Identified Macrosubsystems Involved in the Avionics Use Case

| Macrosubsystem ID Number | Macrosubsystem Name | Number of SPD States | Macrosubsystem Components |
|---|---|---|---|
| 1 | SDR | 2 | Elliptic curve cryptography |
| | | | Smart transmission |
| 2 | Gateway | 1 | S-Gateway |
| 3 | Middleware intrusion detection system | 1 | Middleware intrusion detection system |
| 4 | SHIELD middleware | 1 | Control algorithms |
| | | | OSGi middleware |
| | | | Semantic model |
| | | | Secure discovery |
| | | | Security agent |
| 5 | UAV core | 8 | Reliable avionics (IQ_Engine) |
| | | | OMNIA-IMA |

From a physical perspective, focusing specifically on the avionics system, the identified macrosubsystems are (Figure 6.13):

1. *UAV core (area highlighted in blue)*: It is composed of IQ_Engine and OMNIA. It is physically contained in each UAV considered in the scenario.
2. *Communication network (area highlighted in yellow)*: It is the macrocomponent responsible for the communication between the UAVs and the ground station. This macrosubsystem is physically contained partially in the UAV and partially in the ground station, but only the components contained in the UAV are redundant for the scenario, and therefore contribute to the redundancy functionality of the UAV.
3. *Ground station core (area highlighted in red)*: This macrosubsystem hosts the middleware of the mission ground terminal and influences the ground station SPD functionalities.

The calculation of SPD levels is performed by applying the composition rules to the identified macrosubsystems:

1. The UAV core has eight SPD levels with SPD values varying from 85,464 to a maximum of 89,381. These SPD levels can be characterized by the following considerations:
   a. Considering the UAV redundancy (UAV01, UAV02), it is possible to identify four potential states when two UAVs are available

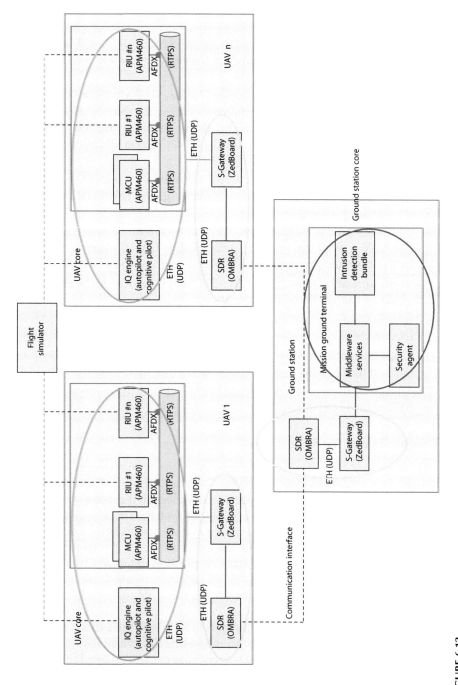

**FIGURE 6.13**
Avionics system macrosubsystems.

(UAV01—patrolling, UAV02—waiting). The states are identified as follows (this list of states is referred to as UAV01):

   i. No alarm.

  ii. (Notification of a hardware failure on OMNIA) or (Notification of the UAV oil overheating) or (Notification of too low fuel level).

 iii. (Notification of a hardware failure on OMNIA) and (Notification of the UAV oil overheating) or (Notification of the UAV oil overheating) and (Notification of too low fuel level) or (Notification of a hardware failure on OMNIA) and (Notification of too low fuel level).

  iv. (Notification of a hardware failure on OMNIA) and (Notification of the UAV oil overheating) and (Notification of too low fuel level).

  b. The same four possible states can also be identified when only one UAV is available (UAV01—unavailable, UAV02—patrolling). In this case, the previous list of states is referred to as UAV02.

2. The communication network has two SPD levels, which can be characterized by the following considerations:

  a. The first state is referred to as a state with no alarm or faults (SPD level: 83,565).

  b. The second state is referred to as a state where there is an interference on the channel used for communication (SPD level: 83,133).

3. The ground station core has a single SPD level, and the corresponding SPD value is 89,161.

**SPD Composition at the System Level**

After the analysis of the SPD states of the avionics macrosubsystems and the calculation of the related SPD levels, the final step is to apply the composition algebra to merge, from the SPD perspective, the identified macrosubsystem into the whole dependable avionics surveillance system. The analysis starts from the composition of the communication network and the ground station core and, subsequently, proceeds with the composition of the results with the UAV core macrosubsystem. The first composition generates two possible states, with calculated values for the SPD levels ranging from 85,175 to 85,502. The second composition step, involving the UAV core, generates the 16 states that characterize the SPD behavior of the whole avionics use case. Table 6.2 illustrates the details of the calculation with reference to the macrosubsystems, their SPD states, and the corresponding SPD values.

The SPD levels derived from SHIELD metrics are expressed by plain numbers (e.g., 84,705) since they are the result of mathematical formulas. In order to make the SPD level easier to understand, specifically for human operators,

**TABLE 6.2**

Avionics Use Case SPD Values

| Avionics Scenario States Ref. No. | (OMNIA + IQ Engine) UAV Core States | | UAV Availability | Communication Network States | | Ground Station Core States | SPD-Level Value | Normal SPD Level |
| --- | --- | --- | --- | --- | --- | --- | --- | --- |
| | Ref. No. | Description | | Ref. No. | Description | Ref. No. | | |
| 1 | 1 | No alarm | 2 Available UAV (UAV01—patrolling, UAV02—waiting) | 1 | No alarm | 1 | 84,95 | 1,00 |
| 2 | 2 | (Notification of an HW failure on OMNIA) or (notification of the UAV oil overheating) or (notification of too low fuel level) | | | | | 84,70 | 0,80 |
| 3 | 3 | (Notification of an HW failure on OMNIA) and (notification of the UAV oil overheating) or (notification of the UAV oil overheating) and (notification of too low fuel level) or (notification of an HW failure on OMNIA) and (notification of too low fuel level) | | | | | 84,37 | 0,54 |
| 4 | 4 | (Notification of an HW failure on OMNIA) and (notification of the UAV oil overheating) and (notification of too low fuel level) | | | | | 84,12 | 0,34 |
| 5 | 5 | No alarm | 1 Available UAV (UAV02—patrolling, UAV01—unavailable) | | | | 84,71 | 0,81 |
| 6 | 6 | (Notification of an HW failure on OMNIA) or (notification of the UAV oil overheating) or (notification of too low fuel level) | | | | | 84,42 | 0,58 |
| 7 | 7 | (Notification of an HW failure on OMNIA) and (notification of the UAV oil overheating) or (notification of the UAV oil overheating) and (notification of too low fuel level) or (notification of an HW failure on OMNIA) and (notification of too low fuel level) | | | | | 84,16 | 0,37 |

*(Continued)*

**TABLE 6.2 (CONTINUED)**

Avionics Use Case SPD Values

| Avionics Scenario States | | (OMNIA + IQ Engine) UAV Core States | | Communication Network States | | Ground Station Core States | | Normal SPD Level |
|---|---|---|---|---|---|---|---|---|
| Ref. No. | Ref. No. | Description | UAV Availability | Ref. No. | Description | Ref. No. | SPD-Level Value | |
| 8 | 8 | (Notification of an HW failure on OMNIA) and (notification of the UAV oil overheating) and (notification of too low fuel level) | | | | | 83,92 | 0,18 |
| 9 | 1 | No alarm | 2 Available UAV (UAV01—patrolling, UAV02—waiting) | 2 | Notification of interference on channel used for communication | | 84,65 | 0,76 |
| 10 | 2 | (Notification of an HW failure on OMNIA) or (notification of the UAV oil overheating) or (notification of too low fuel level) | | | | | 84,37 | 0,54 |
| 11 | 3 | (Notification of an HW failure on OMNIA) and (notification of the UAV oil overheating) or (notification of the UAV oil overheating) and (notification of too low fuel level) or (notification of an HW failure on OMNIA) and (notification of too low fuel level) | | | | | 84,12 | 0,34 |
| 12 | 4 | (Notification of an HW failure on OMNIA) and (notification of the UAV oil overheating) and (notification of too low fuel level) | | | | | 83,88 | 0,15 |
| 13 | 5 | No alarm | 1 Available UAV (UAV02—patrolling, UAV01—unavailable) | | | | 84,42 | 0,8 |

*(Continued)*

**TABLE 6.2 (CONTINUED)**

Avionics Use Case SPD Values

| Avionics Scenario States Ref. No. | (OMNIA + IQ Engine) UAV Core States | | UAV Availability | Communication Network States | | Ground Station Core States | SPD-Level Value | Normal SPD Level |
|---|---|---|---|---|---|---|---|---|
| | Ref. No. | Description | | Ref. No. | Description | Ref. No. | | |
| 14 | 6 | (Notification of an HW failure on OMNIA) or (notification of the UAV oil overheating) or (notification of too low fuel level) | | | | | 84,16 | 0,37 |
| 15 | 7 | (Notification of an HW failure on OMNIA) and (notification of the UAV oil overheating) or (notification of the UAV oil overheating) and (notification of too low fuel level) or (notification of an HW failure on OMNIA) and (notification of too low fuel level) | | | | | 83,92 | 0,18 |
| 16 | 8 | (Notification of an HW failure on OMNIA) and (notification of the UAV oil overheating) and (notification of too low fuel level) | | | | | 83,69 | 0,00 |

**TABLE 6.3**

Normalized SPD ranges

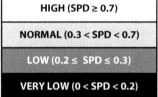

| HIGH (SPD ≥ 0.7) |
| NORMAL (0.3 < SPD < 0.7) |
| LOW (0.2 ≤ SPD ≤ 0.3) |
| VERY LOW (0 < SPD < 0.2) |

a normalization of the SPD level between 0 (lowest relative SPD) and 1 (highest relative SPD) is performed.

The associated SPD ranges and related thresholds are shown in Table 6.3.

The calculation of the SPD levels and the results obtained highlight that the single component or prototype adopted in this scenario has a good balance between the controls implemented and the limitations identified through initial risk management. It is also evident that, according to the SHIELD methodology and, in particular, after the application of its composition rules, when the set of faults and threats became wider with subsystem composition, the SPD level of the resulting scenario or system decreased. Finally, the set of countermeasures adopted in the analyzed use case were adequate to support each other mutually when the system must react to a threat or fault. This is demonstrated by the slight reduction of the SPD levels that the avionics system has in the presence of a threat, just before the implemented countermeasures intended to reestablish the previous SPD level.

## Real Avionics Scenario

This section describes a real scenario that has been identified to test and validate the SHIELD methodology in the avionics domain. The scenario is also intended to demonstrate the dependability and composability features in the context of a complex avionics system.

The identified scenario consists of a mission conceived to monitor a restricted area in order to avoid access of nonauthorized personnel.

The dependable avionics demonstrator is composed of a set of heterogeneous systems that, integrated, are able to cooperate for the execution of one or more tasks. It is composed of prototypes that, when subjected to failure, can recover and restore the normal operative level, exploiting their SDP functionalities. The dependability of the system is achieved in each single layer (node and network), as well as at the system level, thanks to the intervention of the SHIELD middleware and overlay layer.

The dependable avionics demonstrator verifies the SPD performance using the SHIELD methodology and the metric computation algorithm implemented in the demonstrator middleware layer. Validation and verification against relevant high-level and SHIELD system-level requirements is also demonstrated in the context of the avionics scenario, showing one of the relevant applications of SHIELD capabilities in a novel configuration (Figure 6.14).

To accomplish the mission objective, which is focused on keeping an adequate level of security in the restricted area, the borders of the area periodically surveilled by UAVs are equipped with a high-definition camera.

Two UAVs are involved in the surveillance operations: UAV_001 performs the mission patrol of the defined area, and UAV_0002 is off the ground, ready to take off in case UAV_001 fails.

From the SPD perspective, the scenario is automatically monitored by the SHIELD middleware, running in the ground station: the middleware collects the status of the UAV and, depending on the scenario conditions, promptly reacts to different threats or system failure. Figure 6.15 illustrates the mission's logical evolution.

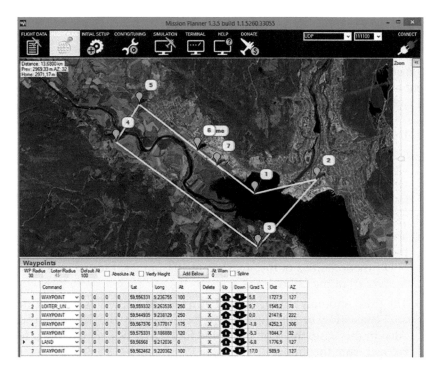

**FIGURE 6.14**
Patrol path to monitor the restricted area.

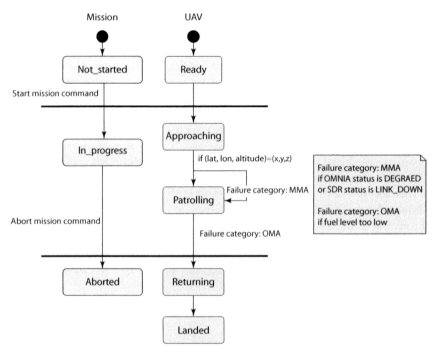

**FIGURE 6.15**
Mission state diagram.

During the mission execution, the avionics system is exposed to four different anomalies consisting of faults or threats, which are intended to test and validate the SHIELD methodology and the avionics system itself:

1. During patrol, a hardware fault is injected into the OMNIA RIU, causing the loss of the GPS signal. A similar event can be easily managed, relying on a spare unit. Thanks to the adoption of the SHIELD methodology, embedded into the OMNIA-IMA components, this fault is identified, isolated, and recovered.

2. The second anomaly is a threat caused by a jammer that creates an interference on the channel used for communication, disabling the link between the UAV and the ground station. The prototype SPD functionalities of the smart transmission layer are exploited to recover from the failure.

3. The third anomaly is represented by a fault identified by IQ_Engine and consisting of the overheating of the oil of the UAV engine. Also in this case, the prototype SPD functionalities are exploited to recover from the failure.

4. The fourth anomaly is generated by the fuel level that became too low to complete the mission. This failure cannot be managed by a

single macrosubsystem; therefore, in this case, the SHIELD overlay layer reacts to compose the system SPD functionalities in a different way to maintain the required SPD level.

The state diagram in Figure 6.16 illustrates the states that the avionics system can assume during the execution of the scenario and the calculated system SPD level.

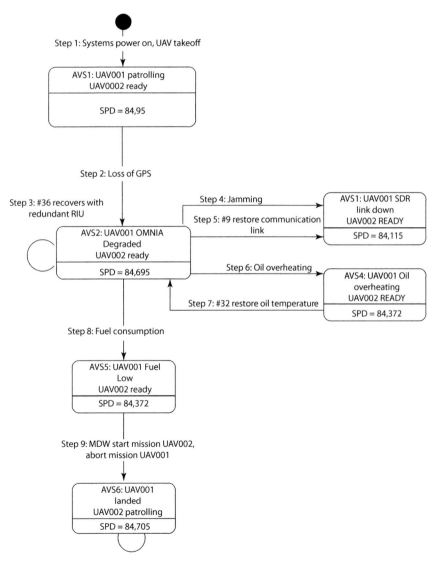

**FIGURE 6.16**
Scenario state diagram.

As shown by the state diagram (Figure 6.16), during the execution of the scenario, the system SPD level assumes different values, as a result of the attack surface metrics composition rules. Each macrosubsystem contributes, with its own internal status, to define the system SPD level.

The messages are exchanged between the UAVs and the ground station on a periodic basis or in an event-driven way. The periodic messages report the status of the UAVs and of the mission. They are generated by the OMNIA subsystem, which collects status information from IQ_Engine and sends it to the ground control station passing through the S-Gateway and the SDR subsystems.

The event-driven messages are transmitted when specific conditions occur (e.g., fault or threats) and represent the commands generated by the SHIELD middleware running on the ground control station and sent to the UAVs.

During mission execution, it is possible to observe that all the system events do not influence the system SPD level. This behavior is due to the nature of normal internal system operations managed, for example, by surveillance personnel: these operations do not influence the SPD level of the scenario because in that particular state, the SPD functionalities are not necessary, but their contribution to the overall scenario SPD level is always present.

The following list provides a detailed description of all the steps of the scenario execution:

T0  An operator set username and password to connect to the ground control station, using the operating system facilities.

T1  All the subsystems of UAV_0001 and UAV_0002 are powered on and perform their built-in tests and application launching. At start-up, the OMNIA power-up software component performs the built-in tests (Figure 6.17).

The SHIELD middleware performs the initialization of the ground control station (Figure 6.18).

T2  The communication between UAVs and the control station is established on the predefined channel, using the soldier broadband waveform (SBW) waveform; therefore, a secure communication channel is established (cryptography activated: CRYPTO_ON status and LINK_OK status). The UAV is ready to take off (Figure 6.19).

OMNIA performs a REQUEST_STATUS command to IQ_Engine. IQ_Engine sends the UAV ID, Mission ID, and MISSION status. This information, passing through OMNIA, S-Gateway, and the SPD-driven smart transmission layer, is delivered to the ground control station.

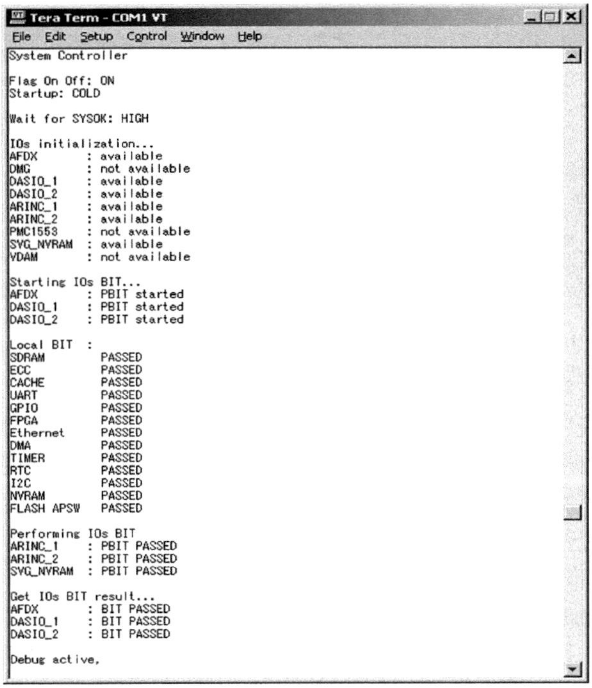

**FIGURE 6.17**
OMNIA power-up sequence.

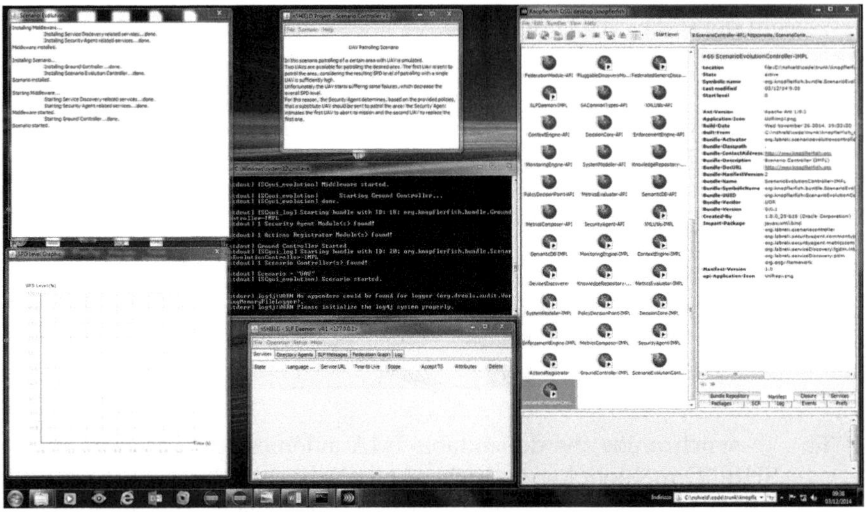

**FIGURE 6.18**
Ground control station initialization.

X-Plane/IQ_Engine                    OMNIA display

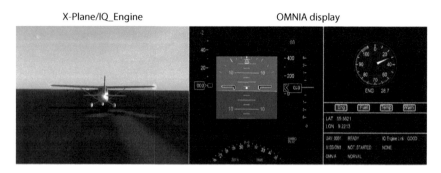

**FIGURE 6.19**
OMNIA display : UAV ready to take-off.

T3    The SHIELD middleware registers UAV_0001 and UAV_0002 as available. The information of the available UAVs with their status is visualized on the ground control station.

UAV_0001:READY

MISSION_001:NOT_STARTED

MISSION_FAILURE_CAT:NONE

OMNIA_STATUS:NORMAL

SDR_STATUS:LINK_OK,CRYPTO_ON

UAV_0002:READY

MISSION_002:NOT_STARTED

MISSION_FAILURE_CAT:NONE

OMNIA_STATUS:NORMAL

SDR_STATUS:LINK_OK,CRYPTO_ON

The system SPD level is calculated and reaches the maximum value of 1.

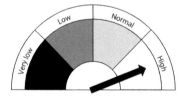

T4    To synchronize the dependable IMA avionics application to the SHIELD methodology, which relies on the evaluation of the scenario's SPD level, the onboard S-Gateway converts these internal parameters to match the correct correspondence with values necessary for the SPD calculation.

The SPD level is calculated by the SHIELD middleware and displayed.

T5 The middleware, evaluating the system status, makes the decision to start the mission, visualizing the START MISSION command. In the ground control station, the operator selects the START MISSION command. The middleware generates the start command on UAV_0001. The S-Gateway sends the command to OMNIA, and OMNIA to IQ_Engine. The SPD level remains at the maximum level.

T6 The mission starts, OMNIA and IQ_Engine actuate the START_MISSION command, and the UAV_0001 takes off. The UAV_0001 is controlled by IQ_Engine, which follows MISSION_001. On the flight simulator display, the UAV taking off is visible (chase view of X-plane). In the ground control station, the following states are visualized (Figure 6.20):

UAV_0001:APPROACHING

MISSION_001:IN_PROGRESS

MISSION_FAILURE_CAT:NONE

OMNIA_STATUS:NORMAL

SDR_STATUS:LINK_OK,CRYPTO_ON

T7 OMNIA continuously collects data from the flight simulator and from IQ_Engine. The OMNIA display provides flight information, including the avionics and aircraft health status and the position of the UAV in terms of latitude and longitude.

T8 According to the request coming from the S-Gateway, UAV_001 sends to the ground station information related to the status of the mission and of the UAV, and the mission failure category. Status data are visualized on the ground station. UAV_001 arrives in the patrolling area (Figure 6.21).

X-Plane/IQ_Engine     OMNIA display

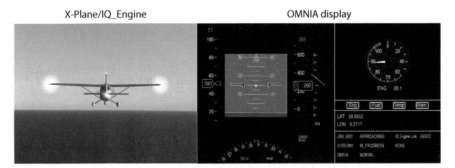

**FIGURE 6.20**
OMNIA display: UAV take-off.

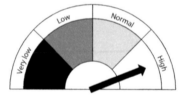

**FIGURE 6.21**
Status of the UAV mission with related failure category.

UAV_0001:PATROLLING

MISSION_001:IN_PROGRESS

MISSION_FAILURE_CAT:NONE

OMNIA_STATUS:NORMAL

SDR_STATUS:LINK_OK,CRYPTO_ON

The SPD level is still at the maximum value.

T9   During the mission, suddenly an internal fault occurs on UAV_001: OMNIA detects the presence of a hardware fault on the GPS sensor (Figure 6.22).

T10  OMNIA sets the mission failure category to mild mission alert (MMA) and sends it to the ground station (Figure 6.23).

UAV_0001:PATROLLING

MISSION_001:IN_PROGRESS

MISSION_FAILURE_CAT:MMA

OMNIA_STATUS:NORMAL

SDR_STATUS:LINK_OK,CRYPTO_ON

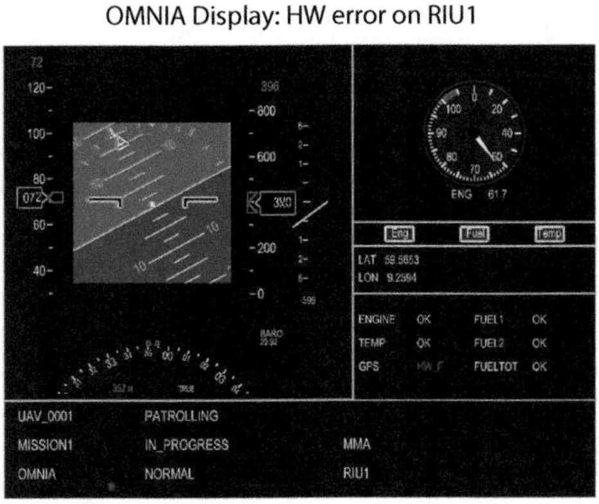

OMNIA Display: HW error on RIU1

**FIGURE 6.22**
OMNIA display: HW fault (GPS) on RIU 1.

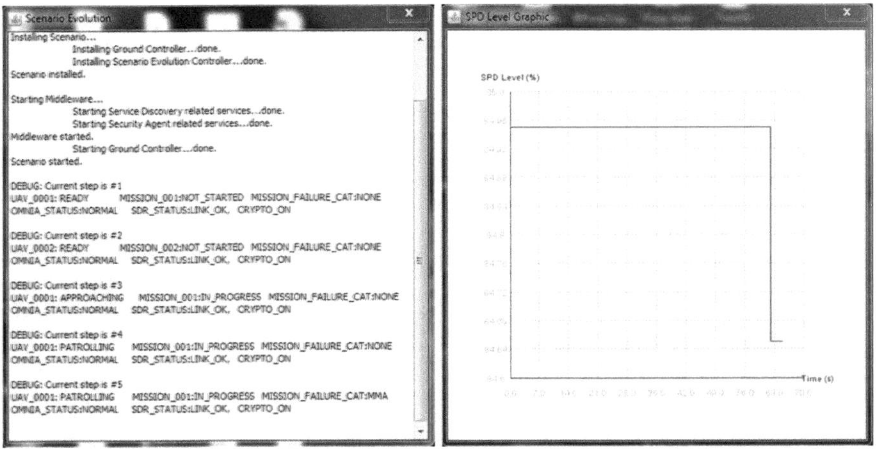

**FIGURE 6.23**
Status of the mission with related failure category.

The system SPD normalized value is 0.6.

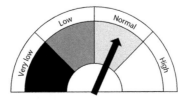

T11   OMNIA automatically recovers the fault exploiting the redundant RIU2. The OMNIA status is now set to DEGRADED (Figure 6.24).

**FIGURE 6.24**
OMNIA display: HW fault (GPS) recovered. OMNIA system DEGRADED.

T12 While OMNIA is performing the fault recovery procedure, an event-driven message about the new status of the UAV_001 is transmitted to the ground station:

UAV_0001:PATROLLING

MISSION_001:IN_PROGRESS

MISSION_FAILURE_CAT:NONE

OMNIA_STATUS:DEGRADED

SDR_STATUS:LINK_OK,CRYPTO_ON

The SHIELD middleware calculates the new SPD level, based on the received new UAV metrics. The middleware will store this new status of the UAV_001, but no alert condition is raised; therefore, it will not take any countermeasure. Consequently, the patrolling mission can proceed.

T13 A second fault occurs: a jammer starts creating interference on the channel used for communication, disabling the link with the UAVs (Figure 6.25).

The normalized SPD level reaches its minimum value: 0.

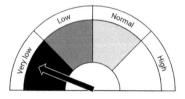

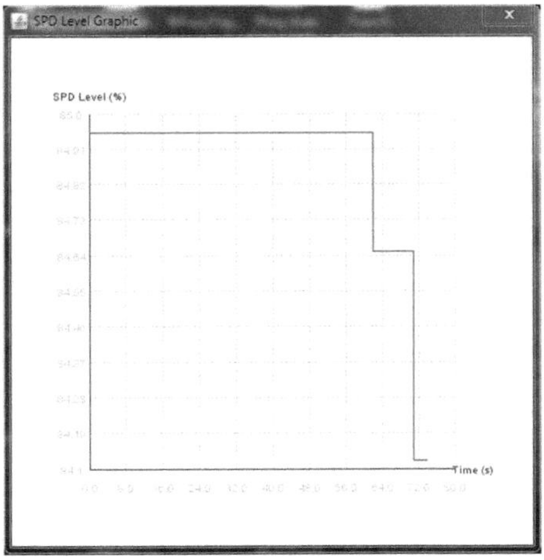

**FIGURE 6.25**
Graphic SPD level: Trend after two failures.

T14  The mission failure category MMA is sent to the SHIELD middleware for the calculation of the new SPD level, through the S-Gateway.

UAV_0001:PATROLLING

MISSION_001:IN_PROGRESS

MISSION_FAILURE_CAT:MMA

OMNIA_STATUS:DEGRADED

SDR_STATUS:LINK_DOWN,CRYPTO_OFF

T15  The SHIELD middleware calculates and visualizes the new SPD
level, based on the received new UAV metrics. The middleware
will store this new status of the UAV_001 but will not take any
counteraction until a new event happens in the system. Thus, the
patrolling mission can proceed (Figure 6.26).

T16  The SPD-driven smart transmission layer detects the occurrence
of jamming and switches the communication channel (channel
surfing) according to a predefined pattern. SDRs are capable of
switching to another licensed predefined channel, thus reinstat-
ing the communication.

T17  The mission failure category NONE is sent to the middleware
for the calculation of the SPD parameters, through the S-Gateway.
The SPD level is recalculated and visualized (Figure 6.27).

UAV_0001:PATROLLING

MISSION_001:IN_PROGRESS

MISSION_FAILURE_CAT:NONE

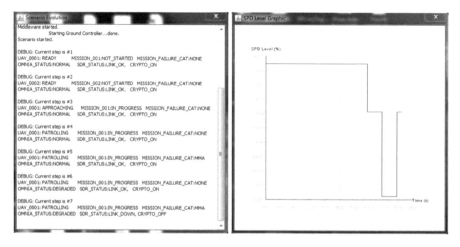

**FIGURE 6.26**
Status of the UAV mission with related failure category.

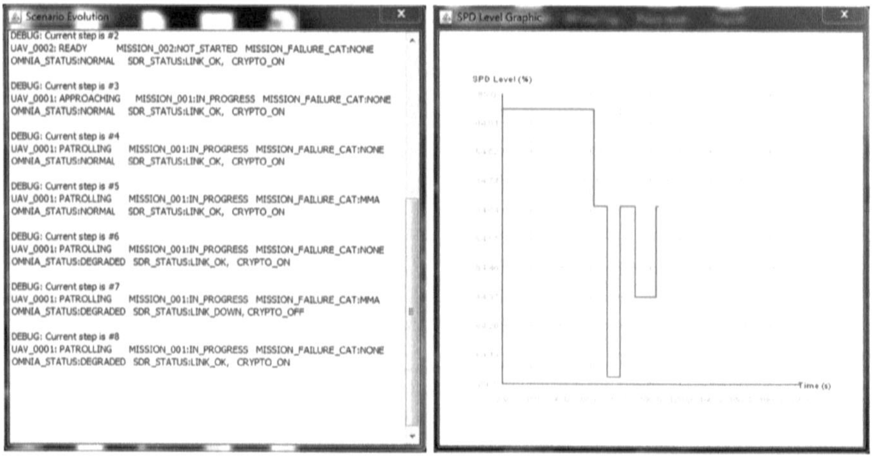

**FIGURE 6.27**
Status of the UAV mission with related failure category.

OMNIA_STATUS:DEGRADED

SDR_STATUS:LINK_OK,CRYPTO_ON

The normalized SPD value is restored to 0.6.

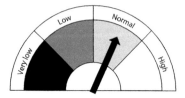

T18  A third fault happens on the UAV_0001: IQ_Engine detects that the oil temperature is too high. OMNIA performs an integrity check on the acquired data and visualizes that the temperature is over the normal range. IQ_Engine sends the MMA status to OMNIA (Figure 6.28).

T19  OMNIA sends new status information to the S-Gateway. The SHIELD middleware calculates the new SPD level, and the mission failure category is updated to MMA.

UAV_0001:PATROLLING

MISSION_001:IN_PROGRESS

MISSION_FAILURE_CAT:MMA

OMNIA_STATUS:DEGRADED

SDR_STATUS:LINK_OK,CRYPTO_ON

The normalized SPD level falls to 0.3.

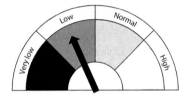

T20  IQ_Engine recovers the fault acting on the "cowl flap." The oil temperature returns to a normal value. OMNIA receives the updated status and sends it to the S-Gateway. The SHIELD middleware calculates the new SPD level. The mission failure category returns to NONE (Figure 6.29).

**FIGURE 6.28**
OMNIA display: UAV mission---failure of oil overheating.

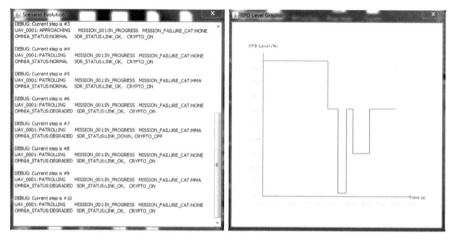

**FIGURE 6.29**
Status of the UAV mission with related failure category.

> UAV_0001:PATROLLING
>
> MISSION_001:IN_PROGRESS
>
> MISSION_FAILURE_CAT:NONE
>
> OMNIA_STATUS:DEGRADED
>
> SDR_STATUS:LINK_OK,CRYPTO_ON

The normalized SPD level returns to 0.6.

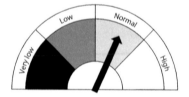

T21  A fourth fault occurs: the fuel tank level is low, due to a long flight. The UAV mission failure category, sent to the ground station, is ordinary mission abort (OMA).

> UAV_0001:PATROLLING
>
> MISSION_001:IN_PROGRESS
>
> MISSION_FAILURE_CAT:OMA
>
> OMNIA_STATUS:DEGRADED
>
> SDR_STATUS:LINK_OK,CRYPTO_ON

T22  This fault generates a situation that requires immediate action, because it could implicate the loss of the UAV_0001 and represent a threat for the entire mission. Due to the low fuel status, the UAV_0001 cannot complete the mission, and a safety condition alert is raised. A countermeasure must be identified quickly: the information about the low fuel level is translated by the S-Gateway in an SPD parameter and sent to the ground station. The ground station recalculates and visualizes the new SPD level of the system based on the received new metrics and compares it against the desirable scenario SPD level (Figures 6.30 and 6.31).

The normalized SPD level falls again at 0.3.

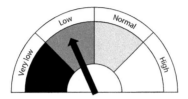

T23 In the ground control station, an alert message is displayed, and the operator is informed that UAV_0002 is ready to take part in the mission. The operator selects the START MISSION command for the UAV002 (Figure 6.32).

UAV_0002:APPROACHING

MISSION_002:IN_PROGRESS

MISSION_FAILURE_CAT:NONE

OMNIA_STATUS:NORMAL

SDR_STATUS:LINK_OK,CRYPTO_ON

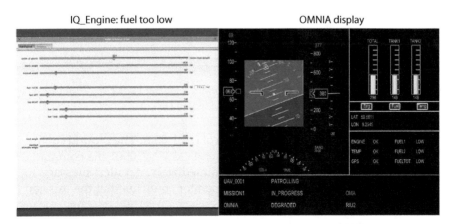

**FIGURE 6.30**
OMNIA display: UAV mission: lack of fuel.

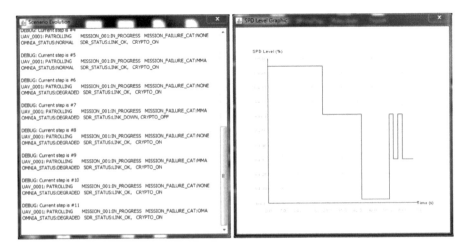

**FIGURE 6.31**
Status of the UAV mission with related failure category.

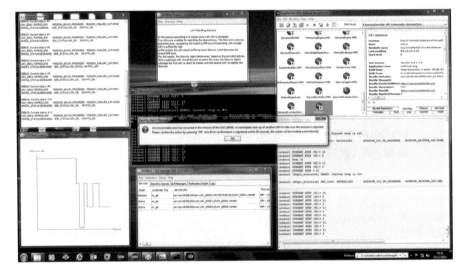

**FIGURE 6.32**
Recovery of the UAV mission: Start mission for the second UAV.

T24  UAV_0002 takes over the mission of UAV_0001. The ground control station through the S-Gateway sends to UAV_0001 the command to abort the mission and land immediately. An alert message to the operator is displayed. The operator confirms that the mission can be aborted.

UAV_0002:PATROLLING

MISSION_002:IN_PROGRESS

MISSION_FAILURE_CAT:NONE

OMNIA_STATUS:NORMAL

SDR_STATUS:LINK_OK,CRYPTO_ON

The normalized SPD level is still at 0.3.

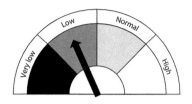

T25 UAV_0002 continues the mission, starting from the position where the first UAV had aborted it, while UAV_0001 lands. UAV_0002 restores the desired SPD level of the scenario by accomplishing the mission (Figures 6.33 and 6.34).

UAV_0001:LANDED

MISSION_001:ABORTED

MISSION_FAILURE_CAT:NONE

OMNIA_STATUS:NORMAL

SDR_STATUS:LINK_OK,CRYPTO_ON

The normalized SPD level is restored to 0.7.

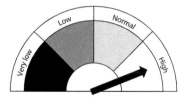

## Final Remarks

The objective of the SHIELD methodology applied to an airborne application is to foster the technological progress of avionics embedded systems through an innovative approach, conceived for critical applications and focusing on the safety requirements.

The proposed solution is able to handle dynamic changes in real-time embedded systems, as well as virtualization of the application, thus enabling faster system reconfiguration with recurring costs and especially nonrecurring costs (cost of development).

**FIGURE 6.33**
OMNIA display: UAV-1 aborted mission.

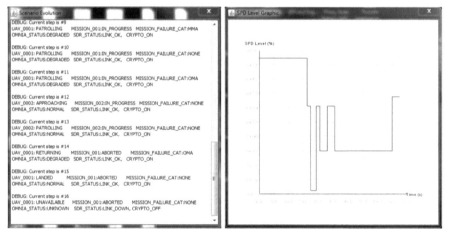

**FIGURE 6.34**
Status of the UAV mission with related failure category.

The SHIELD approach introduces metrics and methodologies to jointly target certain technologies for the use of different avionics applications, and thus facilitate a radical change in the ability to exchange information between heterogeneous embedded systems.

The innovations that led to this radical change include:

- Significant expansion of use in all areas, particularly in areas where performance and security requirements (mixed-criticality environment) are becoming more common

- Better flexibility and optimization in trade-off between costs, security, energy consumption, and reliability
- Better interoperability and connectivity

Methodologies and rules for avionic certification must fulfill stringent reliability requirements, especially the ones that have a direct impact on flight safety (safety or mixed critical computer), the avionics scenario architecture presented in this chapter could be used as a prototype reference architecture in a future assessment of SAE ARP 4761 ("Guidelines and Methods for Conducting the Safety Assessment Process on Civil Airborne Systems and Equipment"), together with other projects, starting from the N-AMMC IMA-based computer actually certified in accordance with DO178B and DO254 level A, for introducing the SPD requirements in a mixed/safety avionic system.

The avionics scenario application consists of a laboratory exercise finalized to demonstrate the dependability and security of the system, composed by 11 SHIELD native prototypes. Each of the 11 prototypes is located at one of the SHIELD architectural levels: node, network, and middleware or overlay.

The ADDP used for the communications between all the IMA computers is a paradigm based on the publisher–subscriber mechanism (RTPS) used for the exchange of information between different nodes in a system. The implementation of the ADDP is not compliant with the standard Object Management Group request for proposal (OMG RFP) "Data-Distribution Service Interoperability Wire Protocol," because of the restriction imposed by the guideline RTCA DO-178/EUROCAE ED-12, which makes the overall OMNIA-IMA software certifiable for civil avionics.

The avionics scenario therefore tries to contribute to standards in the following aspects:

- An infrastructure supporting dynamic changes of software, as well as the dynamic interconnection of different embedded systems and system reconfiguration at runtime. This is not covered by existing standards today. Standardizing such platforms is, however, an important key to ensure interoperability and thus to significantly improve the economic competitiveness and impact.
- Current safety standards do not address dynamic system adaptations—some standards even prohibit necessary concepts, for example, dynamic reconfiguration. The impact of security breaches on safety is only vaguely addressed by safety standards. The results of the solution proposed in this book on safety and security will therefore try to provide important contributions to future safety standardization activities, including security as an important property of system dependability and trust.

## References

1. C.B. Watkins and R. Walter. Transitioning from federated avionics architectures to integrated modular avionics. Presented at Digital Avionics Systems Conference, Dallas, TX, October 2007.
2. Distributed equipment Independent environment for Advanced avioNics Application (DIANA Project); SCAlable & ReconfigurabLe Electronics plaTform and Tools (SCARLETT Project); Avionics Systems Hosted on a distributed modular electronics Large scale dEmonstrator for multiple tYpe of aircraft (ASHLEY Project).

# 7

# Railway Domain

Paolo Azzoni, Francesco Rogo, and Andrea Fiaschetti

## CONTENTS

## Introduction

Modern railway infrastructure has been developed rapidly in the last two decades, including its communication systems. In the past, only wired communication solutions were used for signaling and data communication in the railway industry. Recently, wireless communication systems have emerged as a valid alternative for wired systems. The wireless solutions are currently adopted very frequently as protection systems to monitor assets within a railway infrastructure and ensure reliable, safe, and secure operation. This trend also tries to face common threats associated with freight transportation, from the perspective of improving critical infrastructure protection and railway security. In this context, the detection of anomalous operating situations, of dangerous environmental conditions in wagons, and of threats like burglary, acts of vandalism, or theft represents an example application of great interest for freight train monitoring. In the case of mobile assets monitoring, interesting indicators of the train health include vibrations, smoke, tilt, position, ambient temperature, and humidity in wagons. For instance, temperature monitoring safeguards wagons against fire outbreak, while vibration and tilt monitoring proactively prevents potential accidents, which could be extremely dangerous in the transportation of hazardous materials. In addition, for security reasons, other important indicators are provided by access control devices, allowing us to prevent cargo thefts and manumissions of the material inside the train cars.

The adoption of a wireless monitoring solution presents several issues and challenges, which are addressed by the innovative features provided by SHIELD:

- Since most freight cars are unpowered, there is the need to provide a power-aware and power-autonomous system architecture.
- The railway is a geographically distributed infrastructure, and train cars are mobile entities; therefore, there is the necessity to provide a connection to a central monitoring system through a wireless wide area network (WAN).
- The solution must be low cost, easy to install, and easy to maintain.
- The overall monitoring system is highly heterogeneous in terms of detection technologies, embedded computing power, and communication facilities, with very different hardware–software architectures and very different capacities for providing *security*, *privacy*, and *dependability* (SPD).

---

## Transportation of Hazardous Material: A Case Study

The transportation of hazardous materials deals with the physical management of solid, liquid, or gas substances that can expose the health and safety of individuals to extreme risks and cause serious environmental damage or pollution. The class of hazardous materials includes radioactive materials, flammable materials, explosive materials, corrosive materials, and oxidizing, asphyxiating, toxic, infectious, or allergenic materials. This class also includes compressed gases and, more in general, all the materials that can become dangerous under specific conditions (e.g., certain substances release flammable gases when put in contact with water). Common examples of hazardous materials are oils, paints, batteries, fertilizers, and chemicals.

The shipping of hazardous materials can be performed in different ways and by various means of transportation; each one is characterized by specific risks and security issues. The handling of hazardous materials includes identification, labeling, packaging, storage, transportation, and disposal, and requires the continuous monitoring of the environmental conditions and of the security level of the whole railway infrastructure.

In a similar context, the adoption of the SHIELD methodology for monitoring freight trains transporting hazardous materials represents a significant vertical application to demonstrate the added value of SHIELD in the critical infrastructure protection and railway security domain. More specifically, this case study allows us to validate the whole SHIELD methodology, including the "composition" capabilities, which allows us to orchestrate a complex system of systems from the SPD perspective. The case study is focused on:

- Providing SPD functionalities to off-the-shelf smart sensors, measuring environmental parameters (e.g., temperature and vibration level).
- Developing a monitoring application that detects anomalous operating conditions.
- Measuring and testing the SPD functionalities of the overall monitoring system.

### Norms and Regulations

The international regulations that apply to the transportation of hazardous materials are mainly represented by the "UN Recommendations on the Transport of Dangerous Goods," known as the "Orange Book" and published by ONU in 1957. The objective of this document is to regulate the circulation of hazardous materials, guaranteeing the security of people, environments, and goods during the transportation. The Orange Book represents the baseline for all the regulations regarding the transportation of hazardous materials, at international and European levels. At the international level, the United Nations Economic and Social Council (ECOSOC) and the Transport of Dangerous Goods Subcommittee (TDG Subcommittee) are responsible for defining and applying this regulation. The European Union (EU) guarantees that all countries adopt the ONU regulation for the transportation of hazardous materials, at both European and national levels. In Europe, the transportation of hazardous materials is subject to a specific directive, applicable to road, rail, or maritime transportations. Based on this directive, the transportation is authorized if

- It is compliant with the provisions specified in the RID (Règlement concernant le transport ferroviaire International des marchandises Dangereuses) agreement for the international carriage of dangerous goods by rail

- It respects the European Agreement Concerning the International Carriage of Dangerous Goods by Road (ADR)
- It observes the DNA (Accord Européen relatif au transport international des marchandises Dangereuses par voie de Navigation intérieure), the European agreement concerning the international carriage of dangerous goods by inland waterways

## Threats and Issues

The Hazardous Materials Transportation Act (HTMA), the first U.S. law on the transportation of hazardous materials, has defined "hazardous materials incident" as an unintentional emission of dangerous substances during the transportation, loading and unloading, and storage of the material. The Pipeline and Hazardous Materials Safety Administration (PHMSA) periodically informs of accidents occurring since the entry into force of the HTMA [1]: according to the PHMSA, the number of accidents from 2001 to 2010 varied from 15,000 to 20,000, depending on the year. Most accidents occur during transport by highway (about 87.1%), while air accidents account for 7.9%, the railway accidents are approximately 4.7%, and 0.3% are marine. Most of these accidents are in the transport of petrol tankers. Although most of the reported incidents did not have serious consequences, unfortunately, a still significant number of incidents present casualties or environmental damage, with an important socioeconomic impact.

To increase the safe transport of dangerous goods, it is necessary to monitor the following components:

- Locomotive and freight cars (speed, acceleration, vibration, and inclination). Through these data, it is possible, for example, to detect collisions and derailments and analyze the behavior of the driver (also noting any breaches that may compromise the security of cargo, such as exceeding speed limits on the way).
- Transported goods. If, for example, a liquid is transported, it is important to monitor the pressure, the liquid level (information that allows the detection of losses), and temperature (information needed in case of flammable goods).
- GPS position.

## System for Hazardous Materials Transportation Monitoring

SHIELD-enabled technologies represent a solid baseline for the design and development of a monitoring system that satisfies the requirements of the hazardous materials transportation domain. The system focuses on the monitoring and protection of fixed or mobile assets, including freight cars, against both natural and intentional threats. It is based on a security control center, where a security management system is capable of remotely collecting alarms about some environmental parameters (e.g., temperature, humidity, vibrations, and intrusions) and monitoring the status of assets. The solution proposed the advantages of using self-powered wireless devices, with advanced intelligent capabilities.

Regarding the specific case of freight train monitoring, the main requirements to be fulfilled by the autonomous monitoring system are as follows:

- Secure handling of the critical information of the transported material
- Secure and dependable monitoring of the transport

In the considered scenario, both natural and malicious threats can have an impact on system availability and, indirectly, on safety. Some examples of critical situations related to these threats include (Figure 7.1)

- The mismanagement of hazardous materials that potentially translates into environmental pollution and human health threats
- Lack of control of car integrity that could allow the theft or leakage of hazardous material

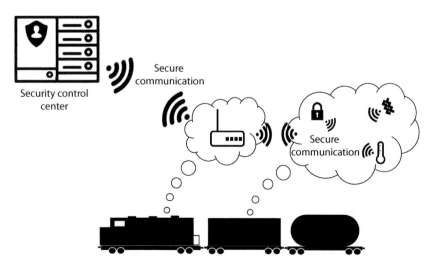

**FIGURE 7.1**
Transportation monitoring system architecture.

The SHIELD-based monitoring solution is based on four subsystems:

- An S-Gateway, consisting of a field-programmable gate array (FPGA)–based SHIELD power node that provides modular system reconfiguration and self-dependability, software security and privacy, and management of power sources.
- A wireless sensor network (WSN) for data collection that provides SPD functions in an integrated embedded sensor.
- The SHIELD semantic model (ontology) and middleware.
- A security control center.

The monitoring system is aimed at the detection of abnormal operating environmental conditions on board vehicles, as well as threats of burglary. Therefore, the proposed solution follows a basic working logic: whenever an abnormal event (e.g., very high temperature or an anomalous vibration pattern) is detected by a wireless sensor, its transmission unit is activated and the related data are received by a local SHIELD gateway for elaboration (S-Gateway; see Chapter 3). If the anomaly is confirmed, the S-Gateway sends an appropriate alert message to a security control center to manage the anomaly.

The objective is to simplify anomaly and fault detection (by considering the same parameters in the same area for sensor measuring), to improve the overall detection reliability, and to make possible complex threats detection (including also distributed heterogeneous sensors).

The self-powered smart sensors send data to the S-Gateway through a secure connection (the information between the sensors and gateway is encrypted and the communication is resilient to jamming), and the S-Gateway transmits data to the control room by means of a secure wireless connection (e.g., Hypertext Transfer Protocol Secure [HTTPS] or Secure Sockets Layer [SSL]).

## S-Gateway

Dependability is one of the most important aspects for many embedded system markets. Usage of hardware redundancy is frequently a way to reach high dependability. But hardware redundancy increases system cost drastically. The usage of FPGA-based embedded systems can drastically solve this issue, because FPGAs are intrinsically redundant (due to their design characteristics).

The concept of runtime reconfiguration is applicable to FPGAs and represents the capability to modify or change the functional configuration of the device during a fault or normal operation, through either hardware or software changes. This capability can be specialized in different ways in order to reduce component count, power consumption, reuse, fault tolerance, and so forth, increasing in this way the global SPD capabilities of the system.

The selection of this technology aims to develop a new approach for an FPGA runtime reconfiguration that is capable of increasing the nodes' dependability. Following this objective, the S-Gateway has been equipped with an FPGA supplied by Xilinx that provides dynamic partial reconfigurability (DPR) features.

The role of the S-Gateway in the selected railway scenario is to collect data from the car of the freight train, preprocess the data, and interact with the security control center to signal and manage potential anomalies, faults, and threats.

In the car, a WSN collects information and sends it to the S-Gateway via frequency-shift keying (FSK) modulation. The S-Gateway receives the signals, demodulates them, decrypts and processes the data, and sends the data to the security control center through the SHIELD network.

The adoption of the S-Gateway in the railway domain, and specifically in the use case described in this chapter, provides the following advantages and benefits:

- *Dependability*: Dependability is ensured by detecting errors in the demodulator, and tolerating them, through FPGA partial reconfiguration. After a fault being detected by the FPGA and the fault affecting the demodulator, an error is generated and the FPGA is partially reprogrammed.

- *Security*: Security is ensured by receiving encrypted data and being able to decrypt it.

- *Self-reconfiguration*: The system can automatically reconfigure itself because it can detect when a different carrier is used in the FM signal, and reconfigure the FPGA to adapt to a new carrier. This situation is forced by switching at runtime the carrier that is being used by the modulator to produce the FM signal.

- *Metrics*: SHIELD metrics can easily be calculated by collecting and providing data such as the number of messages received, errors detected, and so forth.

- *Composability*: The S-Gateway provides discovery and composability information, such as the identification of the modules in the FPGA and its characteristics.

- *High performance*: The S-Gateway provides real-time demodulation, decryption, and preprocessing for all the received data. This feature is fundamental, for example, for intrusion detection.

- *Legacy component integration* with the SHIELD ecosystem is ensured by providing SPD functionalities to a legacy FM demodulator and enabling the collection of metrics and the provision of composability information.

### FSK Modulator

In many industrial, telecommunication, or transportation appliances the transmission of digital information through noisy environments makes use of FSK due to its immunity to "adverse environment" conditions (i.e., electromagnetic interference, noise, surge, and ground loop–ground plane shift problems), its ability to transmit data across commutators or sparking sources (sliding contacts, slip rings, rolling wheels, etc.), and the use of any two conductor wires, shielded or unshielded. Furthermore, it is employed even in wireless communications, such as in a digital cellular communication system (Global System for Mobile Communications [GSM]), using Gaussian minimum shift keying (GMSK), a special type of FSK.

FSK is a frequency modulation scheme in which digital information is transmitted through discrete frequency changes of a carrier wave. The simplest FSK is binary FSK (BFSK). BFSK uses a pair of discrete frequencies to transmit binary (0 and 1) information. With this scheme, the 1 is called the mark frequency and the 0 is called the space frequency. The time domain of an FSK-modulated carrier is illustrated in Figure 7.2.

The context application, used to demonstrate how the implementation of an FSK-based modulator on the S-Gateway could be compliant with the security, dependability, and privacy requirements of a SHIELD system, is illustrated in Figures 7.3 and 7.4.

The FSK modulator has been implemented using an Altera board. The board is the Altera cycloneIII_3c120_dev development kit [2]. The selected

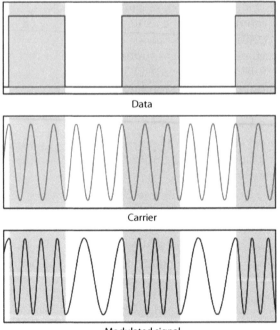

Data

Carrier

Modulated signal

**FIGURE 7.2**
FSK signal example.

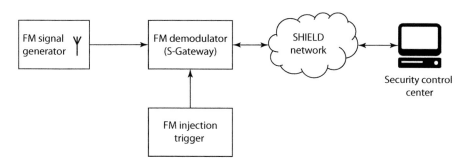

**FIGURE 7.3**
FSK demodulator test and evaluation context.

**FIGURE 7.4**
FSK demodulator laboratory test setup.

FPGA is model EP3C120F780 FPGA, belonging to the Cyclone III Altera FPGA family. This family is a very good compromise between cost and performance.

The 32-bit SRAM and 16-bit NOR flash give the developer very good support for a low-resource system prototype, while the 256 MB of DDR2 RAM is the perfect precondition for an operating system full-featured system prototype.

The Ethernet interface allows the development of a network-based system prototype, while the different graphical interfaces (graphic, characters, and Quad 7 segments) allow the developer to export the output to the external world in many different ways. Finally, the hIgh speed mezzanine card (HSMC) ports are the perfect method to expand the capability on the board.

**TABLE 7.1**

FSK-Modulated Signal Parameters

| Digital Signal to Transmit | Parameter |
|---|:---:|
| FSK rate | Variable between 50 and 100 Hz |
| **FSK-Modulated Signal** | **Values** |
| Space frequency | 968 or 1937 Hz (f1 in Figure 7.3) |
| Mark frequency | 2062 or 1031 Hz (f2 in Figure 7.3) |
| Amplitude | 1 Vpp |
| **Analog-to-digital sampling rate** | 32 or 16 kHz |

### FSK Signal Generator

The FSK signal generator has been implemented using the Altera board. The signal data are saved in a wave file that may be played using simple wave player software running on the PC.

The generated signal consists of audio FSK-modulated signal parameters (Table 7.1).

Figure 7.5 illustrates an FSK signal sample, while Table 7.2 reports the specification of the audio signal.

### FSK Demodulator

The FSK demodulator receives the data samples sent by the signal generator and performs a digital demodulation of the samples. After that, it analyzes the characteristics of the sampled signal, in order to check whether it is compliant with the expected characteristics. If it is compliant, the FSK demodulator decrypts the signal and provides the sample to the security control center, through a web service accessed from the SHIELD network,

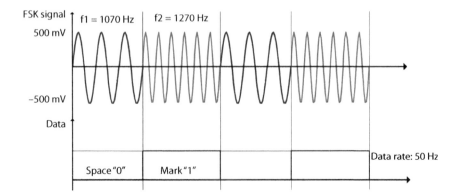

**FIGURE 7.5**
FSK signal sample.

**TABLE 7.2**

FSK Audio Signal Specification

| FSK Data Sources | |
|---|---|
| Telnet console characters | 8 bit; ASCII-coded characters |
| Proximity sensor sampled data | 8 bit; proximity signal level 8 bit quantized |
| General data file | 8-bit characters; ASCII-coded file |

using the Ethernet communication interface. If an error occurs, it generates metrics that highlight the discovered problem.

During normal operation, the system continuously sends *metrics data* to the SHIELD network. They contain information about the health status of each internal module (hardware and software) of the system. This information is processed by the security control center, which is responsible for deciding what action is to be done (reconfigure or recovery) based on the obtained results.

### Fault Injection Trigger

The fault injection trigger is a mechanism that performs a change of the processing algorithm parameters of the demodulator block. It is generated with a very simple trigger event. The scope of this block is to inject a fault into the demodulator process for test and debug purposes.

### SHIELD Control Center

A remote PC, connected to the network via the Ethernet, is used to simulate the security control center. A web browser running on the PC allows a remote user to

- Receive and store the data samples sent by FSK demodulator
- Receive and analyze the metrics of the system
- Send the commands (reconfigure and recover) to the system

### FSK Integration with the S-Gateway

The FSK demodulator has been integrated into the S-Gateway and can be considered a part of the gateway itself. Figure 7.6 illustrates the block diagram of the logical architecture of the integration.

From a hardware point of view, the integration is performed directly interfacing the S-Gateway FPGA with the Altera board, as illustrated in Figure 7.7.

The demodulator implementation is illustrated in Figure 7.8.

The *fmin* is the byte representing the input-modulated sample; the *clk* is the necessary clock signal to process the incoming modulated byte stream. The

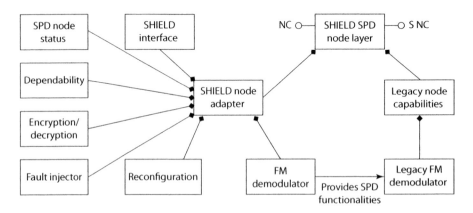

**FIGURE 7.6**
FSK demodulator SPD node layer.

*dmout* word (only 12 bits meaningful) is the demodulated baseband signal samples.

A more detailed description of the demodulator is the given in Figure 7.9.

The *fmin* byte stream is supposed to be fed by the FM modulator data output and clock, while the *fmout* word is supposed to feed a FIFO (first in, first out).

The CPU core is a PowerPC® 440 and provides the following features:

- PowerPC $440 \times 5$ dual issue, superscalar 32 bit
- 32 KB instruction cache, 32 KB data cache example design
- Memory management unit (MMU)
- Crossbar interconnect with nine inputs and two outputs (128 bits wide), implemented in hardware
- 128-bit processor local bus (PLB), version 4.6
- High-speed memory controller interface
- Auxiliary processor unit (APU) controller and interface for connecting the floating-point unit (FPU) or custom coprocessor

The cache, the MMU, and the FPU are configurable; therefore, the user can select to use or not the component, allowing the possibility to save resources of the FPGA. This aspect is essential for the scalability of the proposed solution.

The fault injection trigger is a simple push button: it is polled by the fault injection application, and when asserted, it generates a failure of the demodulator.

When the fault injector trigger is asserted, the demodulator stops to feed the output FIFO. The event of empty-output FIFO is evidence of a malfunctioning demodulator. The fault injector is an artifact put in the system on chip (SoC) to show how the system restores its capability after a fault.

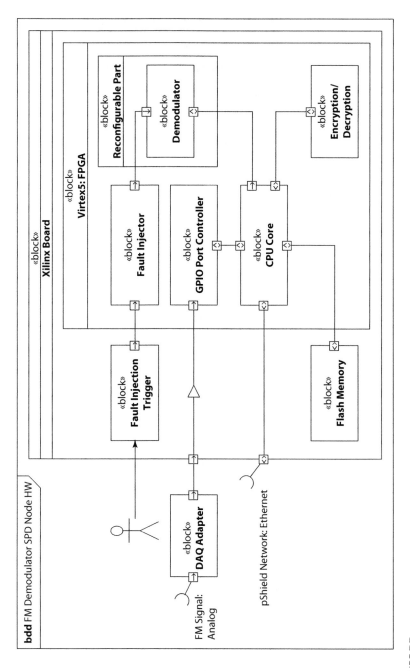

**FIGURE 7.7**
FSK demodulator hardware architecture.

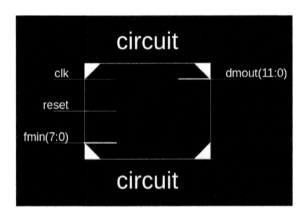

**FIGURE 7.8**
Demodulator implementation block diagram.

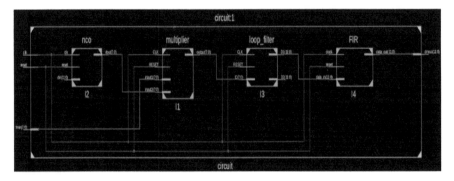

**FIGURE 7.9**
Detailed demodulator diagram.

The encryption and decryption system is implemented at the software level and only through the network layer. The blowfish algorithm is used in both parts of the system (modulator and demodulator).

The flash memory is implemented using a compact flash (CF) card. It is used to store the status of the system and SPD level of the node.

The partial reconfiguration feature is provided on the S-Gateway by the Internet content adaptation port (ICAP) port of the onboard FPGA. This is a proprietary Internet Protocol (IP) core made available by Xilinx Corporation to manage the partial reconfiguration at the device driver level. The ICAP port IP core used is hwicap 6.01.

## Monitoring Application

The monitoring application that implements the use case logic has been written in C language without the use of any general-purpose operating system.

This choice was made to prove the real scalability available on very basic devices, working inside an SPD distributed system.

The application is essentially an infinite loop managing

- A very light IP stack
- An FAT32 file system
- A blowfish encrypting algorithm
- A partial reconfiguration device

The application (main loop) can be described as follows:

- IP application
  - Reading the data structure to send
  - Sending the data structure to the network
  - Reading the data from the stack
  - Filling the data structure to use inside the application
- File system application
  - Building the application files
  - Managing the files for ensuring correct access
- Encryption application
  - Getting clear data
  - Encrypting clear data
  - Filling the data structure to be sent
  - Receiving the encrypted data
  - Decrypting data
  - Filling the private data structure with clear received data
- Reconfiguration application
  - Monitoring the fault injector
  - If the fault event occurs, getting the fresh bit stream from the flash and reloading it into the partial reconfigurable area to restore the proper work of the demodulator

## FPGA Partial Reconfiguration

The FPGA partial reconfiguration is a built-in feature of some Xilinx FPGA chips. The partial reconfiguration allows us to divide all the FPGA chips into two parts: the static configured part and the dynamic configurable part. The configuration of the static part is permanently defined at design time. When required, the static part of the system can load onto the partial reconfigurable part a bit stream implementing a particular hardware function.

The development method to reach such a result needs a particular workflow. Both parts (static and reconfigurable) need to be built from the source codes in the form of net lists.

The system will give a full-featured bit stream file as a result, allowing us to run the default full system configuration, and several partial bit stream files (one for each different synthesized configuration of the system).

The static part can use the ICAP hardware resource of the FPGA to load from a memory support the required partial bit stream file and store it to the partially reconfigurable region, setting up such a region to work the specified way.

It is clearly evident that the partial reconfiguration feature allows the system to be more dependable. In fact, if the designers identify a part of the system absolutely vital for the system itself, they may put such a part into a reconfigurable region and reprogram it just in case of failure of this part.

From the integration point of view, the components delivering the SPD compliance are the microprocessor, the FPGA, the Ethernet chip, and the flash memory. The microprocessor is hardwired into the FPGA; therefore, it ensures robustness and reliability. Furthermore, it implements the decryption of the demodulated data.

The FPGA hosts several crucial IPs (the demodulator, the FIFO data buffer, and the MAC part of the Ethernet equipment) and the native feature of dynamic partial reconfiguration, the core of the dependability feature of the system.

The flash memory allows us to store the dynamic information generated by the system and the several required partial bit streams.

The fault injector subsystem acts on the demodulator. It masks the incoming clock source to the demodulator. This action freezes the demodulated data sending, so the main system recognizes that the demodulator is failing and can start to reconfigure it.

This is one of the possible techniques to force demodulator failure, and it has been chosen because it is relatively simple to add to and remove from the project, allowing the designer to pass from a development environment to the production one with a minimum risk of undesirable side effects.

## Wireless Sensor Network

The environmental conditions inside and outside the train cars, as well as intrusion detection information, are collected by a WSN. The WSN is composed of environmental sensors, which measure physical quantities, including temperature, humidity, and vibration, and an intrusion detection system composed of remote proximity sensors that continuously measure the distance to the surrounding objects. The WSN nodes communicate with the concentrator using a secure connection and the WSN concentrator, and the WSN concentrator communicates with the S-Gateway through the Altera board, which is equipped with a data encryption module and an FSK modulator.

Two different WSNs have been adopted:

1. The WSN in charge of the environmental monitoring is composed of Telos-based sensor nodes, which measure temperature, humidity, and pressure.
2. The second WSN is responsible for the detection of vibration, position, and intrusions and is composed of MICA-based sensor nodes.

### SHIELD Middleware

The SHIELD middleware is responsible for monitoring the SPD behavior of the components of the transportation monitoring system and ensures that the appropriate actions are executed in the case of threats or faults, trying to recover the optimal SPD level. At runtime, depending on the information provided by the components of the monitoring system and considering the current SPD state, the middleware identifies the next SPD state and communicates the new state to the components of the monitoring system. The application logic adopted for monitoring the train cars and the transported hazardous materials allows the middleware to calculate the next SPD state starting from the current SPD state. This logic defines the behavior of the monitoring system when it is exposed to a fault or a menace and allows us to identify the reaction consisting of specific countermeasures. Every component of the system receives the new SPD state and adopts the operation required to "enter" the suggested SPD state. The SHIELD methodology adopted in this dynamic process and the metrics calculation are described in Chapters 4 and 5.

Figure 7.10 illustrates the basic architecture of the SHIELD middleware. The middleware core services and the semantic representation allow the

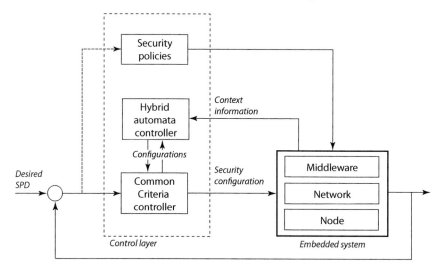

**FIGURE 7.10**
Rationale of the SHIELD middleware.

overlay to discover the system. The overlay is able to compose them to achieve the desired SPD level in two separate ways:

1. By adopting policies
2. By following the Common Criteria composition approach, combined with the context-aware hybrid automata approach

From the component perspective, the prototype of the monitoring system at the middleware level is based on the following elements:

- The middleware core services for discovery and composition of SHIELD components (developed using OSGi)
- A web ontology language (OWL) file representing the SHIELD ontology that is used, in conjunction with the SHIELD middleware, to make the composition possible
- The reasoner for Common Criteria–compliant composition of SPD metrics
- The architectural design and performance analysis of a policy-based approach by which the middleware composition could be driven
- A MATLAB simulation and theoretical formalization of a hybrid automata approach to drive the SPD composition in a context-aware way

In Figure 7.11, the architecture of Figure 7.10 is translated into a real software prototype: the core SPD services (based on the OSGi framework), the SHIELD ontology (OWL), and the Common Criteria reasoner have been implemented in a Java software environment, while the policy-based management and the hybrid automata controller have been designed and simulated using MATLAB Simulink.

## Integrated Monitoring System

The hazardous materials transportation monitoring solution is a system of systems resulting from the integration of the hardware and software components described in the previous sections. This solution involves components covering all the SHIELD layers: node, network, middleware, and overlay. Figure 7.12 illustrates the monitoring system integrated architecture: the gray boxes indicate hardware components, while the black boxes represent software components. The black dashed lines represent logical connections (interfaces between components), while the gray lines represent wireless connections (Figure 7.12).

As anticipated, the wireless sensors collect environmental data from the freight car and detect intrusions, sending to the S-Gateway, through a

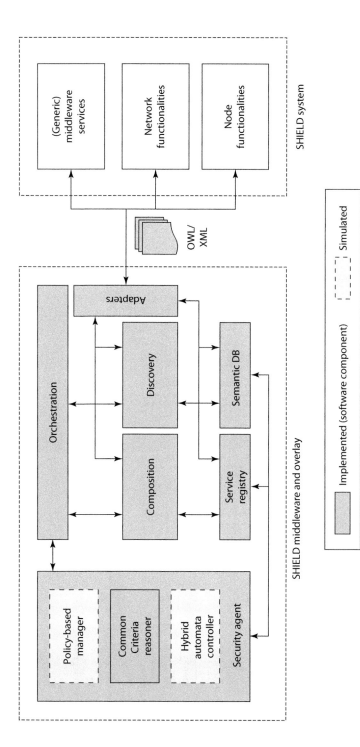

**FIGURE 7.11**
Middleware component integration.

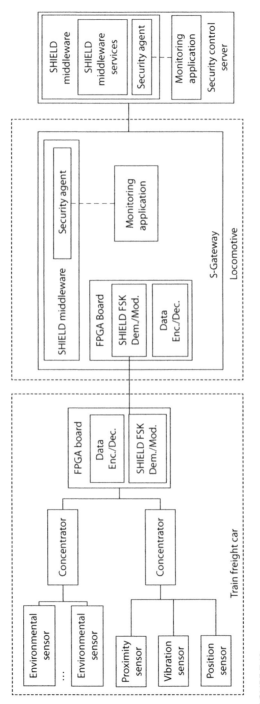

**FIGURE 7.12**
Prototype-integrated architecture.

secure connection, all the information related to anomalous situations. The S-Gateway hosts the monitoring application that is in charge of data preprocessing and anomaly identification, and provides reconfiguration capabilities that ensure communication robustness. The SHIELD middleware runs on both the S-Gateway and the security control center, with specific security agents. The security agent running on the S-Gateway is responsible for faults and threats that could alter the SPD level of the remote monitoring application, while the security agent running on the security control server is responsible for the server-side counterpart of the monitoring application. The part of the SHIELD middleware that runs on the security control server is responsible for the SHIELD middleware services, managing the semantic model and providing secure discovery, orchestration, and composition. The interface offered by the middleware is based on the SLP messages that the S-Gateway and the monitoring application exchange with their own security agents.

## SPD Metrics Calculation

The SPD metrics calculation approach has been based on the system attack surface metrics approach [3], which has been applied at both the subsystem and system levels. With this approach, the composition phase of the metrics calculation generated 89 different SPD states that characterize the hazardous materials transportation monitoring system.

The following sections illustrate the results obtained with the system attack surface metrics method for SPD metrics calculation, applied first to each of the subsystems and subsequently (by composition) to the entire identification system. The operative behavior of each subsystem is analyzed, trying to identify potential faults or threats that could negatively influence the SPD level.

### S-Gateway SPD Levels

The S-Gateway hosts parts of the SHIELD middleware and is responsible for collecting the status and metrics that allow the overlay to decide on system composition. The following metrics are collected:

- SPD power node identification and status
- FM signal generator identification and status
- Demodulator identification and status, including carrier frequency
- Decryptor identification and status
- Received data samples from the signal generator, with statistics
  - Sample ID and time stamp
  - Number of valid and invalid samples

- Decryption errors (could be intrusion attempts in the connection between the FM signal generator and the SPD power node)
- Demodulation errors
- Self-reconfiguration (software, partial FPGA reconfiguration, or full FPGA reconfiguration)
- Error recovery (software, partial FPGA reconfiguration, or full FPGA reconfiguration)

The S-Gateway hosts a web server, with a web service providing information about node identification, capabilities, and status of the device. The control center accesses this web page through a web browser and can collect all the data received from the S-Gateway, including data related to the FSK demodulator.

Table 7.3 and Figure 7.13 report an example of the data collected with the metrics and sent to the security control center. The communication is based on HTTP and supported by a web server build in to the S-Gateway.

**TABLE 7.3**

Control Center User Interface–Provided Data

| Identification | | Responses | |
|---|---|---|---|
| ID | SPD_PN_01 | Distance | 123 |
| Name | FSK demodulator 1 | Intrusion | Not detected |
| Status | | Metrics | |
| Node status | Running | | |
| SPD level | 2 | | |
| Watchdog timer | On | Errors detected | 0 |
| | | Error recovery | 1 |
| | | Total failures | 1 |
| Decryption | On | Decryption errors | 1 |
| | | Decrypted frames | 20 |
| Commands | On | Reconfiguration requests | 2 |
| | | Reconfiguration failures | 1 |
| | | Service requests | 30 |
| | | Not recognized requests | 3 |
| Demodulation | On | Samples/s | 32,000 |
| | | Demodulator faults | 1 |
| | | Demodulator errors | 5 |
| | | Signal bandwidth | Neighborhood area network (NaN) |
| Capabilities | | | |
| CPU model | | Error detection | Watchdog timer |
| CPU frequency | | Error recovery | Software restart |
| RAM size | | Error recovery | FPGA reconfiguration |

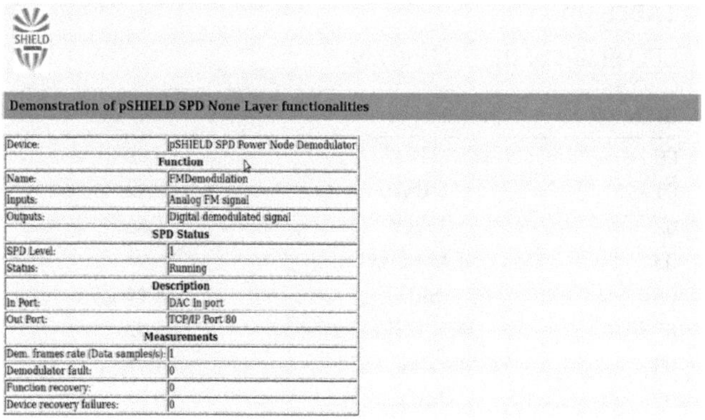

**FIGURE 7.13**
Security control center web status interface.

The analysis of the S-Gateway role, of the previous metrics, and of the collected data allows the identification of a set of faults and threats that are significant for the S-Gateway:

*DEM*: A demodulation error occurred. The WSN periodically generates data collected from the sensors. These data are encrypted, modulated, and sent to the S-Gateway. The S-Gateway demodulates the signal, but an error occurs.

*DEC*: A decryption error occurred. The WSN periodically generates data collected from the sensors. These data are encrypted, modulated, and sent to the S-Gateway. The S-Gateway demodulates the signal and tries to decrypt it in order to process the data and detect anomalous situations, but the decryption process fails.

*SEL*: Self-reconfiguration failed. The FSK modulator switches to a different carrier. The S-Gateway detects a demodulation error, and the demodulator is automatically reconfigured to this new carrier, by a partial reconfiguration of the FPGA. The reconfiguration fails.

*NET*: The S-Gateway is not able to establish a connection with the security control center.

Considering this set of faults and threats, the SPD-level calculation generated 11 states, as reported in Table 7.4. State 00 represents normal operations.

## Wireless Sensor Network SPD Levels

The analysis of the WSN role, of the previous metrics, and of the collected data allows the identification of a set of faults and threats that are significant for the WSN:

**TABLE 7.4**

SPD States of the S-Gateway

| State ID | Faults/Threats | Actual SPD Level |
|---|---|---|
| 00 | — | 90,02 |
| 01 | DEM | 87,42 |
| 02 | DEC | 87,86 |
| 03 | DEM+DEC | 86,91 |
| 04 | SEL | 85,14 |
| 05 | DEC+SEL | 85,02 |
| 06 | DEM+DEC+SEL | 84,90 |
| 07 | NET | 86,87 |
| 08 | DEM+NET | 86,30 |
| 09 | DEC+NET | 86,30 |
| 10 | SEL+NET | 86,20 |

*POW:* The battery charge of one or more sensors is under 5% of the total capacity. The wireless network is still capable of guaranteeing the minimum set of functionalities, but a maintenance intervention is required as soon as possible. This is a serious event that compromises the dependability of the entire subsystem.

*NOD:* One or more nodes of the WSN has stopped working. This event covers hardware faults, software faults, or battery fully discharged. Also, in this case a maintenance activity is mandatory. This is a serious event that compromises the dependability of the entire subsystem.

*JAM:* A tentative jamming is ongoing or, more generally, a fault that prevents one or more nodes from communicating has occurred.

*TEM:* The temperature inside the freight car is critical.

*HUM:* The humidity level inside the freight car is critical.

*VIB:* The vibration level indicates an anomalous situation.

*INT:* An intrusion has been detected.

From this set of faults and threats, the SPD level calculation generated 10 different states, as reported in Table 7.5. State 00 represents normal operations.

## SPD-Level Composition at the System Level

The final step of the SPD metrics calculation consists of the composition of the SPD metrics obtained for the single components of the monitoring system. This step is fundamental for the identification of the resulting SPD at the system level.

**TABLE 7.5**

SPD States of the WSN

|    | Faults/Threats | Actual SPD Level |
|----|----------------|------------------|
| 00 | —              | 89,11            |
| 01 | TEM            | 88,22            |
| 02 | HUM            | 88,00            |
| 03 | VIB            | 87,91            |
| 04 | INT            | 86,01            |
| 05 | JAM            | 86,52            |
| 06 | TEM + HUM      | 87,26            |
| 07 | INT + JAM      | 85,77            |
| 08 | POW            | 86,90            |
| 09 | NOD            | 85,40            |

The result of the composition is a set of SPD states that define how the SPD levels evolve when the monitoring system is exposed to a fault and/or threat and reacts with the corresponding countermeasure.

Every fault and/or threat considered for the calculation of the SPD metrics at the component level is also considered in the composition phase, in this way providing full coverage of these faults and menaces at the system level.

Table 7.6 provides the list of all the SPD states that characterize the system at runtime. The table reports the results obtained from the composition of the SPD values calculated for the single components. For each SPD state, the table reports

- A reference number of the state at the scenario level
- For each of the four prototypes involved in the demonstrator,
  - A reference number for the state involved in the composition
  - A description of the state
- The SPD value obtained by the composition at the use case level
- A normalization of the SPD value obtained by the composition at the use case level

The normalized values have been introduced to simplify the use of the SPD information in a real environment and unify the range of SPD values between 0 and 1 across different domains, heterogeneous systems, and applications. With this approach, it is possible to manage the SPD composition at both the system level and the system of systems level.

The normalization considers the minimum and maximum values of the SPD level obtained by the composition at the use case level and maps them over the range [0, 1].

**TABLE 7.6**

Hazardous Materials Transportation Monitoring System SPD States and Corresponding SPD Values

| Hazardous Materials Transportation Scenario | Wireless Sensor Network | | S-Gateway | | SHIELD Middleware | | SPD-Level Value | Normal SPD Level |
|---|---|---|---|---|---|---|---|---|
| Ref. No. | Ref. No. | Description | Ref. No. | Description | Ref. No. | Description | | |
| 0 | 0 | Normal operations | 0 | Normal use | 0 | Normal use | 89,57 | 1,000 |
| 1 | | | 1 | DEM | | | 83,89 | 0,664 |
| 2 | | | 2 | DEC | | | 84,09 | 0,676 |
| 3 | | | 3 | DEM+DEC | | | 83,66 | 0,651 |
| 4 | | | 4 | SEL | | | 74,35 | 0,099 |
| 5 | | | 5 | DEC+SEL | | | 74,31 | 0,097 |
| 6 | | | 6 | DEM+DEC+SEL | | | 74,27 | 0,094 |
| 7 | | | 7 | NET | | | 74,09 | 0,084 |
| 8 | | | 8 | DEM+NET | | | 73,90 | 0,072 |
| 9 | | | 9 | DEC+NET | | | 73,90 | 0,072 |
| 10 | | | 10 | SEL+NET | | | 73,86 | 0,070 |
| 11 | 1 | TEM | 0 | Normal use | 0 | Normal use | 83,36 | 0,632 |
| 12 | | | 1 | DEM | | | 77,51 | 0,286 |
| 13 | | | 2 | DEC | | | 77,70 | 0,297 |
| 14 | | | 3 | DEM+DEC | | | 77,07 | 0,260 |
| 15 | | | 4 | SEL | | | 73,76 | 0,064 |
| 16 | | | 5 | DEC+SEL | | | 73,71 | 0,061 |
| 17 | | | 6 | DEM+DEC+SEL | | | 73,66 | 0,058 |

*(Continued)*

**TABLE 7.6 (CONTINUED)**

Hazardous Materials Transportation Monitoring System SPD States and Corresponding SPD Values

| Hazardous Materials Transportation Scenario Ref. No. | Wireless Sensor Network | | S-Gateway | | SHIELD Middleware | | SPD-Level Value | Normal SPD Level |
|---|---|---|---|---|---|---|---|---|
| | Ref. No. | Description | Ref. No. | Description | Ref. No. | Description | | |
| 18 | | | 7 | NET | | | 74,45 | 0,105 |
| 19 | | | 8 | DEM+NET | | | 74,22 | 0,091 |
| 20 | | | 9 | DEC+NET | | | 74,22 | 0,091 |
| 21 | | | 10 | SEL+NET | | | 74,18 | 0,089 |
| 22 | 2 | HUM | 0 | Normal use | 0 | Normal use | 83,73 | 0,654 |
| 23 | | | 1 | DEM | | | 77,18 | 0,267 |
| 24 | | | 2 | DEC | | | 77,38 | 0,278 |
| 25 | | | 3 | DEM+DEC | | | 76,96 | 0,254 |
| 26 | | | 4 | SEL | | | 72,78 | 0,006 |
| 27 | | | 5 | DEC+SEL | | | 72,73 | 0,003 |
| 28 | | | 6 | DEM+DEC+SEL | | | 72,68 | 0,000 |
| 29 | | | 7 | NET | | | 73,47 | 0,047 |
| 30 | | | 8 | DEM+NET | | | 73,24 | 0,033 |
| 31 | | | 9 | DEC+NET | | | 73,24 | 0,033 |
| 32 | | | 10 | SEL+NET | | | 73,20 | 0,031 |
| 33 | 3 | VIB | 0 | Normal use | 0 | Normal use | 84,34 | 0,691 |
| 34 | | | 1 | DEM | | | 77,37 | 0,78 |
| 35 | | | 2 | DEC | | | 77,56 | 0,289 |
| 36 | | | 3 | DEM+DEC | | | 77,15 | 0,265 |

*(Continued)*

**TABLE 7.6 (CONTINUED)**

Hazardous Materials Transportation Monitoring System SPD States and Corresponding SPD Values

| Hazardous Materials Transportation Scenario Ref. No. | Wireless Sensor Network | | S-Gateway | | SHIELD Middleware | | SPD-Level Value | Normal SPD Level |
|---|---|---|---|---|---|---|---|---|
| | Ref. No. | Description | Ref. No. | Description | Ref. No. | Description | | |
| 37 | | | 4 | SEL | | | 73,84 | 0,068 |
| 38 | | | 5 | DEC+SEL | | | 73,79 | 0,066 |
| 39 | | | 6 | DEM+DEC+SEL | | | 73,74 | 0,063 |
| 40 | | | 7 | NET | | | 73,66 | 0,058 |
| 41 | | | 8 | DEM+NET | | | 73,44 | 0,045 |
| 42 | | | 9 | DEC+NET | | | 73,44 | 0,045 |
| 43 | | | 10 | SEL+NET | | | 73,40 | 0,042 |
| 44 | 4 | INT | 0 | Normal use | 0 | Normal use | 77,67 | 0,296 |
| 45 | | | 1 | DEM | | | 77,54 | 0,288 |
| 46 | | | 2 | DEC | | | 77,71 | 0,298 |
| 47 | | | 3 | DEM+DEC | | | 76,90 | 0,250 |
| 48 | | | 4 | SEL | | | 75,78 | 0,184 |
| 49 | | | 5 | DEC+SEL | | | 75,74 | 0,181 |
| 50 | | | 6 | DEM+DEC+SEL | | | 75,69 | 0,178 |
| 51 | | | 7 | NET | | | 75,58 | 0,172 |
| 52 | | | 8 | DEM+NET | | | 75,37 | 0,159 |
| 53 | | | 9 | DEC+NET | | | 75,37 | 0,159 |
| 54 | | | 10 | SEL+NET | | | 75,33 | 0,157 |
| 55 | 5 | JAM | 0 | Normal use | 0 | Normal use | 77,74 | 0,299 |

*(Continued)*

**TABLE 7.6 (CONTINUED)**

Hazardous Materials Transportation Monitoring System SPD States and Corresponding SPD Values

| Hazardous Materials Transportation Scenario Ref. No. | Wireless Sensor Network | | S-Gateway | | SHIELD Middleware | | SPD-Level Value | Normal SPD Level |
|---|---|---|---|---|---|---|---|---|
| | Ref. No. | Description | Ref. No. | Description | Ref. No. | Description | | |
| 56 | 6 | TEM+HUM | 1 | DEM | | | 76,94 | 0,252 |
| 57 | | | 2 | DEC | | | 77,12 | 0,263 |
| 58 | | | 3 | DEM+DEC | | | 76,73 | 0,240 |
| 59 | | | 4 | SEL | | | 75,15 | 0,146 |
| 60 | | | 5 | DEC+SEL | | | 75,11 | 0,144 |
| 61 | | | 6 | DEM+DEC+SEL | | | 75,06 | 0,141 |
| 62 | | | 7 | NET | | | 74,98 | 0,136 |
| 63 | | | 8 | DEM+NET | | | 74,75 | 0,123 |
| 64 | | | 9 | DEC+NET | | | 74,75 | 0,123 |
| 65 | | | 10 | SEL+NET | | | 74,72 | 0,121 |
| 66 | 6 | TEM+HUM | 0 | Normal use | 0 | Normal use | 84,21 | 0,683 |
| 67 | | | 1 | DEM | | | 76,85 | 0,247 |
| 68 | | | 2 | DEC | | | 77,47 | 0,284 |
| 69 | | | 3 | DEM+DEC | | | 76,65 | 0,235 |
| 70 | | | 4 | SEL | | | 75,50 | 0,167 |
| 71 | | | 5 | DEC+SEL | | | 75,46 | 0,164 |
| 72 | | | 6 | DEM+DEC+SEL | | | 75,41 | 0,162 |
| 73 | | | 7 | NET | | | 75,33 | 0,157 |
| 74 | | | 8 | DEM+NET | | | 75,11 | 0,144 |

*(Continued)*

**TABLE 7.6 (CONTINUED)**

Hazardous Materials Transportation Monitoring System SPD States and Corresponding SPD Values

| Hazardous Materials Transportation Scenario Ref. No. | Wireless Sensor Network Ref. No. | Description | S-Gateway Ref. No. | Description | SHIELD Middleware Ref. No. | Description | SPD-Level Value | Normal SPD Level |
|---|---|---|---|---|---|---|---|---|
| 75 | | | 9 | DEC+NET | | | 75,11 | 0,144 |
| 76 | | | 10 | SEL+NET | | | 75,07 | 0,141 |
| 77 | 7 | INT+JAM | 0 | Normal use | 0 | Normal use | 77,65 | 0,198 |
| 78 | | | 1 | DEM | | | 77,46 | 0,138 |
| 79 | | | 2 | DEC | | | 77,64 | 0,148 |
| 80 | | | 3 | DEM+DEC | | | 77,24 | 0,126 |
| 81 | | | 4 | SEL | | | 74,80 | 0,125 |
| 82 | | | 5 | DEC+SEL | | | 74,75 | 0,123 |
| 83 | | | 6 | DEM+DEC+SEL | | | 74,70 | 0,120 |
| 84 | | | 7 | NET | | | 74,62 | 0,115 |
| 85 | | | 8 | DEM+NET | | | 74,40 | 0,102 |
| 86 | | | 9 | DEC+NET | | | 74,40 | 0,102 |
| 87 | | | 10 | SEL+NET | | | 74,36 | 0,099 |
| 88 | 8 | POW | 0 | Normal use | 0 | Normal use | 84,04 | 0,673 |
| 89 | | | 1 | DEM | | | 77,56 | 0,289 |
| 90 | | | 2 | DEC | | | 77,74 | 0,300 |
| 91 | | | 3 | DEM+DEC | | | 77,35 | 0,276 |
| 92 | | | 4 | SEL | | | 75,33 | 0,157 |
| 93 | | | 5 | DEC+SEL | | | 75,29 | 0,154 |

*(Continued)*

**TABLE 7.6 (CONTINUED)**

Hazardous Materials Transportation Monitoring System SPD States and Corresponding SPD Values

| Hazardous Materials Transportation Scenario Ref. No. | Wireless Sensor Network Ref. No. | Description | S-Gateway Ref. No. | Description | SHIELD Middleware Ref. No. | Description | SPD-Level Value | Normal SPD Level |
|---|---|---|---|---|---|---|---|---|
| 94 | | | 6 | DEM+DEC+SEL | | | 75,24 | 0,151 |
| 95 | | | 7 | NET | | | 75,16 | 0,147 |
| 96 | | | 8 | DEM+NET | | | 74,93 | 0,134 |
| 97 | | | 9 | DEC+NET | | | 74,93 | 0,134 |
| 98 | | | 10 | SEL+NET | | | 74,90 | 0,131 |
| 99 | 9 | NOD | 0 | Normal use | 0 | Normal use | 77,47 | 0,284 |
| 100 | | | 1 | DEM | | | 76,84 | 0,247 |
| 101 | | | 2 | DEC | | | 77,03 | 0,257 |
| 102 | | | 3 | DEM+DEC | | | 76,63 | 0,234 |
| 103 | | | 4 | SEL | | | 74,62 | 0,115 |
| 104 | | | 5 | DEC+SEL | | | 74,57 | 0,112 |
| 105 | | | 6 | DEM+DEC+SEL | | | 74,53 | 0,109 |
| 106 | | | 7 | NET | | | 74,44 | 0,104 |
| 107 | | | 8 | DEM+NET | | | 74,22 | 0,091 |
| 108 | | | 9 | DEC+NET | | | 74,22 | 0,091 |
| 109 | | | 10 | SEL+NET | | | 74,18 | 0,089 |

**TABLE 7.7**

Normalized SPD Ranges

| High (SPD≥0.7) |
|---|
| Normal (0.3<SPD<0.7) |
| Very low (0<SPD<0.2) |

Furthermore, to provide visual feedback of the status of the system in terms of SPD, four different macro SPD levels (high, normal, low, and very low) have been introduced. They are represented by four colors and give a quick indication of the status of the system. The macro SPD levels are defined in Table 7.7.

## Hazardous Materials Transportation: A Real Scenario

This section illustrates the storyboard that has been selected to validate and verify the adoption of the SHIELD methodology in the hazardous materials transportation scenario. The main objectives of the storyboard are:

- Demonstrate the use of the monitoring system.
- Identify a sequence of execution steps that include the faults and threats considered for the SPD metrics calculation.
- Validate and verify the use of the system attack surface metrics.

The storyboard is based on four main events that significantly describe the use of the monitoring system during the transportation of hazardous material. The real scenario consists of continuously monitoring the trains and railway infrastructure, trying to achieve the main objectives:

1. Detecting any unusual conditions, such as high temperature, anomalous vibration patterns, and unexpected movement.
2. Transferring such information to different actors (i.e., train operator, train infrastructure owner, and consumer) involved in the rail system, both automatically and in a request or response demand-based passive mode.

When the railway convoy is planning to move from one station to another, the desired SPD level can change during the travel, due to different environmental conditions, faults, or threats. During the travel, and especially while the train stops at intermediate stations, the freight cars must be continuously monitored.

The train is equipped with the pervasive component of the monitoring system, which is in charge of collecting information from the environment and from the train, processing it, and identifying potential menaces and anomalous situations. The monitoring system has been installed on a real freight car made available by the Italian Railway Authority (RFI/Trenitalia) at the Roma Smistamento railway infrastructure. Eight sensors have been installed on a freight car and grouped in two networks: the WSN can monitor temperature and humidity, acceleration, vibrations, position, and detect intrusions. On the locomotive, the S-Gateway has been installed and is connected to the security control center via wireless communication. The networks are based on the TinyOS operating system and implement the cognitive radio network to enforce confidentiality and integrity requirements at the WSN communication level.

The network that measures humidity and temperature was installed inside the freight cars (Figure 7.14). The second network, which measures acceleration, vibration, and position and detects intrusions, was installed outside the car, as showed in Figures 7.15 and 7.16. As illustrated in the figures, the

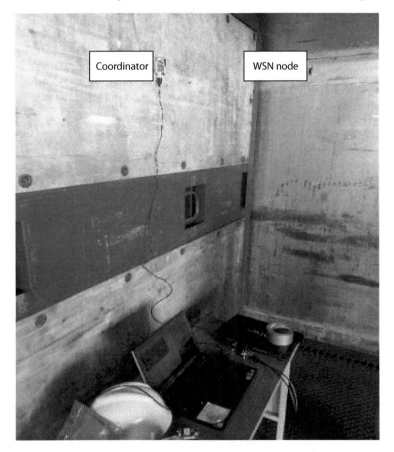

**FIGURE 7.14**
Setup of the WSN in charge of environmental monitoring.

**FIGURE 7.15**
Setup of vibration monitoring nodes.

test setup is comparable to a laboratory test: a real industrial solution would require tamper-proof rugged devices, installed in a safe and reliable way.

The storyboard starts with the setup of the monitoring system, which illustrates the operations required to set up every single component of the monitoring system, register them into the SHIELD middleware, and start the normal operations. For normal operations, we intend an idle state in which the system is already monitoring the train. Focusing on a single freight car, in the idle state the WSN is collecting data and the S-Gateway analyzes the data to identify anomalous situations.

The first event is the train that starts moving, but no anomalous events are detected.

During the travel, the second event occurs: the WSN detects an anomalous vibration pattern, probably due to the instability of the railway. In this situation, the monitoring system alerts the train driver, suggesting a reduction

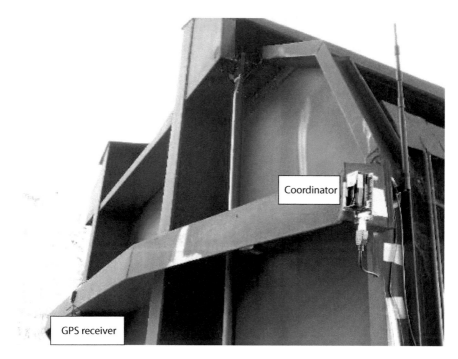

**FIGURE 7.16**
Setup of the coordinator and GPS receiver of the second WSN.

in speed in order to prevent damage to the freight car and the transported material.

The instable section of the railway is passed, but after few kilometers, the third event occurs. It is raining and water is entering from a fissure created by the previous vibrations: an old seal is no longer capable of containing the water that enters from the door of one of the freight cars. The monitoring system detects an anomalous humidity level and alerts the train driver. The train is approaching the next station, and the train drivers call the maintenance service. After the maintenance activities, the monitoring system returns to the idle state.

While the convoy is waiting for departure, the fourth event occurs: someone is trying to disrupt the monitoring system communication and an intrusion is detected. The monitoring system promptly reacts, and the police intervene, resolving the theft tentative.

Figure 7.17 illustrates the storyboard.

Each main event of the storyboard is composed of one or more execution steps that, in turn, represent the various operations that the system performs when the event occurs. Four different situations have been considered in the scenario, in order to try to cover both correct operations and situations that present faults and menaces:

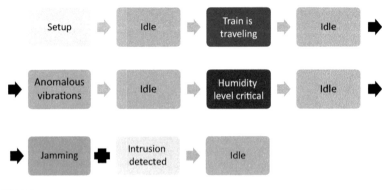

**FIGURE 7.17**
Storyboard of the monitoring scenario.

- Transportation is ongoing safely.
- Instable railway section.
- Fault to a freight car.
- Theft tentative.

The following list reports all the steps of the scenario execution:

T0  The train is staying at the departing station waiting to leave. The monitoring system is set up, started, and enters the idle state. The system is characterized by the highest SPD level.

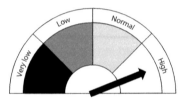

T1  Whenever the train is stationary or traveling, the driver can use a manager panel to check the status of the monitoring system. The monitoring panel shows the graphical structure of the convoy network and contains a section, which becomes visible when selecting a node from the network graph, that allows us to see the sensor status and the related alert messages (Figures 7.18 and 7.19).

T2  The train leaves the station and starts to travel. The monitoring system is in the idle state, and the SPD is still at the maximum level.

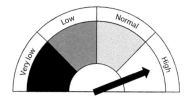

T3 The WSN detects an anomalous vibration pattern due to the instability of the railway section that the train is covering. The S-Gateway sends an alert to the security control center, which alerts the driver to reduce the speed as much as possible (Figure 7.20).

The SHIELD middleware calculates the new SPD level: 0,691.

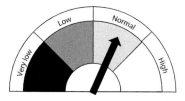

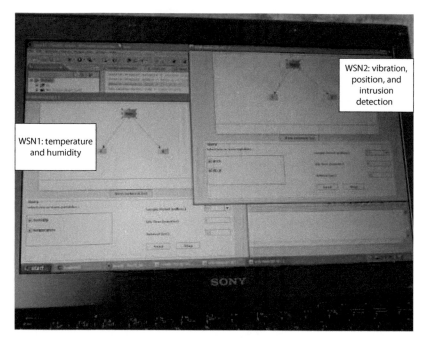

**FIGURE 7.18**
Security operator panel.

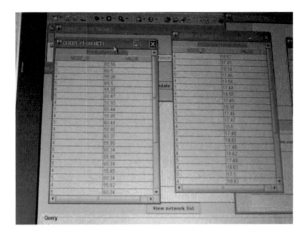

**FIGURE 7.19**
Humidity and temperature data visualization.

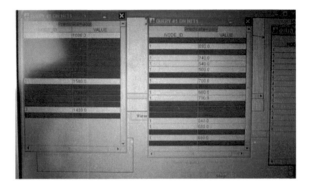

**FIGURE 7.20**
Anomalous vibration pattern detected.

T4 The train passes the instable railway section, and the anomalous situation disappeared. The highest level of SPD is reestablished. The monitoring system is in the idle state.

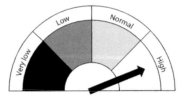

T5 A few kilometers before the next station, the system detects an anomalous humidity temperature in one of the train cars: it is raining and water is entering from a fissure in the sealing of the car's door (Figure 7.21).

| QUERY #1 ON NET1 | | QUERY #1 ON NET1 | |
| --- | --- | --- | --- |
| **Predicate=humidity** | | 4 | 4.0 |
| NODE_ID | VALUE | 5 | 6.0 |
| 5 | 3.0 | 4 | 8.0 |
| 4 | 5.0 | 5 | 10.0 |
| 5 | 7.0 | 4 | 12.0 |
| 4 | 9.0 | 5 | 14.0 |
| 5 | 11.0 | 4 | 16.0 |
| 4 | 13.0 | 5 | 18.0 |
| 5 | 15.0 | 4 | 20.0 |
| 4 | 17.0 | 5 | 22.0 |
| 5 | 19.0 | 4 | 24.0 |
| 4 | 21.0 | 5 | 26.0 |
| 5 | 23.0 | 4 | 28.0 |
| 4 | 25.0 | 5 | 30.0 |
| 5 | 27.0 | 4 | 32.0 |
| 4 | 29.0 | 5 | 34.0 |
| 5 | 31.0 | 4 | 36.0 |
| 4 | 33.0 | 5 | 38.0 |
| 5 | 35.0 | 4 | 40.0 |
| 4 | 37.0 | 5 | 42.0 |
| 5 | 39.0 | 4 | 44.0 |
| 4 | 41.0 | 5 | 46.0 |

**FIGURE 7.21**
Humidity alert has been detected.

The monitoring system alerts the drivers and schedules a maintenance intervention at the next station. The SPD level is degraded and the SHIELD middleware calculates the new value: 0,654.

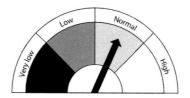

T6 The train reaches the next station, and the damaged car undergoes maintenance. At the end of the maintenance activity, the monitoring system is in the idle state and the train is ready to depart. The SPD is back to the maximum level.

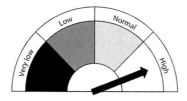

T7 While the train is waiting to leave the station, someone tries to access the car. With a jammer, the WSN communications are disrupted and the SHIELD middleware promptly degrades the SPD level: 0,299.

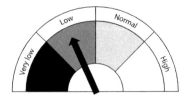

T8 The monitoring application on the S-Gateway reacts to the SPD-level reduction and, trying to contrast the jamming tentative, applies the countermeasures based on FSK and cognitive technologies. The SHIELD middleware is informed and informs the S-Gateway to reconfigure the FSK demodulator as a countermeasure.

T9 While the SHIELD middleware is managing the jamming attack, the monitoring system detects an intrusion tentative. The middleware decreases the SPD value to the lowest level, and the police are called. The SPD decreases to 0,198.

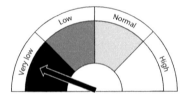

T10 After the intervention of the police, the normal SPD level is back to the highest level and the monitoring system enters the idle state.

---

## References

1. Pipeline and Hazardous Materials Safety Administration. Incident reporting. Washington, DC: U.S. Department of Transportation. https://www.phmsa.dot.gov/incident-report.
2. Intel. Cyclone III FPGA development kit. Santa Clara, CA: Intel. https://www.altera.com/products/boards_and_kits/dev-kits/altera/kit-cyc3.html.
3. P.K. Manadhata and J.M. Wing. An attack surface metric. *IEEE Transactions on Software Engineering*, 37(3): 371–386, 2011.

# 8

## Biometric Security Domain

**Paolo Azzoni, Konstantinos Rantos, Luca Geretti,**
**Antonio Abramo, and Stefano Gosetti**

### CONTENTS

### Introduction

The security of a large public infrastructure is currently managed by traditional systems based on security cameras connected to a security control center. The cameras are remotely controlled by the security personnel, and the acquired video streams are saved on a data server, without any processing or analysis. The most important limitation of this approach is the complete absence of real-time automatic processing of the video stream, which could provide important information directly when the monitored events are occurring. Real-time video analysis performed by the security personnel is not reliable, and the reliability level drops proportionally with the number of video streams: humans are not "designed" for long-term focus and attention. The human brain is very efficient when reacting to stimulation, but it is not efficient in keeping constant control over long periods of time, especially when nothing interesting is happening. The attention of a professional operator drops on average by 60% only after 30 min of video streaming surveillance. Attention is further lost when an event occurs and the security operators are focused on that single event, with the risk of missing the identification of other important events.

Existing solutions allow the analysis of the video stream after the events occurred, only during investigations, and rely entirely on the security personnel expertise. Very frequently, security companies incur high expenses just to provide unreliable surveillance.

The idea of adopting biometric security technologies that are natively compliant with the SHIELD framework introduces a new secure paradigm in surveillance that is based on the concept of prevention, which is implemented using a proactive video stream analysis that allows near-real-time event identification.

## Biometric-Based People Identification: A Case Study

The security of large public infrastructures, such as airports, stations, and stadiums, strongly depends on the surveillance and identification of people. This domain offers several excellent use cases that allow us to demonstrate and exploit the security, privacy, and dependability (SPD) functionalities and capabilities of the SHIELD approach, when applied to a real environment.

Large public infrastructure security is currently managed, in the large part of this specific market, by traditional systems based on security camera circuits connected to a security control center. With this approach, the security of the monitored infrastructure largely depends on the professional profiles and the efficiency of the security operators that are responsible for the identification of dangerous situations, tumults, people identification, access control, and so forth. Furthermore, access to a large public infrastructure is still based on tickets, and it is difficult to find more advanced solutions for people identification and access control.

Biometrics offers some important enabling technologies that enhance the potentialities of surveillance systems, facilitating people identification, authentication, access control, and monitoring. Biometrics can be further supported by the use of tamper-resistant devices, such as smart cards, that can host the user biometrics profile and provide a robust solution for authenticating and registering people at entrance points. Increasing the autonomy of the surveillance systems simplifies and improves the reliability of the identification of events and the accuracy of people identification, during or immediately after the events are happening.

From a technology point of view, this approach is based on face recognition algorithms that are executed directly on a security camera that, in this way, is capable of identifying people and situations autonomously. This solution provides information when the monitored event is happening and does not require the presence of security personnel.

The solution is enforced by the adoption of smart card–based access delegation services and by a dependable framework for distributed computing. The use of smart card provides an important alternative source of information related to the identity of people. A dependable framework for distributed computing introduces several advantages on the technical side, allowing the implementation of a distributed system of smart cameras that are optimized in terms of efficiency, security, dependability, and costs.

The reliability of the solution is guaranteed in terms of SPD by the native adoption of the SHIELD approach, which allows us to manage any potential threat with the appropriate strategy, ensuring always the maximum level of SPD during operation.

Following the paradigm based on prevention and adopting the SHIELD methodology, the implementation of the proposed solution covers, for example, the security aspects related to people identification both at the entrance of the public infrastructure and inside the infrastructure during a public event. At the entrance, at the turnstile, the identity of people is checked to deny access to criminals and unwanted and untrusted people. Inside the public infrastructure, the system dynamically identifies people that are responsible for security-related events.

## System for People Identification

The first security barrier against threats in a public infrastructure is located at the entrance, where people identification represents a fundamental preventive "filter" that acts directly at the source of the security issue: people access to the infrastructure. Figure 8.1 illustrates the basic building blocks of a distributed solution, conceived to securely manage people access to a public infrastructure with multiple entrances that must be surveilled simultaneously.

The solution is based on a smart turnstile equipped with an intelligent camera and a smart card reader, connected to a security control center: the smart camera is responsible for biometric-based people identification in the video stream, while the smart card reader is used to retrieve the biometric profile stored in the smart card. Depending on the dimensions and the number of entrances of the public infrastructure, the smart turnstile becomes the basic element of a distributed identification system that, with a client–server architecture, provides near-real-time entrance automated surveillance and

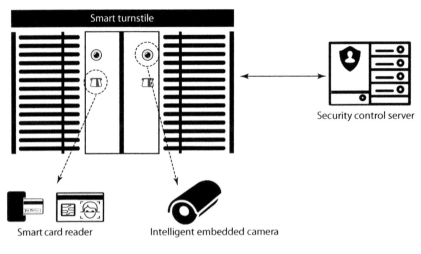

**FIGURE 8.1**
Smart turnstile.

access control. At this level, SPD is based on turnstile redundancy and is managed by the SHIELD middleware and by a framework for dependable distributed computation (Figure 8.2).

The SHIELD middleware is integrated in the identification system and ensures SPD support when the system is exposed to active and passive menaces, like camera obfuscation, noncollaborative user behavior, face

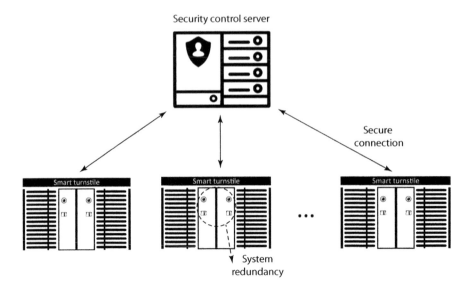

**FIGURE 8.2**
Distributed identification system.

alteration or obfuscation, and technical issues, like camera malfunctions and an absence of communication with the security control center.

## Identification Process

The identification process at the entrance of a public infrastructure begins at the turnstile, where the identity of a person is checked twice using two different technologies: face recognition and smart card secure services. The identification is based in both cases on the use of a biometric profile that is associated unequivocally to a person.

The first level of identification based on face recognition is adopted to dynamically acquire the biometric profile of a person from the video stream acquired by the intelligent embedded high-definition (HD) camera that is installed in the turnstile. The security system asks each person to position and stand in front of the camera for face recognition.

The face recognition algorithm automatically detects the presence of a face in the video stream, calculates the best image in the stream, and extracts the biometric profile of the person. During this phase, a collaborative approach from the user is required, to avoid situations where the face is obfuscated by clothes or voluntarily altered, trying to hide the real identity. The level of collaboration required in real conditions is very limited and is determined by the ICAO standard [1–3]: the image selected in the video stream that is used to extract the biometric profile is the one that better satisfies the requirements of the ICAO standard. For security reasons, if during the sampling interval the face recognition algorithm is not capable of identifying an image containing a photo of the face that satisfies the ICAO standard, the turnstile does not open. In this situation, the identification system asks the person to remove any object or clothes obfuscating the face and reposition in front of the camera. The procedure is repeated three times, and in the case it fails, the identification process is directly assisted by the security operators.

During this phase, the system is exposed to active and passive menaces, like camera obfuscation, noncollaborative user behavior, face alteration or obfuscation, and technical issues, like intelligent camera malfunctions and an absence of communication with the security control center.

The second technology adopted to cross-check people's identity is based on the security services offered by the secure messaging and by the intrinsic features of smart cards. This solution provides a second source of information for the biometric profile: a smart card is used as an electronic ID card that contains the biometric profile of the user. The secure messaging service (see Chapter 3) guarantees that this information is read from the smart card in a secure way. Furthermore, this service could be exploited to ensure that, after entrance, visitors are allowed to access only areas that they are authorized to. This functionality can also help people move in different areas, avoiding the possibility of getting lost or accessing restricted areas.

Also, in this stage of the identification process, the system is exposed to potential menaces, like smart card use by unauthorized people and exchange of smart cards between people, as well as technical issues, like smart card failure and smart card reader failure.

With this approach, the identification of a person is based on two independent sources of information that are used to cross-check the biometric profile acquired by the camera and provided with the smart card. This double check increases the security and dependability of the identification process, because the two technologies adopted for the acquisition of the biometric profile enforce themselves reciprocally. The weak aspects of face recognition are enforced by the support offered by the smart card, and the possibility that the smart card is not used by the real owner is excluded or confirmed by face recognition.

After the biometric profile has been acquired from the two sources, the first check consists of comparing the two versions to understand if they match. If a first matching is found, the biometric profile is stored in a temporary archive. If the profiles do not match, access to the public infrastructure premises is refused and the security personnel takes charge of the issue.

The biometric profile stored in the temporary archive is then used to find a matching profile in the central database containing the biometric profiles of trusted, untrusted, and unwanted people. Access to the premises is granted only in the case of matching with the profile of a trusted person. An alternative solution, which can be adopted, for example, for occasional visitors, is the use of a ticket (and a ticket reader) as a substitute for the smart card. The security and dependability of this solution are lower than those of the previous one, but they are still acceptable. With this solution, it is possible to ensure that the owner of the ticket is the person that has been identified by the camera.

Figure 8.3 illustrates the procedure adopted for the identification of a person at the turnstile.

Finally, to increase the dependability of the system and reduce the overall costs, the Dependable Distributed Computation Framework (DDCF) is used to physically partition and distribute the software components of the people identification system. Three modules compose the face recognition system: the face finder module (FF), the ICAO module (ICAO), and the face recognition module (FR). The DDCF is capable of executing and managing the face finder and ICAO modules directly on the camera, while the face recognition module is executed and managed on the central server. The role of the DDCF is to monitor every software component of the identification system and manage any possible fault that can compromise the correct operation of the system.

During the identification process the SHIELD middleware continuously monitors the SPD level of the people identification system and reacts to faults and threats changing the SPD level, allowing the components of the identification system to apply appropriate countermeasures to reestablish the optimal SPD level.

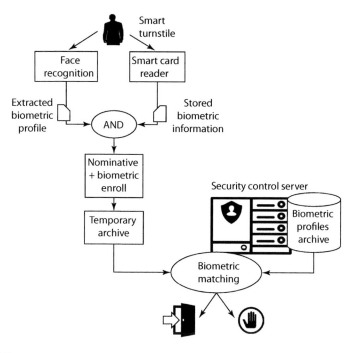

**FIGURE 8.3**
Identification process.

## Hardware Components

The smart turnstile is based on two embedded devices that physically provide the biometric information required for people identification: an intelligent camera and a smart card reader.

The intelligent camera is a SHIELD node that provides automatic people identification based on face recognition and includes intrinsic SPD features. The node is a stand-alone unit that can perform the identification autonomously and provides rugged features, which make it suitable for installation in a harsh environment. The camera is a custom board based on a system on chip (SoC) manufactured by Texas Instruments and based on ARM architecture: the DM816x DaVinci system on a chip. The board has been conceived specifically for video processing and provides native hardware encoders and a digital signal processor (DSP) (Figure 8.4).

The key elements of the DM816x DaVinci are three HD video and imaging coprocessors (HDVICP2). Each coprocessor can perform a single 1080p60 H.264 encode or decode or multiple lower resolution encodes and decodes. It also provides the TI C674x VLIW floating-point DSP core, HD video and imaging coprocessors, and multichannel HD-to-HD or HD-to-standard definition (SD) transcoding, along with multicoding. This solution has been

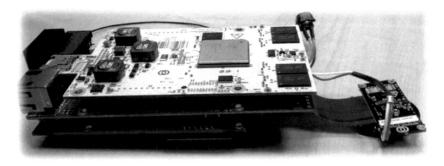

**FIGURE 8.4**
Intelligent camera based on TI SoC.

conceived specifically for demanding HD video applications. The architecture of the board is described in Figure 8.5.

The intelligent camera prototype integrates a high-quality complementary metal-oxide semiconductor (CMOS) charge-coupled device (CCD) sensor. This approach allows us to directly access the raw data acquired by the sensor, avoiding the compression and decompression of the images acquired by the camera connected, for example, via USB. Furthermore, the board allows us the possibility of using different type of sensors, depending on the requirements of the application scenario. It will be possible to select the sensor depending on the resolution, the pixel size, the frame rate, and so forth. Currently, the prototype supports two kinds of sensors:

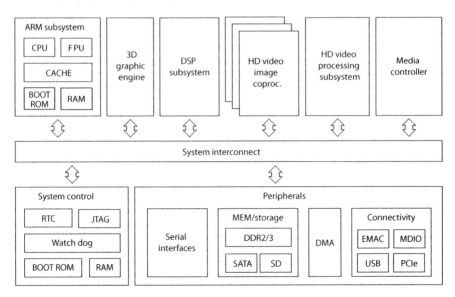

**FIGURE 8.5**
Intelligent camera architecture.

1. *Aptina AR0331*: An image sensor that integrates a small, high-performance global shutter technology for high-speed image capture into a 1/3-inch optical format HD device. It provides a 3.75-micron global shutter pixel with low-light performance that can stop action without the artifacts typically associated with conventional rolling shutter pixels.

2. *Aptina MT9P001I12STC*: An HD CMOS sensor with a pixel size of 2.2 µm, low light sensitivity, and low noise level. This sensor enables high-speed image capture capabilities and includes variable functions, such as gain, frame rate, and exposure, while maintaining low power consumption.

Optionally, if the application context requires high-quality images, the board is able to also support the Aptina MT9M031: this is a full HD, 3.1-megapixel sensor based on the new 2.2-micron pixel. The AR0331 targets the mainstream 1/3-inch optical format surveillance camera market with excellent image quality. It provides full HD video with wide dynamic range (WDR) capability and built-in adaptive local tone mapping. The new sensor is capable of working up to 1080p at 60 fps, adopting a technique to enable the sensor's sub-1-lux low-light performance.

The intelligent camera is integrated in the turnstile, but can also be used in other configurations of the system as a stand-alone device. This is possible because the board has been designed with a removable enclosure that provides all the rugged features of a standard surveillance camera for industrial and defense application.

Depending on the distributed configuration of the face recognition software, the board hosts the entire recognition software or a subset of modules managed remotely by the DDCF.

The smart card reader represents the second source of biometric information: it is integrated in the turnstile, and it is connected to the intelligent camera via USB. The primary role of this device is to read the biometric profile of a person from a smart card, but it is also used as a tamper-resistant device for storing cryptographic keys and performing sensitive cryptographic computations for the needs of secure channel establishment between the camera and the security control server. It allows the camera to have access to a directly attached trusted platform to rely on for transmitting and receiving data securely. This is particularly important to ensure the protection of the user's privacy regarding the images taken from the camera and being sent to the control server and receiving authenticated commands from it (Figure 8.6).

Prior to deploying a card to be used with a specific camera, it must be personalized for the camera and the required set of keys using the camera serial number as a parameter must be generated. As a result, the card is loaded with the node master encryption key and the node master authentication key. The personalized card can then be deployed and used by the corresponding camera.

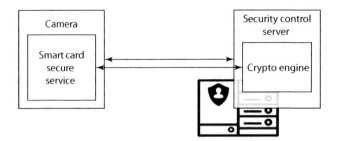

**FIGURE 8.6**
Smart card–based secure communications.

The secure channel functionality is implemented by a smart card applet. The commands that the host system can issue to the card for the needs of node authentication and establishment of a secure channel are defined in Table 8.1.

## Software Components

The software architecture of the biometric system for people identification is composed of the following components:

- *Face recognition software*: It is responsible for people face recognition and biometric-based people identification.
- *Smart card security services*: This component provides services for biometric profile encryption, storage, and secure sessions management.
- *Dependable distributed computation framework*: It manages and monitors the distributed execution of the face recognition software.

**TABLE 8.1**

Secure Channel Applet Commands

| Command | CLA | INS | P1 | P2 | Description |
|---|---|---|---|---|---|
| ENCRYPT | 0×80 | 0×76 | 0×02 | 0×00 | Creates new session keys that will be used for the encryption or decryption of data exchanged with the security control server. It returns the random number generated by the card and used for the creation of session keys. |
| ≫ | 0×80 | 0×76 | 0×04 | 0×00 | Encrypt and compute master authentication key (MAC) for the data sent to the card using the precomputed session keys. |
| ≫ | 0×80 | 0×76 | 0×06 | 0×00 | Compute MAC for the data sent to the card. |
| DECRYPT | 0×80 | 0×78 | 0×04 | 0×00 | Decrypt and verify MAC. |
| ≫ | 0×80 | 0×78 | 0×06 | 0×00 | Verify MAC. |

*Note:* CLA: class of instruction; INS: instruction code; P1: instruction parameter 1; P2: instruction parameter 2.

- *SHIELD middleware*: It is responsible for monitoring the SPD behavior of the components of the identification system and ensuring that the appropriate actions are executed in the case of threats or faults.

### Face Recognition Software

The face recognition software is a modular application that implements a technique called the eigenface method for face recognition [4,5]. This method is based on the idea of extracting the basic features of the face: the objective is to reduce the recognition problem to a lower dimension, maintaining, at the same time, the level of dependability required for such an application context. This approach has been theoretically studied during the nineties and has been recently reconsidered because it provides a good ratio between the required resources and the quality of the results, and it is well dimensioned for embedded systems. Today, it is becoming the most credited method for face recognition. The face recognition software is based on three main modules:

1. Face finder module (FF)
2. ICAO module (ICAO)
3. Face recognition module (FR)

These modules cooperate to implement the recognition and identification procedure illustrated in Figure 8.7.

The face recognition application can operate in two different modes: "enroll" mode and "transit" mode. The enroll mode is used to populate the

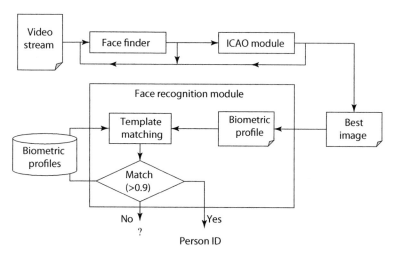

**FIGURE 8.7**
Face recognition and user identification process.

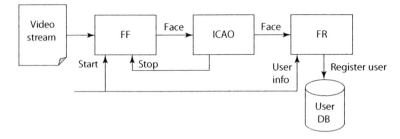

**FIGURE 8.8**
Enroll mode. DB, database.

database with the biometric profiles of the people that will be accepted by the system. The transit mode is used to dynamically recognize and identify the people that pass (transit) in front of the camera. Figures 8.8 and 8.9 illustrate the operations performed in these two working modes.

Every single module can run autonomously, giving the identification software the possibility of running entirely on the intelligent cameras or being distributed on multiple nodes, depending on the system configuration. In both cases, the DDCF controls and monitors the modules, in order to manage potential failures.

The modules can be remotely controlled and communicate with each other using an application programming interface (API) based on standard remote procedure call (RPC). All the face recognition modules can be configured using a simple configuration file or through a web interface.

The face finder component is responsible for the acquisition of the video stream, for the analysis of the video stream, and for the generation of the output messages when a human face is found in the input stream. Every time a face is detected in the video stream, the extracted features of the face are passed to the ICAO component that is responsible for the selection of the best image in the set of images identified by the face finder module. Once the module identifies a face with an ICAO score that allows identification, the best result of this detection phase is sent to the face recognition module.

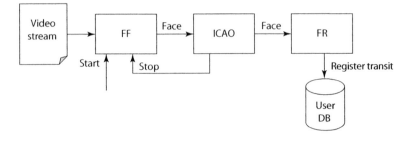

**FIGURE 8.9**
Transit mode. DB, database.

The face recognition module is responsible for the extraction of the biometric profile and for the identification of a matching profile in the database. During the enroll phase, the face recognition module stores the detected biometric profile and personal information of a specific person in the security control server database. In the transit mode, the face recognition module searches the database for the biometric profile that matches the one extracted from the video stream.

### Smart Card–Based Authentication

Smart cards have been adopted to enforce the reliability of people identification as secure components for storing user's biometric credentials and as tamper-resistant devices that facilitate key management and secure channel establishment.

During the identification process, the smart card represents the user's security token for storing his or her biometric credentials and granting access to the public infrastructure. Once the person is authenticated, the system can make decisions regarding his or her access to the premises and control attempts to access unauthorized places.

Use of smart cards for storing biometrics credentials typically involves two phases: the registration phase and the matching phase. During the registration phase, a proper image of the subject is captured by the intelligent camera, processed to check compliance to the ICAO standard and extract the biometric profile, and finally, stored in the smart card. After the registration phase, the smart card information is stored in the security control center and can be used as a security token (Figure 8.10).

When such a security token is presented for identification and authentication purposes, the identification phase is initiated. During this phase, the user biometric profile extracted by the intelligent camera in the video stream is compared with the profile that was stored in the card during the registration phase (Figure 8.11).

The functionalities required during the registration and identification phase are offered by a smart card applet. The commands that the applet can accept in the context of the above functionality are described in Table 8.2.

The image is stored in a two-dimensional array. The user selects the array size during the installation phase with SET_SIZE command, where P1 is rows and P2 is columns. The maximum size of the array is limited to $128 \times 128$ bytes $= 16$ K.

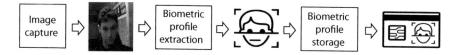

**FIGURE 8.10**
Registration phase.

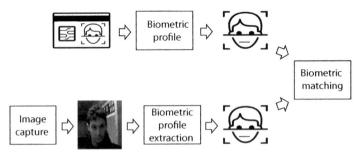

**FIGURE 8.11**
Identification phase.

### *Dependable Distributed Computation Framework*

Computation in a distributed environment introduces several complications compared with a "centralized" approach. In particular, it becomes necessary to account for faults within the array of resources allocated to the execution of a given application. Sometimes this requirement translates into the need to reconfigure the flow of execution, in order to guarantee that a task will correctly run to completion.

The model-based DDCF addresses these problems in the design and execution of distributed applications (see Chapter 3). The main objective of the framework is to provide a safe way to describe distributed applications, which makes interactions between subsections of code explicit; the framework also obviously provides a distributed runtime environment that supports such an execution model. The framework interacts with the SHIELD middleware through a configurable log service that logs the data flowing through the components of the identification software. This enables auditing for security

**TABLE 8.2**

Smart Card Biometric Applet Commands

| Command | CLA | INS | P1 | P2 | Description |
|---|---|---|---|---|---|
| BIO_SET | 0×80 | 0×74 | 0×XX. | 0×00 | It is used for uploading an image to the card. It must be called as many times as the number of rows of the array. The P1 value indicates the current row for uploading. |
| BIO_GET | 0×80 | 0× | 0×XX. | 0×00 | It is used for downloading the image from the card. As above, it must be called as many times as the number of the rows. The P1 value informs which row we want the card to send. |
| GET_SIZE | 0×80 | 0×78 | 0×00 | 0×00 | It is used to get the stored image size. |
| SET_SIZE | 0×80 | 0×78 | 0×00 | 0×00 | It is used for informing the system about the stored image size. |

purposes (during operation) or debugging reasons (during the development phase). In addition, the encryption of data is configurable. Disabling both these features allows the optimization of interprocess communication, which can be a significant benefit for applications with high workload.

The framework defines the hardware or software architecture for the runtime, which supports a distributed execution of an application. Such an application is described using a domain language that exploits composability and reuse. The DDCF is designed to enhance both the development and production phases of an application. The development phase is concerned with dependable design, while the production phase is concerned with dependable execution. Here, attention is placed on the production phase, but many concepts (such as the application metamodel) also directly impact the development phase.

In the context of the biometric-based identification system, the DDCF covers the need to deploy versioned code in an automatic way, and to avoid single-point-of-failure situations as much as possible. This enhancement to dependability originates from two features:

1. *The application can be explicitly designed to be distributed.* If multiple cameras for identification are available, we can tolerate temporary downtimes or permanent failures of several cameras without compromising the operation of the system.

2. *We can introduce redundancy in some resources.* Apart from the presence of multiple cameras, there may be other computational nodes related to communication or persistence that are redundant. The middleware is responsible for handling failures and preserving the execution of the application.

### SHIELD Middleware

The SHIELD middleware is responsible for monitoring the SPD behavior of the components of the identification system, and ensuring that the appropriate actions are executed in the case of threats or faults, trying to recover the optimal SPD level. At runtime, depending on the information provided by the components of the identification system and considering the current SPD state, the middleware identifies the next SPD state and communicates the new state to the components themselves. A set of rules that describe the application logic of the people identification process allow the middleware to calculate the next SPD state starting from the current SPD state. These rules describe the behavior of the identification system when it is exposed to a fault or a menace, and define the reaction consisting of specific countermeasures. Every component of the system receives the new SPD state and adopts the operation required to "enter" the suggested SPD state. The SHIELD methodology adopted in this dynamic process and the metrics calculation are described in Chapters 4 and 5.

### Identification System Integrated Architecture

The biometric-based identification system is an integrated solution that combines the hardware and software components described in the previous sections, orchestrated and managed from the SPD perspective by the SHIELD framework. The block diagram in Figure 8.12 illustrates the architecture of a single smart turnstile and the integration map of its components. The gray boxes indicate hardware components, while the black boxes represent software components. The black dashed lines represent logical connections (interfaces between components), while the gray lines represent physical connections (USB or Ethernet). The blue elements represent DDCF descriptors, the configuration files that are used by the DDCF during the start-up of the system.

The intelligent embedded camera is responsible for automatic face recognition and for the initial phase of people identification at the entrance of the public infrastructure premises. The intelligent cameras are connected to the security control server, hosting the DDCF, and the SHIELD middleware via the Ethernet: the biometric profile obtained during the face recognition process is used by identification software running on the security control server for people identification. The interfaces provided by the intelligent camera for integration are the RPC functions offered by the face finder module, the ICAO module, and the face recognition module. These functions allow remote control of the modules.

Each intelligent camera is connected to a smart card reader via USB. The role of smart card–based security services is twofold: smart cards are tamper-resistant devices that are used to store a visitor's biometric profile, and at the same time, the security service is used to check the real identity of the embedded nodes of the identification system. The node identity can be checked using the secure messaging API.

The DDCF defines a data flow model to design distributed applications. In this context, applications are automatically deployed or updated to remote nodes through the network. The role of the DDCF is to monitor the instances of the face recognition software running on the various intelligent cameras integrated in the smart turnstiles. Transient data can be logged for auditing purposes, faults of the face recognition software are notified through specific events, and automatic recovery is supported.

The DDCF runs on the security control server, and is connected to the intelligent cameras via the Ethernet. The DDCF prototype incorporates routines in the form of libraries with supporting descriptor files. The actual interface between the routines and the framework is made through well-defined access functions. The framework then becomes responsible for spawning processes that employ these libraries, and for data communication and synchronization between the nodes running such processes.

The SHIELD middleware runs both on the intelligent cameras and on the security control server with specific security agents. The security agent

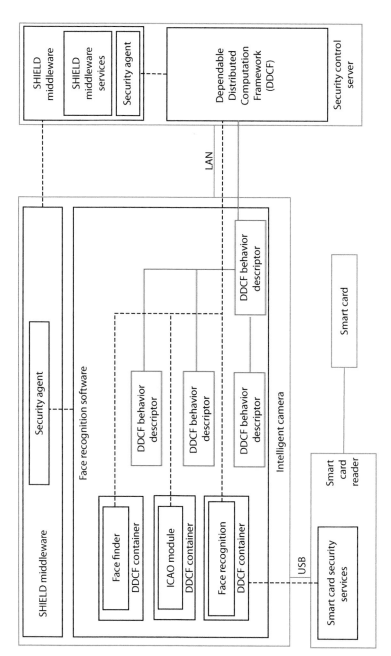

**FIGURE 8.12**
Prototype architecture. Gray lines, hardware components/interfaces; black lines, software components/interfaces; blue lines, files required for the initial setup.

running on the cameras is responsible for faults and threats that could alter the SPD level of the face recognition process, while the security agents running on the security control server interact with the DDCF. The SHIELD security services run on the security control server, managing the semantic model and providing secure discovery. The interface offered by the middleware is based on the Service Location Protocol (SLP) messages that the intelligent cameras and the DDCF exchange with their own security agents.

The face recognition software provides an API that allows remote configuration and control and is used for system integration.

The face finder module is responsible for the acquisition of the video stream and generates a message when a human face is found in the input stream. It provides the following RPC functions:

| | |
|---|---|
| face_detection.start | Start the face detection process |
| face_detection.stop | Stop the face detection process |
| face_detection.kl | Send a standard keep-alive message |
| face_detection.send.image | Send the image containing a face |
| face_detection.klm(msg) | Send a keep-alive message with the specific message "msg" |

Every time a face is detected in the video stream, the extracted features are passed to the ICAO module, which accepts them as input data and elaborates them with an appropriate function.

The ICAO module is responsible for the selection of the best image in the set of images identified by the face finder module. The module provides the following RPC functions:

| | |
|---|---|
| icao.compute.score | Calculate the ICAO parameters for the images identified in the video stream |
| icao.send.image | Send the image of the face with the correct ICAO score |
| icao.kl | Send a standard keep-alive message |
| icao.klm(msg) | Send a keep-alive message with the specific message "msg" |

When the module finds a face with an ICAO score that allows the identification, the stop function of the face finder module is called and the best result of this detection phase is sent to the face recognition module.

The face recognition module is responsible for the extraction of the biometric profile and for the identification of a matching profile in the database. The module provides the following RPC functions:

| | |
|---|---|
| facerec.bio.extract | Extract the biometric profile of the face contained in the selected image |
| facerec.bio.store | During the enroll phase, store the biometric profile in the database |
| facerec.bio.identify | In transit mode, query the database to find a biometric match |
| facerec.kl | Send a standard keep-alive message |
| facerec.klm(msg) | Send a keep-alive message with the specific message "msg" |

The smart card module is responsible for reading the encrypted biometric profile of a person from his or her smart card. The module provides the following RPC functions:

| | |
|---|---|
| smartcard.secure.open | Open a secure connection with the smart card reader |
| smartcard.secure.close | Close a secure connection with the smart card reader |
| smartcard.secure.getkey | Get the secure key used in the connection |
| smartcard.bio.store | If a secure connection is open, store a biometric profile in the smart card |
| smartcard.bio.load | If a secure connection is open, read the biometric profile in the smart card |
| smartcard.kl | Send a standard keep-alive message |
| smartcard.klm(msg) | Send a keep-alive message with the specific message "msg" |

The communication with the SHIELD middleware is based on the SLP. The SLP is a packet-oriented protocol defined in RFCs 2608 and 2609 [6]. The protocol is used to transmit, in the Transmission Control Protocol/User Datagram Protocol (TCP/UDP) packet, the information of the use case, in terms of both status and commands.

Each identification subsystem composed of an intelligent camera integrated into the turnstile and hosting the identification software and of a smart card reader is associated with a unique identifier. The unique identifier is a URL formatted as follows:

```
service:SHIELDDevice:Cam _ XXX.://Cam _ XXX..shield
```

Also, the commands that the security agent sends to the identification subsystems and to the DDCF are formatted using a URL. The messages are unique in the SLP service URLs, and for this reason, different commands sent to a specific identification subsystem or to the DDCF are registered, by the service agent, using the SLP in the following form:

```
service:SHIELDAction.Cam _ XXX.:yyyy://Cam _ XXX..shield
service:SHIELDAction.DDCF:yyyy://DDCF.shield
```

In the URLs previously mentioned, the fields have the following meaning:

- *Cam_XXX.*: This is a unique identifier (XXX. is an integer with three digits) that identifies a specific identification subsystem in the use case.
- *yyyy*: This is a four-digit time stamp used to discriminate the different messages received from the security agent.

The face recognition software and the DDCF send the status information to the security agent using the SLP. The status is described in terms of SPD metrics, and the messages contain the following information:

- The values of the SPD metrics expressed in XML, as follows:

```xml
<?xml version="1.0" encoding="UTF-8" standalone="yes"?>
<metrics>
  <vulnerabilities>
    <basic></basic>
    <e_basic></e_basic>
    <moderate></moderate>
    <high></high>
    <beyond_high></beyond_high>
  </vulnerabilities>
  <limitations>
    <anomalies></anomalies>
    <concerns></concerns>
    <exposures></exposures>
    <weaknesses></weaknesses>
  </limitations>
  <classA>
    <authentication></authentication>
    <identification></identification>
    <resilience></resilience>
    <subjugation></subjugation>
    <continuity></continuity>
  </classA>
  <classB>
    <non_Repudiation></non_Repudiation>
    <confidentiality></confidentiality>
    <privacy></privacy>
    <integrity></integrity>
    <alarm></alarm>
  </classB>
  <complexity></complexity>
  <trust></trust>
  <accessesList>
    <dp></dp>
    <ef></ef>
    <num></num>
  </accessesList>
</metrics>
```

- The ID of the intelligent camera, in the form Cam_XXX., or the code DDCF
- The status of the identification subsystem (see the "Embedded Face Recognition SPD Levels" section)

- The status of the DDCF (see the "Dependable Distributed Computation Framework SPD Levels" section)

The intelligent cameras and the DDCF periodically check the SLP service registry for any new message sent from the security agent. The procedure required to retrieve the messages is based on two different steps: send a request to receive all the URLs containing the messages from the security agent and send an SLP request in order to obtain the actual commands for the different identification subsystems or for the DDCF.

## SPD Metrics Calculation

The calculation of the SPD metrics has been performed following the SHIELD methodology described in Chapters 4 and 5. The background experience in the security field, the information obtained by the behavioral analysis of the system component, and the set of identified failures, threats, and menaces are the main information that can be used to compute the SPD levels associated with the system component, according to proper metrics calculation methods (e.g., medieval castle and multimetrics). One of the strengths of the SHIELD approach is indeed its modularity and independence with respect to any specific metrics computation method.

The SPD metrics calculation for the people identification system has been calculated following the medieval castle method (see Chapter 5), which has been applied at both the subsystem and system levels. With this approach, the composition phase of the metrics calculation generated 143 different SPD states that characterize the biometric-based people identification system.

The following sections illustrate the results obtained following the SHIELD method for SPD metrics calculation, applied first to each of the subsystems and subsequently by composition to the entire identification system. The operative behavior of each subsystem is analyzed, trying to identify potential faults or threats that could negatively influence the SPD level. A state diagram illustrating the operative behavior is defined, and for each state of the state diagram, the SPD level is calculated.

### Embedded Face Recognition SPD Levels

The identification of the SPD states and the calculation of the related SPD values are based on the analysis of the operative behavior of the intelligent camera and of the face recognition software, during normal operation and in the presence of a faults and/or menaces.

Figure 8.13 illustrates the functional behavior of the intelligent camera prototype during the people identification process.

The identified states correspond to the main steps of the face recognition and people identification process:

- *ST*: The face recognition software registers on the DDCF and performs the setup.
- *W*: This is the waiting state, in which the software analyzes the video stream, looking for a face.
- *I*: After a face has been identified, the best image in the stream is selected.
- *B*: The recognition software extracts the biometric profile from the face found in the selected image.
- *P*: The recognition software searches for a match in the biometric profile database.
- *A*: Depending on the results of step P, the turnstile is opened or security is called.

For each state, the diagram reports the module of the face recognition software involved in that particular phase: FF is the face finder module, ICAO is the ICAO module, FR is the face recognition module, and AM is the access management module.

Two types of failures and menaces have been considered:

- Physical failures and menaces
- System failure

An example of a physical menace is the camera obfuscation: the CCD sensor integrated into the intelligent camera can be obfuscated trying to break the reliability of the identification process or simply for an act of vandalism. In this case, depending on the specific installation of the identification system and on the type of turnstile adopted, the security and dependability of the identification process are compromised. The countermeasure adopted for this kind of issue is to call security personnel.

A second example is the noncollaborative behavior of the person that has to be identified. Typically, this situation happens when a person hides or alters his or her face, fully or partially, in order to try to avoid or confuse the identification process. In this situation, the user interface gives to the person a set of suggestions aimed at removing any possible source of obfuscation from the face and positioning it in the correct way. Noncollaborative behavior can be easily transformed into collaborative

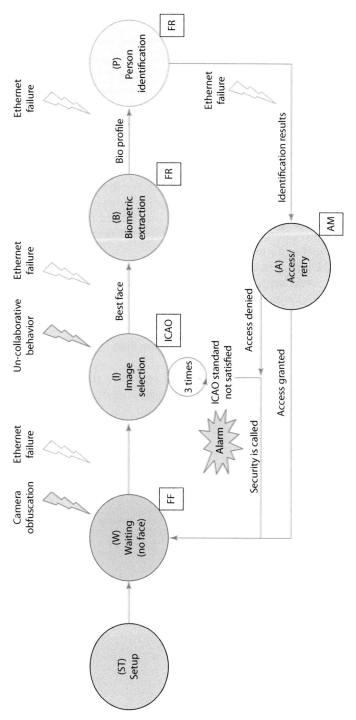

**FIGURE 8.13**
Functional state diagram of the intelligent camera.

behavior, and if it is due to criminal activity, it can be solved by calling the security personnel.

A system failure could be an example of the absence of the Ethernet link on the local area network that connects the intelligent cameras with the security control server. This failure can be caused by a hardware issue of the intelligent camera, a fault in a switch on the local area network, a hardware fault on the security control server side, and so forth. This kind of failure is a severe issue, because it completely stops the identification activities and the flow of people entering the stadium. It requires the immediate intervention of maintenance and has an important impact on the SPD level.

Background experience in the security field, the information obtained by the analysis of the state diagram, and the set of failures and menaces selected have been used to apply the medieval castle method for the calculation of the metrics. The output of this phase is the characterization, in terms of SPD, of the states described in Figure 8.13.

From the SPD characterization of the states, it is possible to calculate all the real SPD states: considering a state in the functional block diagram and the possible failures and menaces that affect that state, one or more different SPD levels for the system can be calculated. Each different level of SPD defines a single SPD state.

Table 8.3 provides a list of the SPD states identified for the face recognition prototype and the related SPD values.

The table contains a set of acronyms that represent the faults or menaces considered for the intelligent camera prototype. The acronyms have the following meanings:

**TABLE 8.3**

SPD Levels of the Intelligent Camera

| State ID | Faults and Threats | Actual SPD Level |
|---|---|---|
| 00 | SEC | 90,63 |
| 01 | 3 ETH + SEC | 87,27 |
| 02 | SEC + OBF | 88,96 |
| 03 | 3 ETH + SEC + OBF | 86,61 |
| 04 | SEC + 3T | 89,24 |
| 05 | 4 ETH + SEC + 2 3T | 86,74 |
| 06 | SEC + OBF + 3T | 88,10 |
| 07 | 4 ETH + SEC + OBF + 2 3T | 86,17 |
| 08 | 3 SEC + 2 3T | 88,30 |
| 09 | 3 SEC + OBF + 2 3T | 87,42 |
| 10 | 5 ETH + 4 SEC + 3 3T | 86,28 |
| 11 | 5 ETH + 4 SEC + OBF + 3 3T | 85,78 |

- *ETH*: Failure in the Ethernet connection.
- *OBF*: The camera has been obfuscated or the person does not collaborate for the identification.
- *3T*: An alarm occurred during the ICAO phase and a warning to the user has been generated (three times).
- *SEC*: An alarm that requires the intervention of security (after 3T warnings) has been generated.

Multiple threats and menaces of the same type may occur, so in the table, each acronym is associated with a number representing the multiplicity of the specific menace in the laboratory test.

### Smart Card Secure Service SPD Levels

This section describes the SPD metrics calculation for the smart card secure services. The most significant faults and threats influence this component of the identification system when it establishes a secure channel between the smart card and the security control server for the protection of messages exchanged between the two entities. The necessity of a secure channel might be the result of a need to transmit sensitive data over an exposed or unprotected communication channel or an identified attack. The system changes the state of the smart card component, which uses the preinstalled cryptographic keys to establish a secure channel with the remote counterpart, which in this case is the intelligent camera. Finally, if requested by the intelligent camera, the system passes the session keys back to the intelligent camera.

Given the required level of protection for the exchanged messages, the system can provide maximum message protection, for example, confidentiality, data integrity, and message authentication (high SPD level), or simply message integrity and authentication (medium SPD level). State changes can be the result of suspected or identified attacks on the network resources, transmission errors, or message manipulation and injection attacks.

The functional state diagram of the smart card prototype is defined in Figure 8.14, which depicts the following functional states of the prototype:

- *ST*: The software components are set up.
- *W*: The card waits for the issuance of the appropriate command from the host system (e.g., camera).
- *S*: Upon system request, the card can use its preconfigured cryptographic keys to establish a secure channel with a remote party (security control server) and, optionally, return session keys.
- *P*: Upon system request, the card can protect the messages for the host system using the session keys (confidentiality, integrity, and authentication).

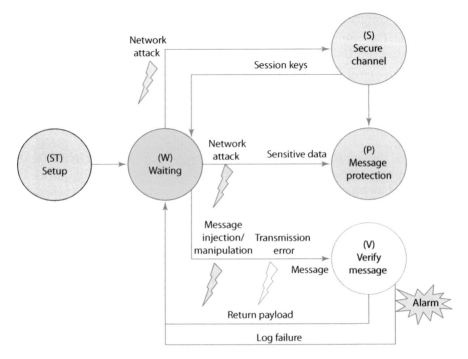

**FIGURE 8.14**
Functional state diagram of the smart card security service.

- *V*: Upon system request, the card can use the session keys to verify the integrity and message origin and, if necessary, decrypt incoming messages.

The system is designed to react to the following failures and threats:

- Transmission errors
- Basic network attacks (eavesdropping and message injection or manipulation)

Transmission errors can be the result of networking component malfunction, interference, and environmental conditions that can lead to data corruption. The robust integrity mechanisms used by the smart card secure service can be used to identify erroneous messages even if typical error detection codes are deployed for these types of errors.

Other networking attacks can occur by eavesdroppers or intruders who want to either intercept communications and have access to sensitive data or manipulate exchanged data in order to bypass security controls. For example, consider the case where an intruder replaces the biometric profile exchanged between the intelligent camera and the security control server in

**TABLE 8.4**

SPD States of the Smart Card

| State ID | Faults and Threats | Actual SPD Level |
|----------|--------------------|------------------|
| 00 | MAN+MON | 72,62 |
| 01 | MON | 83,25 |
| 02 | 0 | 96,59 |

order to masquerade as a legitimate and authorized person. Communication interception is also a threat to the system, as it can disclose private data to unauthorized entities, such as images captured by the cameras. The secure channel established by the smart card and the secure messaging services can help nullify those threats.

Furthermore, certain controls can be bypassed if the attacker manages to manipulate the host's interaction with the smart card to intentionally reduce the required level of security. This can be typically achieved by sending the wrong command, with the effect of requesting lower-level protection for the messages or bypassing the additional security provided by the smart card.

Table 8.4 lists the SPD states for the smart card security service and the related SPD values. The table contains a set of acronyms that represent the faults or menaces:

- *MON*: The network connection is subject to unauthorized monitoring.
- *MAN*: The communicated data are subject to data manipulation by an adversary.

The SPD states reflect the following situations:

- *00*: No issue is present.
- *01*: Confidentiality is not chosen for the protection of messages, and therefore they are subject to interception and private data can be exposed.
- *02*: Integrity is not chosen for the protection of messages, and therefore they are subject to data manipulation.

## Dependable Distributed Computation Framework SPD Levels

This section describes the SPD metrics calculation for the DDCF subsystem: the identification of the SPD states and the calculation of the related SPD values are based on the analysis of the operative behavior of the DDCF, during normal operation and in the presence of a fault or menace.

Figure 8.15 illustrates the functional behavior of the DDCF prototype during the execution of a generic application.

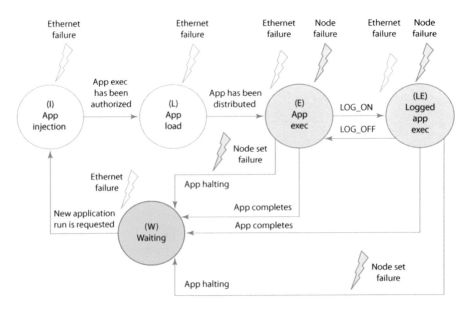

**FIGURE 8.15**
Functional state diagram of the DDCF.

The identified states are as follows:

- *I*: The application is injected by requesting the insertion of a new instance of the application into the cloud of nodes.
- *L*: The application is actually loaded by the nodes, which download the application code.
- *E*: The application is executed in a distributed way.
- *LE*: The application is executed in a distributed way, while logging of the exchanged data is performed.
- *W*: The application is not running anymore, due to either a critical failure or completion. The runtime is waiting for a new application injection.

Three types of failures have been considered in this state diagram:

- Ethernet failure
- Single-node failure
- Node set failure

An Ethernet failure implies that the connection between one node and the computation cloud is unavailable. This is considered a transient situation, and it is not specifically handled unless it results in the failure of a node set. In fact, Ethernet failures in a distributed replicated system do not necessarily cause a failure of the whole system, but rather a reduction of the functionality of the system or a loss of observability from the perspective of the user.

If for some reason (including a loss of connectivity) a single node becomes unreachable, a node failure is recognized. A node failure is not necessarily permanent, and its impact on the execution may be tolerated, but it is still useful to provide an alarm for this event.

On the other hand, if all nodes from a node set (i.e., a "role") fail, then the DDCF runtime cannot process the execution further. This is a critical situation, which again results in an alarm and also causes the execution to halt. After the issues are addressed through human intervention, a new application instance can be injected to resume the execution.

The background experience in the security field, the information obtained by the analysis of the state diagram, and the set of failures and menaces selected have been used to apply the medieval castle method for the calculation of the metrics. The output of this phase is the characterization in terms of SPD of the states described in Figure 8.15. From the SPD characterization of states, it is possible to calculate all the real SPD states: considering a state in the functional block diagram, and the possible failures and menaces that affect that state, one or more different SPD levels for the system can be calculated. To each different SPD state, a single SPD level is associated.

Table 8.5 provides a list of the SPD states identified for the face recognition prototype and the related SPD values. The table contains a set of acronyms that represent the faults or menaces considered for the DDCF prototype. The acronyms have the following meanings:

- *ETH*: Failure in the Ethernet connection
- *FAU*: All nodes with a given role are unavailable due to any kind of fault

Multiple threats and menaces of the same type may occur, so in the table, each acronym is associated with a number representing the multiplicity of the specific menace in the laboratory test.

**TABLE 8.5**

SPD States of the DDCF

| State ID | Faults and Threats | Actual SPD Level |
| --- | --- | --- |
| 01 | ETH | 89,18 |
| 02 | 2 ETH | 87,16 |
| 03 | 2 ETH+FAU | 86,00 |
| 04 | 2 ETH+4 FAU | 84,85 |

The SPD states reflect the following situations:

- *01*: No issue is present.
- *02*: At least one failed node is present within the system.
- *03*: Logging is unavailable due to the failure of all nodes.
- *04*: The execution of the application is halted due to the failure of at least one node set.

### SPD Composition at System Level

The final step of the SPD metrics calculation consists in the composition of the SPD metrics obtained for the single components of the people identification system. This step is fundamental for the identification of the SPD behavior at the system level. According to the previous steps of the metrics calculation, the composition follows the approach defined by the medieval castle method.

The result of the composition is a set of SPD states that define how the SPD levels evolve when the people identification system is exposed to a menace or fault and reacts with the corresponding countermeasure.

Every fault and/or menace considered for the calculation of the SPD metrics at the component level is also considered in the composition phase, in this way providing full coverage of these faults and menaces at the system level.

Table 8.6 provides a list of all the SPD states that characterize the system at runtime. The table reports the results obtained from the composition of the SPD values calculated for the single components. For each SPD state, the table reports

- A reference number of the state at the scenario level
- For each of the four prototypes involved in the demonstrator,
  - A reference number for the state involved in the composition
  - A description of the state
- The SPD value obtained by the composition at the use case level
- A normalization of the SPD value obtained by the composition at the use case level

The normalized values have been introduced to simplify the use of the SPD information in a real environment and unify the range of SPD values between 0 and 1 across different domains, heterogeneous systems, and applications. With this approach, it is possible to manage the SPD composition at both the system level and the system of systems level.

**TABLE 8.6**

People Identification System SPD States and Corresponding SPD Values

| People Identification Scenario Ref. Num. | Smart Card | | DDCF | | Intelligent Camera + Facial Recognition SW | | MW | | SPD-Level Value | Normal SPD Level |
|---|---|---|---|---|---|---|---|---|---|---|
| | Ref. Num. | Description | Ref. Num. | Description | Ref. Num. | Description | Ref. Num. | Description | | |
| 00 | 00 | Low | 00 | All is working | 00 | Normal use | 0 | Normal use | 76,371 | 0,049 |
| 01 | | | | | 01 | ETH | | | 75,964 | 0,017 |
| 02 | | | | | 02 | OBF | | | 76,349 | 0,047 |
| 03 | | | | | 03 | OBF+ETH | | | 75,942 | 0,015 |
| 04 | | | | | 04 | 3T | | | 76,348 | 0,047 |
| 05 | | | | | 05 | ETH+3T | | | 75,941 | 0,015 |
| 06 | | | | | 06 | OBF+3T | | | 76,325 | 0,045 |
| 07 | | | | | 07 | OBF+ETH+3T | | | 75,919 | 0,014 |
| 08 | | | | | 08 | 3T+SEC | | | 76,324 | 0,045 |
| 09 | | | | | 09 | OBF+3T+SEC | | | 76,302 | 0,043 |
| 10 | | | | | 10 | ETH+3T+SEC | | | 75,918 | 0,014 |
| 11 | | | | | 11 | OBF+ETH+3T+SEC | | | 75,896 | 0,012 |
| 12 | | | 01 | System has at least one failed node | 00 | Normal use | | | 76,324 | 0,045 |
| 13 | | | | | 01 | ETH | | | 75,918 | 0,014 |
| 14 | | | | | 02 | OBF | | | 76,302 | 0,043 |
| 15 | | | | | 03 | OBF+ETH | | | 75,896 | 0,012 |
| 16 | | | | | 04 | 3T | | | 76,301 | 0,043 |

*(Continued)*

**TABLE 8.6 (CONTINUED)**

People Identification System SPD States and Corresponding SPD Values

| People Identification Scenario | Smart Card | | DDCF | | Intelligent Camera + Facial Recognition SW | | MW | | SPD-Level Value | Normal SPD Level |
|---|---|---|---|---|---|---|---|---|---|---|
| Ref. Num. | Ref. Num. | Description | Ref. Num. | Description | Ref. Num. | Description | Ref. Num. | Description | | |
| 17 | | | | | 05 | ETH+3T | | | 75,896 | 0,012 |
| 18 | | | | | 06 | OBF+3T | | | 76,279 | 0,042 |
| 19 | | | | | 07 | OBF+ETH+3T | | | 75,874 | 0,010 |
| 20 | | | | | 08 | 3T+SEC | | | 76,278 | 0,042 |
| 21 | | | | | 09 | OBF+3T+SEC | | | 76,256 | 0,040 |
| 22 | | | | | 10 | ETH+3T+SEC | | | 75,873 | 0,010 |
| 23 | | | | | 11 | OBF+ETH+3T+SEC | | | 75,851 | 0,008 |
| 24 | | | 02 | Logging is unavailable due to failure of all P nodes | 00 | Normal use | | | 76,279 | 0,042 |
| 25 | | | | | 01 | ETH | | | 75,874 | 0,010 |
| 26 | | | | | 02 | OBF | | | 76,257 | 0,040 |
| 27 | | | | | 03 | OBF+ETH | | | 75,852 | 0,008 |
| 28 | | | | | 04 | 3T | | | 76,256 | 0,040 |
| 29 | | | | | 05 | ETH+3T | | | 75,851 | 0,008 |
| 30 | | | | | 06 | OBF+3T | | | 75,888 | 0,011 |
| 31 | | | | | 07 | OBF+ETH+3T | | | 75,829 | 0,007 |
| 32 | | | | | 08 | 3T+SEC | | | 76,233 | 0,038 |

*(Continued)*

**TABLE 8.6 (CONTINUED)**

People Identification System SPD States and Corresponding SPD Values

| People Identification Scenario Ref. Num. | Smart Card Ref. Num. | Description | DDCF Ref. Num. | Description | Intelligent Camera + Facial Recognition SW Ref. Num. | Description | MW Ref. Num. | Description | SPD-Level Value | Normal SPD Level |
|---|---|---|---|---|---|---|---|---|---|---|
| 33 | | | | | 09 | OBF+3T+SEC | | | 76,211 | 0,036 |
| 34 | | | | | 10 | ETH+3T+SEC | | | 75,829 | 0,007 |
| 35 | | | | | 11 | OBF+ETH+3T+SEC | | | 75,807 | 0,005 |
| 36 | | | 03 | Execution halted due to failure of at least one node set | 00 | Normal use | | | 76,212 | 0,036 |
| 37 | | | | | 01 | ETH | | | 75,808 | 0,005 |
| 38 | | | | | 02 | OBF | | | 76,190 | 0,035 |
| 39 | | | | | 03 | OBF+ETH | | | 75,787 | 0,003 |
| 40 | | | | | 04 | 3T | | | 76,190 | 0,035 |
| 41 | | | | | 05 | ETH+3T | | | 75,786 | 0,003 |
| 42 | | | | | 06 | OBF+3T | | | 76,168 | 0,033 |
| 43 | | | | | 07 | OBF+ETH+3T | | | 75,765 | 0,002 |
| 44 | | | | | 08 | 3T+SEC | | | 76,167 | 0,033 |
| 45 | | | | | 09 | OBF+3T+SEC | | | 76,145 | 0,031 |
| 46 | | | | | 10 | ETH+3T+SEC | | | 75,764 | 0,002 |
| 47 | | | | | 11 | OBF+ETH+3T+SEC | | | 75,743 | 0,000 |

*(Continued)*

**TABLE 8.6 (CONTINUED)**

People Identification System SPD States and Corresponding SPD Values

| People Identification Scenario Ref. Num. | Smart Card Ref. Num. | Smart Card Description | DDCF Ref. Num. | DDCF Description | Intelligent Camera + Facial Recognition SW Ref. Num. | Intelligent Camera + Facial Recognition SW Description | MW Ref. Num. | MW Description | SPD-Level Value | Normal SPD Level |
|---|---|---|---|---|---|---|---|---|---|---|
| 48 | 01 | Medium | 00 | All is working | 00 | Normal use | 0 | Normal use | 85,743 | 0,776 |
| 49 | | | | | 01 | ETH | | | 84,451 | 0,676 |
| 50 | | | | | 02 | OBF | | | 85,160 | 0,731 |
| 51 | | | | | 03 | OBF+ETH | | | 84,280 | 0,662 |
| 52 | | | | | 04 | 3T | | | 85,154 | 0,730 |
| 53 | | | | | 05 | ETH+3T | | | 84,275 | 0,662 |
| 54 | | | | | 06 | OBF+3T | | | 84,965 | 0,716 |
| 55 | | | | | 07 | OBF+ETH+3T | | | 84,111 | 0,649 |
| 56 | | | | | 08 | 3T+SEC | | | 84,959 | 0,715 |
| 57 | | | | | 09 | OBF+3T+SEC | | | 84,780 | 0,701 |
| 58 | | | | | 10 | ETH+3T+SEC | | | 84,107 | 0,649 |
| 59 | | | | | 11 | OBF+ETH+3T+SEC | | | 83,950 | 0,637 |
| 60 | | | 01 | System has at least one failed node | 00 | Normal use | | | 84,959 | 0,715 |
| 61 | | | | | 01 | ETH | | | 84,107 | 0,649 |
| 62 | | | | | 02 | OBF | | | 84,780 | 0,701 |
| 63 | | | | | 03 | OBF+ETH | | | 83,950 | 0,637 |
| 64 | | | | | 04 | 3T | | | 84,774 | 0,701 |

*(Continued)*

**TABLE 8.6 (CONTINUED)**

People Identification System SPD States and Corresponding SPD Values

| People Identification Scenario Ref. Num. | Smart Card | | DDCF | | Intelligent Camera + Facial Recognition SW | | MW | | SPD-Level Value | Normal SPD Level |
| --- | --- | --- | --- | --- | --- | --- | --- | --- | --- | --- |
| | Ref. Num. | Description | Ref. Num. | Description | Ref. Num. | Description | Ref. Num. | Description | | |
| 65 | | | | | 05 | ETH+3T | | | 83,946 | 0,636 |
| 66 | | | | | 06 | OBF+3T | | | 84,603 | 0,687 |
| 67 | | | | | 07 | OBF+ETH+3T | | | 83,796 | 0,625 |
| 68 | | | | | 08 | 3T+SEC | | | 84,598 | 0,687 |
| 69 | | | | | 09 | OBF+3T+SEC | | | 84,435 | 0,674 |
| 70 | | | | | 10 | ETH+3T+SEC | | | 83,792 | 0,624 |
| 71 | | | | | 11 | OBF+ETH+3T+SEC | | | 83,648 | 0,613 |
| 72 | | | 02 | Logging is unavailable due to failure of all nodes | 00 | Normal use | | | 84,603 | 0,687 |
| 73 | | | | | 01 | ETH | | | 83,796 | 0,625 |
| 74 | | | | | 02 | OBF | | | 84,440 | 0,675 |
| 75 | | | | | 03 | OBF+ETH | | | 83,652 | 0,614 |
| 76 | | | | | 04 | 3T | | | 84,435 | 0,674 |
| 77 | | | | | 05 | ETH+3T | | | 83,648 | 0,613 |
| 78 | | | | | 06 | OBF+3T | | | 83,984 | 0,639 |
| 79 | | | | | 07 | OBF+ETH+3T | | | 83,509 | 0,603 |
| 80 | | | | | 08 | 3T+SEC | | | 84,274 | 0,662 |

(Continued)

**TABLE 8.6 (CONTINUED)**

People Identification System SPD States and Corresponding SPD Values

| People Identification Scenario Ref. Num. | Smart Card | | DDCF | | Intelligent Camera + Facial Recognition SW | | MW | | SPD-Level Value | Normal SPD Level |
|---|---|---|---|---|---|---|---|---|---|---|
| | Ref. Num. | Description | Ref. Num. | Description | Ref. Num. | Description | Ref. Num. | Description | | |
| 81 | | | | | 09 | OBF+3T+SEC | | | 84,124 | 0,650 |
| 82 | | | | | 10 | ETH+3T+SEC | | | 83,505 | 0,602 |
| 83 | | | | | 11 | OBF+ETH+3T+SEC | | | 83,371 | 0,592 |
| 84 | | | 03 | Execution halted due to failure of at least one node set | 00 | Normal use | | | 84,134 | 0,651 |
| 85 | | | | | 01 | ETH | | | 83,379 | 0,593 |
| 86 | | | | | 02 | OBF | | | 83,990 | 0,640 |
| 87 | | | | | 03 | OBF+ETH | | | 83,250 | 0,582 |
| 88 | | | | | 04 | 3T | | | 83,985 | 0,640 |
| 89 | | | | | 05 | ETH+3T | | | 83,246 | 0,582 |
| 90 | | | | | 06 | OBF+3T | | | 83,846 | 0,629 |
| 91 | | | | | 07 | OBF+ETH+3T | | | 83,121 | 0,572 |
| 92 | | | | | 08 | 3T+SEC | | | 83,842 | 0,628 |
| 93 | | | | | 09 | OBF+3T+SEC | | | 83,708 | 0,618 |
| 94 | | | | | 10 | ETH+3T+SEC | | | 83,117 | 0,572 |
| 95 | | | | | 11 | OBF+ETH+3T+SEC | | | 82,996 | 0,563 |

*(Continued)*

**TABLE 8.6 (CONTINUED)**

People Identification System SPD States and Corresponding SPD Values

| People Identification Scenario Ref. Num. | Smart Card | | DDCF | | Intelligent Camera + Facial Recognition SW | | MW | | SPD-Level Value | Normal SPD Level |
|---|---|---|---|---|---|---|---|---|---|---|
| | Ref. Num. | Description | Ref. Num. | Description | Ref. Num. | Description | Ref. Num. | Description | | |
| 96 | 02 | High | 00 | All is working | 00 | Normal use | 0 | Normal use | 88,631 | 1,000 |
| 97 | | | | | 01 | ETH | | | 86,731 | 0,853 |
| 98 | | | | | 02 | OBF | | | 87,813 | 0,936 |
| 99 | | | | | 03 | OBF+ETH | | | 86,414 | 0,828 |
| 100 | | | | | 04 | 3T | | | 87,799 | 0,935 |
| 101 | | | | | 05 | ETH+3T | | | 86,405 | 0,827 |
| 102 | | | | | 06 | OBF+3T | | | 87,410 | 0,905 |
| 103 | | | | | 07 | OBF+ETH+3T | | | 86,116 | 0,805 |
| 104 | | | | | 08 | 3T+SEC | | | 87,398 | 0,904 |
| 105 | | | | | 09 | OBF+3T+SEC | | | 87,051 | 0,877 |
| 106 | | | | | 10 | ETH+3T+SEC | | | 86,107 | 0,804 |
| 107 | | | | | 11 | OBF+ETH+3T+SEC | | | 85,841 | 0,783 |
| 108 | | | 01 | System has at least one failed node | 00 | Normal use | | | 87,398 | 0,904 |
| 109 | | | | | 01 | ETH | | | 86,107 | 0,804 |
| 110 | | | | | 02 | OBF | | | 87,051 | 0,877 |
| 111 | | | | | 03 | OBF+ETH | | | 85,841 | 0,783 |
| 112 | | | | | 04 | 3T | | | 87,039 | 0,876 |

*(Continued)*

**TABLE 8.6 (CONTINUED)**

People Identification System SPD States and Corresponding SPD Values

| People Identification Scenario | Smart Card | | DDCF | | Intelligent Camera + Facial Recognition SW | | MW | | SPD-Level Value | Normal SPD Level |
|---|---|---|---|---|---|---|---|---|---|---|
| Ref. Num. | Ref. Num. | Description | Ref. Num. | Description | Ref. Num. | Description | Ref. Num. | Description | | |
| 113 | | | | | 05 | ETH+3T | | | 85,833 | 0,783 |
| 114 | | | | | 06 | OBF+3T | | | 86,725 | 0,852 |
| 115 | | | | | 07 | OBF+ETH+3T | | | 85,585 | 0,764 |
| 116 | | | | | 08 | 3T+SEC | | | 86,715 | 0,851 |
| 117 | | | | | 09 | OBF+3T+SEC | | | 86,427 | 0,829 |
| 118 | | | | | 10 | ETH+3T+SEC | | | 85,578 | 0,763 |
| 119 | | | | | 11 | OBF+ETH+3T+SEC | | | 85,347 | 0,745 |
| 120 | | | 02 | Logging is unavailable due to failure of all P nodes | 00 | Normal use | | | 86,725 | 0,852 |
| 121 | | | | | 01 | ETH | | | 85,585 | 0,764 |
| 122 | | | | | 02 | OBF | | | 86,437 | 0,830 |
| 123 | | | | | 03 | OBF+ETH | | | 84,712 | 0,696 |
| 124 | | | | | 04 | 3T | | | 86,427 | 0,829 |
| 125 | | | | | 05 | ETH+3T | | | 85,347 | 0,745 |
| 126 | | | | | 06 | OBF+3T | | | 85,892 | 0,787 |
| 127 | | | | | 07 | OBF+ETH+3T | | | 85,130 | 0,728 |
| 128 | | | | | 08 | 3T+SEC | | | 86,153 | 0,808 |

*(Continued)*

**TABLE 8.6 (CONTINUED)**

People Identification System SPD States and Corresponding SPD Values

| People Identification Scenario Ref. Num. | Smart Card | | DDCF | | Intelligent Camera + Facial Recognition SW | | MW | | SPD-Level Value | Normal SPD Level |
|---|---|---|---|---|---|---|---|---|---|---|
| | Ref. Num. | Description | Ref. Num. | Description | Ref. Num. | Description | Ref. Num. | Description | | |
| 129 | | | | | 09 | OBF+3T+SEC | | | 85,907 | 0,789 |
| 130 | | | | | 10 | ETH+3T+SEC | | | 85,124 | 0,728 |
| 131 | | | | | 11 | OBF+ETH+3T+SEC | | | 84,919 | 0,712 |
| 132 | | | 03 | Execution halted due to failure of at least one node set | 00 | Normal use | | | 85,923 | 0,790 |
| 133 | | | | | 01 | ETH | | | 84,931 | 0,713 |
| 134 | | | | | 02 | OBF | | | 85,691 | 0,772 |
| 135 | | | | | 03 | OBF+ETH | | | 84,737 | 0,698 |
| 136 | | | | | 04 | 3T | | | 85,684 | 0,771 |
| 137 | | | | | 05 | ETH+3T | | | 84,732 | 0,697 |
| 138 | | | | | 06 | OBF+3T | | | 85,467 | 0,754 |
| 139 | | | | | 07 | OBF+ETH+3T | | | 84,547 | 0,683 |
| 140 | | | | | 08 | 3T+SEC | | | 85,459 | 0,754 |

*(Continued)*

**TABLE 8.6 (CONTINUED)**

People Identification System SPD States and Corresponding SPD Values

| People Identification Scenario Ref. Num. | Smart Card Ref. Num. | Description | DDCF Ref. Num. | Description | Intelligent Camera + Facial Recognition SW Ref. Num. | Description | MW Ref. Num. | Description | SPD-Level Value | Normal SPD Level |
|---|---|---|---|---|---|---|---|---|---|---|
| 141 | | | | | 09 | OBF+3T+SEC | | | 85,255 | 0,738 |
| 142 | | | | | 10 | ETH+3T+SEC | | | 84,542 | 0,683 |
| 143 | | | | | 11 | OBF+ETH+3T+SEC | | | 84,367 | 0,669 |

**TABLE 8.7**

SPD Macrostates and
Corresponding SPD Ranges

| High (SPD $\geq 0.7$) |
|---|
| Normal ($0.3 <$ SPD $< 0.7$) |
| Very low ($0 <$ SPD $< 0.2$) |

The normalization considers the minimum and maximum values of the SPD level obtained by the composition at the use case level and maps them over the range [0, 1].

Furthermore, to provide visual feedback of the status of the system in terms of SPD, four different macro SPD levels (high, normal, low, and very low) have been introduced. They are represented by four gray tones and give a quick indication of the status of the system. The macro SPD levels are defined as shown in Table 8.7.

## Real Scenario for People Identification

This section illustrates the storyboard that has been selected to validate and verify the adoption of the SHIELD methodology in the people identification scenario. The main objectives of the storyboard are:

- Demonstrate the use of the people identification system
- Identify a sequence of execution steps that include the faults and menaces considered for the SPD metrics calculation
- Validate and verify the use of the system attack surface metrics

The storyboard is based on four main events that significantly describe the use of the people identification system at the entrance of a public infrastructure premises, e.g., a stadium.

The storyboard starts with the setup of the identification system, which illustrates the operations required to set up every single prototype, register them into the SHIELD middleware, and start the normal operations. For normal operations, we intend an idle state in which the system is waiting for people to be identified. Focusing on a single turnstile, the idle state means that the face recognition software is waiting to find a face in the video stream.

The first event of the storyboard illustrates how the identification system correctly identifies a person at the turnstile and grants him or her the access to the stadium. This event describes the normal usage of the identification system and does not consider any fault or menace.

The second main event illustrates how the identification system reacts when a person in front of the turnstile must be identified but does not collaborate with the identification process. This event starts stressing the SHIELD-based system and illustrates how the prototypes react to an anomalous situation. The people identification system tries to guide the person during the identification with warning messages and practical suggestions. The person does not collaborate, and after three warnings, the security personnel are called.

The third event illustrates what happens in the case of an act of vandalism: a person has obfuscated the camera of one of the intelligent cameras with a colored spray. In this case, an alarm is generated and the security personnel are called. After the intervention of maintenance, the identification system in that turnstile is back in an operative state.

Finally, the fourth event simulates the consequences of a hardware fault at the network level: the switch of the local area network of the identification system has a fault. In this situation, the face recognition system at the turnstile loses connection with the security control server and stops working. In this case, the security and maintenance personnel are called, and after the switch is replaced, the identification system is reinitialized and starts working. Figure 8.16 illustrates the storyboard.

Each main event of the storyboard is composed of one or more execution steps that, in turn, represent the various operations that the system performs when the event occurs. Four different situations have been considered in the scenario, in order to try to cover both correct operations and situations that present faults and menaces:

- Correct identification of a person
- Identification failed due to a noncollaborative behavior of a person
- Camera obfuscation
- Fault in the LAN Ethernet switch that compromises the communications

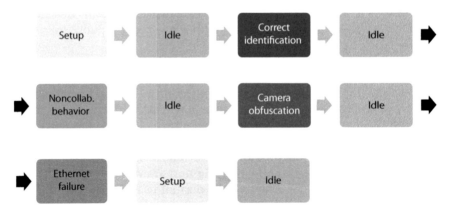

**FIGURE 8.16**
Storyboard of the people identification scenario.

```
ddcf> log runtime -n 2
[2014-12-21 12:35:40:448] Zookeeper ensemble: 192.168.1.121
[2014-12-21 12:35:40:452] Runtime is ready
ddcf> log registry -n 4
[2014-12-21 12:35:41:763] 192.168.1.111 (eth1) registered
[2014-12-21 12:35:41:880] 192.168.1.113 (eth3) registered
[2014-12-21 12:35:42:209] 192.168.1.114 (eth4) registered
[2014-12-21 12:35:42:439] 192.168.1.112 (eth2) registered
```

**FIGURE 8.17**
DDCF logging information.

The following list reports all the steps of the scenario execution:

T0  An operator powers on the system. The intelligent camera registers to the DDCF. The prototypes register to the middleware. The prototypes are in the basic SPD level (Figure 8.17).

The SPD normalized value is

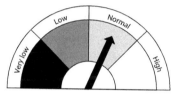

T1  The operator authenticates and logs into the DDCF (Figure 8.18).

The SPD normalized value is

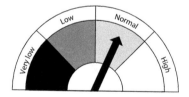

```
lgeretti@earth:~$ ddcf -u admin -h test_ensemble
Password: ***********
Connected to 192.168.1.121 as 'admin'
ddcf>
```

**FIGURE 8.18**
DDCF authentication.

```
ddcf> inject git https://192.168.1.131/gitrepos/ ddcf/apps/vfr
Injected instance of 'Voice Face Recognition Algorithm' with id 1
ddcf> log runtime -n 2
[2014-12-21 12:35:50:751] Requested injection of: git https://192.168.1.131/gitrepos/ at ddcf/apps/vfr, revision latest
[2014-12-21 12:35:51:581] Injection of 'Voice Face Recognition Algorithm' successful (id: 1)
```

**FIGURE 8.19**
DDCF application injection.

```
ddcf> inject git https://192.168.1.131/gitrepos/ ddcf/apps/vfr
Injected instance of 'Voice Face Recognition Algorithm' with id 1
ddcf> log runtime -n 2
[2014-12-21 12:35:50:751] Requested injection of: git https://192.168.1.131/gitrepos/ at ddcf/apps/vfr, revision latest
[2014-12-21 12:35:51:581] Injection of 'Voice Face Recognition Algorithm' successful (id: 1)
```

**FIGURE 8.20**
DDCF application loading.

T2  The DDCF performs the injection of the face recognition software modules (Figure 8.19).

The SPD normalized value is

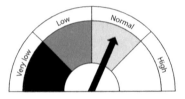

T3  The intelligent camera downloads the face recognition software modules from the security control server (Figure 8.20).

The SPD normalized value is

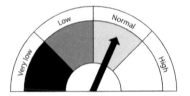

T4  The face recognition software is launched, and the identification system goes into the waiting state (the system is ready to identify a person and is waiting to find a face in the video stream). Every component of the system is executing its functionalities correctly, and the SPD value reaches its maximum level (Figure 8.21).

The SPD normalized value is

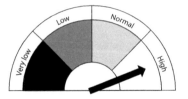

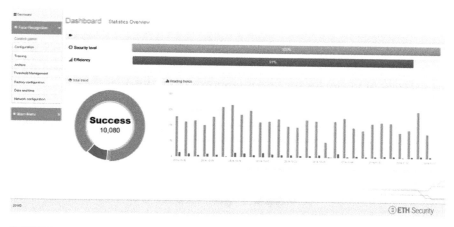

**FIGURE 8.21**
The identification system is in idle state.

T5  The face finder module finds a face in the video stream and passes the control to the ICAO module. A smart card in inserted in the smart card reader. The SPD level is unchanged (Figure 8.22).

The SPD normalized value is

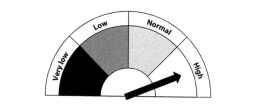

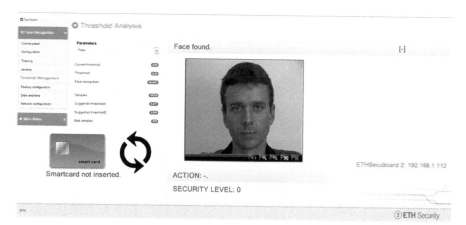

**FIGURE 8.22**
A face is identified in the video stream.

T6  The ICAO module calculates the ICAO scores for the images of the face found in the video stream. It identifies a set of images that satisfy the required thresholds and passes the control to the face recognition module. The SPD level is unchanged (Figure 8.23).

The ICAO score for the face is is shown in Figure 8.24

The SPD normalized value is

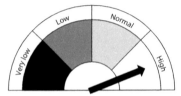

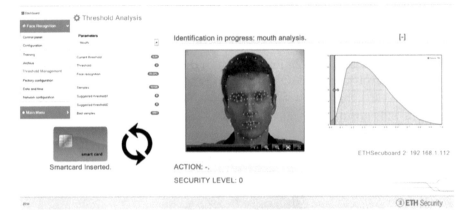

**FIGURE 8.23**
ICAO module calculates the ICAO scores for the identified face.

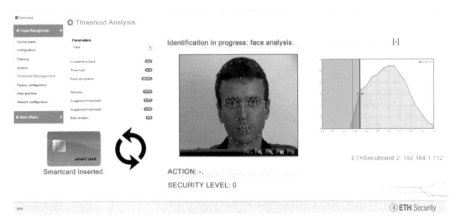

**FIGURE 8.24**
ICAO score for the identified face

![Threshold Analysis interface showing biometric profile extraction with a face being analyzed, a smartcard inserted, and a histogram graph. Text reads "Identification in progress: biometric profile extraction." "Smartcard Inserted." "ACTION: -." "SECURITY LEVEL: 0"]

**FIGURE 8.25**
Biometric profile exctraction.

T7  The face recognition module extracts the biometric profile from the selected images containing the face and passes the biometric profile to the person identification module. The SPD level is unchanged (Figure 8.25).

The SPD normalized value is

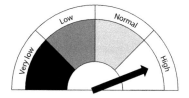

T8  The person identification module loads the biometric profile from the smart card and verifies that it matches the biometric profile extracted from the images. The module queries the biometric database and finds a matching profile. The person is identified and the control passes to the turnstile access module. The SPD level is unchanged (Figure 8.26).

The SPD normalized value is

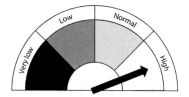

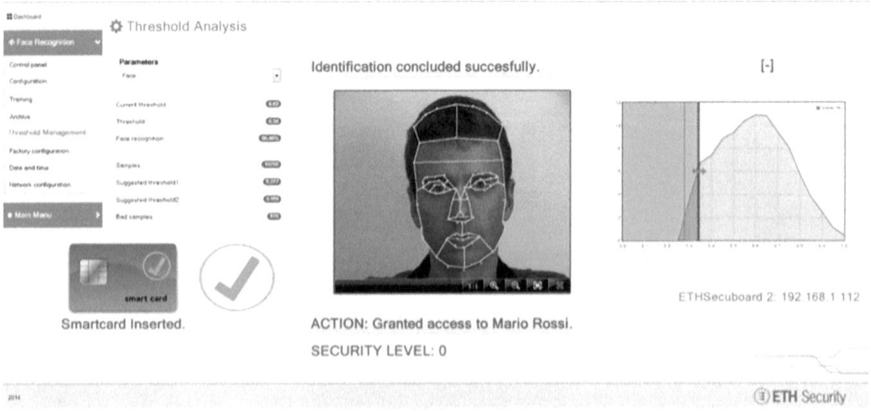

**FIGURE 8.26**
The person has been succesfully identified.

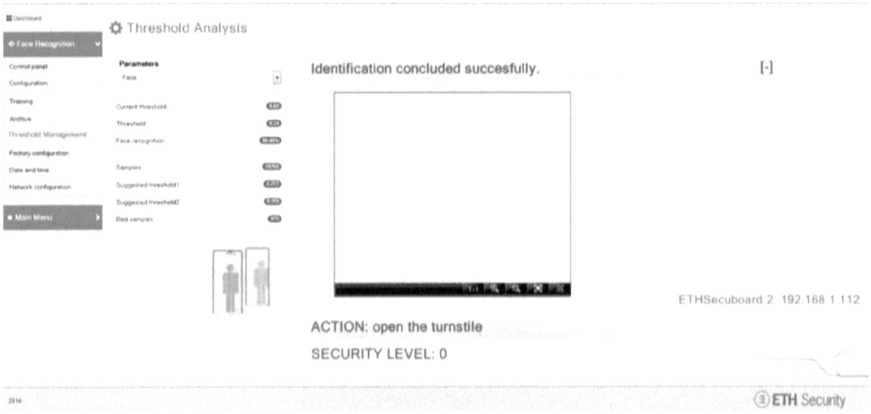

**FIGURE 8.27**
The turnstile opens and the person can access the building.

T9  The access module opens the turnstile. The SPD level is unchanged (Figure 8.27).

The SPD normalized value is

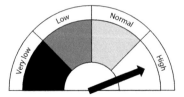

T10 The identification system goes into a waiting state. The SPD level is unchanged.

T11 The face finder module finds a face in the video stream and passes the control to the ICAO module. A smart card is inserted in the smart card reader. The SPD level is unchanged.

T12 The ICAO module calculates the ICAO scores, but it is not able to identify in the video stream a suitable image that respects the ICAO standard. The module detects that the person is wearing black glasses and advises the person to take them off. The person does not follow the indications and an alarm of noncollaborative behavior is generated. The SPD level decreases; the SHIELD middleware calculates the new level and suggests a countermeasure (Figure 8.28).

The SPD normalized value is

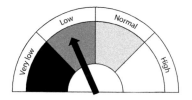

T13 After three warnings, the security personnel are called. The ICAO module passes the control to the face finder module. The middleware calculates the new SPD level that returns to the maximum value (Figure 8.29).

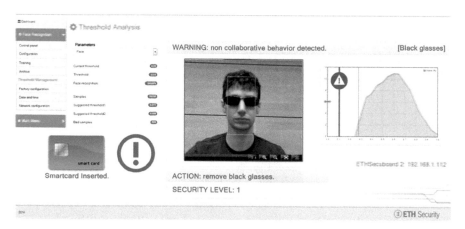

**FIGURE 8.28**
A person with noncollaborative behavior is detected.

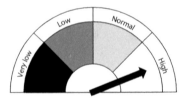

**FIGURE 8.29**
People identification SPD level: High.

The SPD normalized value is

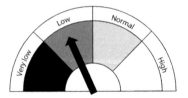

T14  The identification system goes into a waiting state. The SPD level is unchanged.

T15  The face finder module detects that the intelligent camera has been obfuscated. The corresponding alarm is generated. The SPD level decreases; the SHIELD middleware calculates the new level and suggests a countermeasure (Figure 8.30).

The SPD normalized value is

T16  The security and maintenance personnel are called, and the obfuscation is removed. The middleware calculates the new SPD level, which returns to the maximum value.

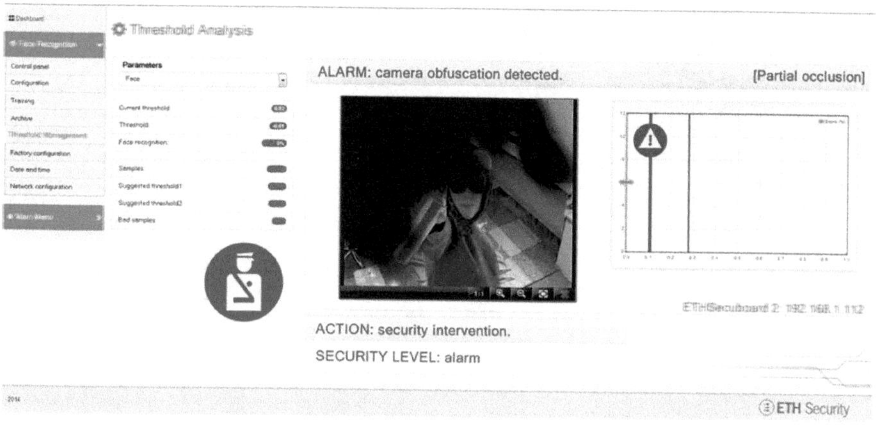

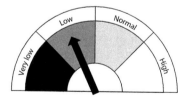

**FIGURE 8.30**
The intelligent camera has been obfuscated.

T17 The identification system goes into a waiting state. The SPD level is unchanged.

T18 The Ethernet connection is compromised due to a fault in the Ethernet switch of the local area network. All the components of the system are not receiving the keep-alive messages. The DDCF and the security control server have lost the connection to the intelligent camera. The security and maintenance personnel intervene. The SPD level decreases; the SHIELD middleware calculates the new level and suggests a countermeasure (Figure 8.31).

The SPD normalized value is

T19 The maintenance personnel substitute the switch and the local area network is reactivated. The face recognition software is closed; the

```
[2014-12-21 12:36:15:926] eth1 has left the network
[2014-12-21 12:36:16:12] eth2 has left the network
[2014-12-21 12:36:16:41] eth4 has left the network
[2014-12-21 12:36:16:134] eth3 has left the network
[2014-12-21 12:36:16:193] Instance #1 cannot progress: suspended
[2014-12-21 12:36:16:208] No wnode is available
```

**FIGURE 8.31**
DDCF log of an Ethernet connection failure event.

```
[2014-12-21 12:37:14:354] eth4 has re-joined the network
[2014-12-21 12:37:14:372] Instance #1 was suspended: cleaning eth4 of dangling tasks
[2014-12-21 12:37:14:432] eth2 has re-joined the network
[2014-12-21 12:37:14:444] Instance #1 was suspended: cleaning eth2 of dangling tasks
[2014-12-21 12:37:14:520] eth1 has re-joined the network
[2014-12-21 12:37:14:538] Instance #1 was suspended: cleaning eth1 of dangling tasks
[2014-12-21 12:37:14:621] Cleaned eth4 of dangling tasks
[2014-12-21 12:37:14:628] eth3 has re-joined the network
[2014-12-21 12:37:14:642] Instance #1 was suspended: cleaning eth3 of dangling tasks
[2014-12-21 12:37:14:695] Cleaned eth1 of dangling tasks
[2014-12-21 12:37:14:704] Cleaned eth2 of dangling tasks
[2014-12-21 12:37:14:731] Cleaned eth3 of dangling tasks
```

**FIGURE 8.32**
DDCF log of re-joining nodes.

intelligent camera registers to the DDCF. The SHIELD middleware sets the new SPD level at the basic level (Figure 8.32).

The SPD normalized value is

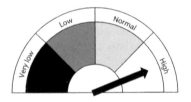

T20  The operator authenticates and logs in to the DDCF.

T21  The DDCF performs the injection of the face recognition software modules.

T22  The face recognition software is launched, and the identification system goes into the waiting state (the system is ready to identify a person and is waiting to find a face in the video stream). Every component of the system is executing its functionalities correctly, and the SPD value reaches its maximum level.

The execution steps T0–T22 can be mapped on the functional states of the single subsystems that compose the people identification system. The sequence diagrams of the subsystems have been introduced in the "SPD Metrics Calculation" section, and the mapping between execution steps and functional states is shown in Table 8.8.

**TABLE 8.8**

Scenario Execution Steps and Subsystems States

| Step | Face Recognition | DDCF | Smart Card Reader |
|------|------------------|------|-------------------|
| [T0] | ST | W | ST |
| [T1] | ST | W | ST |
| [T2] | ST | I | ST |
| [T3] | ST | L | ST |
| [T4] | W | E | W |
| [T5] | W | E | W |
| [T6] | I | E | S |
| [T7] | B | E | S |
| [T8] | P | E | P, V |
| [T9] | A | E | S |
| [T10] | W | E | W |
| [T11] | W | E | W |
| [T12] | I | E | S |
| [T13] | W | E | W |
| [T14] | W | E | W |
| [T15] | W | E | W |
| [T16] | W | E | W |
| [T17] | W | E | W |
| [T18] | W | W | W |
| [T19] | ST | W | ST |
| [T20] | ST | W | ST |
| [T21] | ST | I | ST |
| [T22] | W | E | W |

# References

1. ETSI TS 33.401 V12.10.0 (2013–2012): 3rd Generation Partnership Project; technical specification group services and system aspects; 3GPP system architecture evolution (SAE); security architecture (Release 12).
2. 3GPP TS 33.203 V12.4.0 (2013–2012): 3rd Generation Partnership Project; technical specification group services and system aspects; 3G security; access security for IP-based services (Release 12).
3. 3GPP TS 33.210 V12.2.0 (2012–2012): 3rd Generation Partnership Project; technical specification group services and system aspects; 3G security; network domain security (NDS); IP network layer security (Release 12).
4. Kyong, H., I. Chang, K. W. Bowyer, and P. J. Flynn, An evaluation of multimodal 2d+3d face biometrics, *IEEE Transactions on Pattern Analysis and Machine Intelligence*, 27 (4), pp. 619–624, 2005.
5. Moon, H. and P. J. Phillips, Computational and performance aspects of PCA-based face-recognition algorithms, *Perception*, 30(3), pp. 303–321, 2001.
6. RFC-Base, Service Location Protocol, Version 2, http://www.rfc-base.org/rfc-2608.html.

# 9

## Perspectives in Secure SMART Environments

Josef Noll, Iñaki Garitano, Christian Johansen,
Javier Del Ser, and Ignacio Arenaza-Nuño

**CONTENTS**

## Introduction

Our society is driven by digital services, empowered through embedded systems (ESs) and the Internet of Things (IoT). Chapter 2 provided several aspects of digitization, and reported from a recent McKinsey study analyzing the technology potential of automation (Chui et al., 2016). The main conclusion of the study was that more than half of the working time of employees in the United States has high potential for automation, between 64% for data collection and 78% for predicable physical work. Thus, the question is not if, but when, automation will enter the various domains.

Currently automated systems are often monitored from a control room or a control application, with a wide spread of security measures. Operation centers for the mobile networks, or control centers for payment transactions are highly secured, in terms of both physical security and information technology (IT) security.

However, analyzing the trends of smart environments as smart homes and smart cities, the focus is on interoperability and novel services, rather than on security. Most smart homes are controlled through mobile apps, with a minimum of physical security in mind. In addition, devices for the smart home are highly competitive; thus, the focus is on functionality rather than security. Devices in the smart home should perform a certain task, for example, video streaming, and with price-sensitive consumers, the "invisible" security often lags behind. The best known example is the Mirai attack, which first occurred in October 2016, exploiting mainly unchanged passwords in IoT devices (IoTSec, 2016). Devices like video recorders and closed-circuit television (CCTV) cameras were captured through the Mirai bot net and used for a distributed denial of service (DDoS) attack. The attack of some 90,000 devices reached a network load of 1.2 Tbit/s, and thus over-loaded the web services from Dyn, thereby effectively taking down IT companies like Amazon, LinkedIn, and PayPal. The tool kit was rather simple, trying a small set of typical username–password combinations, such as root–root, admin–admin, and root–1234, for gaining access to home devices. As most devices in the home segment come with rather simple default user-name–password combinations, the attack was easy and successful.

Smart homes will provide a variety of services, including temperature regulations, access and alarm functions, health services, and software updates for smart vehicles. Given the weak security of smart home devices, exploited devices might become the entry point for attacking other parts of the home infrastructure, for example, the router or the smart grid gateway.

The average user in the smart home will only look at functionality, and will not be able to recognize an infected device or network. Given that the smart home will run in an automated way, control of security requirements, as well as monitoring of attacks of the home, will become a major task of Internet service providers (ISPs).

This chapter focuses on the security requirements, and demonstrates the applicability of the measurable security, privacy, and dependability (SPD) for two smart environments: (1) the vehicle and (2) the smart grid. We focus our analysis on the application goals for services, for example, which level of security is required for monitoring a home versus the level of security required for controlling the home.

## Privacy in Smart Vehicles: A Case Study

This chapter focuses on the multimetrics (MM) approach described in Chapter 5, and applies it to a SMART mobility use case. This use case, previously described by Garitano et al. (2015), presents three scenarios of a smart vehicle where the privacy level for the actions of the vehicle is calculated. In each scenario, the privacy level is measured in terms of user location privacy.

However, instead of analyzing just the privacy on its own, the result-ing level is associated with security and dependability levels. The result of combining privacy, security, and dependability will show the necessary commitment to reach an overall SPD level for the system. In addition, we discuss how to evaluate the SPD level of single components to finally join all of them and obtain an overall SPD level for the system (in the form of an SPD dashboard).

A graphical description of the scenario can be seen in Figure 9.1.

The smart vehicle consists of three different subsystems in order to fulfill the requirements from the use case:

- A backend system (BE), building the interface from the motorbike to the end user
- An ES, mounted on the vehicle, to monitor the conditions of the motorcycle (MC)
- The ES-BE communication, a mobile link between the ES and BE

This use case envisages the privacy aspects of motorbike riding (MC). A young driver is allowed to use the MC, given that he is not speeding ($v \leq 80$ km/h). His parents ensure that he has full privacy while keeping within the agreed speed limit (Scenario 1), but they will be informed if he exceeds the

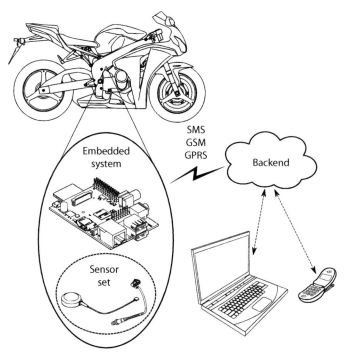

**FIGURE 9.1**
Smart vehicle topology, consisting of the ES with sensors, communication, and the BE.

speed limit (Scenario 2). In Scenario 2, the parents will receive an SMS (Short Message Service) about the speed and location of the vehicle. We also add an emergency scenario in the case of accidents (Scenario 3), where an SMS alert is sent to both parents and emergency services, and where relevant information from the MC is provided to a BE.

Communication is performed through either General Packet Radio Service (GPRS) or SMS in the case of notices to parents or emergency units. The analysis performed focuses on just the communication subsystem (to ease understanding of the overall methodology).

The overall system may operate in nine configurations, addressing decreasing grades of privacy protection:

- *Configuration A*: The ES does not send any SMSs; GPRS data are encrypted with a 128-bit key. The ES accepts remote configuration from the BE.
- *Configuration B*: Same as above, except the ES sends a keep-alive message to the BE every 120 s.
- *Configuration C*: Same as above, except the BE sends messages to the ES and the last one replies every 60 s.
- *Configuration D*: The ES sends an SMS to parents; GPRS data to the BE are encrypted with a 64-bit key. The ES accepts remote configuration from the BE.
- *Configuration E*: Same as above, except the ES sends location and speed information to the BE every 9 s.
- *Configuration F*: Same as above, except the BE sends messages to the ES and the last one replies with location and speed information every 5 s.
- *Configuration G*: The ES sends one SMS to parents and another to emergency services. Unencrypted data about the status of the MC are sent from the ES to the BE. The ES accepts remote configuration from the BE.
- *Configuration H*: Same as above, except the ES sends location and speed information to the BE every 2 s.
- *Configuration I*: Same as above, except the BE sends messages to the ES and the last one replies with location and speed information every 0.5 s.

## Parameter Identification and Weighting

The MM approach has been applied to the ES and BE communication subsystem. The following five metrics have been defined to measure the criticality level of the four components that make up the evaluated subsystem:

1. Port metric
2. Communication channel metric
3. GPRS message rate metric
4. SMS message rate metric
5. Encryption metric

The port metric is described in detail to explain the four metric definition phases. For the other four metrics, we just present the metric and its respective weight.

The port metric evaluates the impact of the port component within the ES-BE communication subsystem. Notice that all presented values refer only to *privacy evaluation*.

- *Parameter identification*: The analysis of the port component and the ES-BE communication subsystem ends up with the identification of four protocols or port classes: SSH, Simple Network Management Protocol (SNMP), SNMP traps, and SMS.

- *Parameter weighting*: The formulation of the weight, by an expert in the field, provides the values shown in Table 9.1.

- *Component–metric integration*: The identification of all possible port values for a given system configuration provides a considerable amount of options.

- *Metric running*: Table 9.2 shows the values obtained from the evaluation of configurations A, B, and C within Scenario 1.

**TABLE 9.1**

Multimetrics Ports Metrics Parameter Weighting

| Parameter | Criticality Level of Privacy | Characteristics |
|---|---|---|
| SNMP (UDP) 161 in the ES | 40 | Usage: BE checks if ES is alive<br>Handled data: IP addresses, ports |
| SNMP trap (UDP) 162 in the BE | 60 | Usage: ES sends keep-alive message to BE<br>Handled data: IP addresses, ports<br>Usage: ES sends observation data to BE<br>Handled data: Speed, location<br>Usage: ES sends alert data in accident case to BE<br>Handled data: Alert, location |
| SSH (TCP) 23 in the ES | 30 | Usage: BE sends configuration to ES<br>Handled data: ES configuration |
| SMS | 80 | Usage: ES sends vehicle owner observation data<br>Handled data: Location and speed of the vehicle<br>Usage: ES sends vehicle owner alert in accident case<br>Handled data: Alert, location, sender phone number, receiver phone number |

**TABLE 9.2**

Multimetrics Ports' Metrics Integration and Running

| Configuration | Metric Parameters | Criticality Level of Privacy |
|---|---|---|
| **Configuration A** | SSH | 30 |
| **Configuration B** | SSH + SNMP trap | 61 |
| **Configuration C** | SSH + SNMP | 41 |

The communication channel metric measures the SPD level offered by the two possible channels used by the ES-BE communication subsystem: GPRS data transmission and SMS message. Table 9.3 shows the parameters of the communication channel metric and their weights.

The GPRS message rate metric evaluates the period in which data are transmitted over the GPRS channel. This metric considers the amount of information that could be jeopardized in case the communication channel is compromised. Table 9.4 shows the parameters of the GPRS message rate metric and their weights.

The SMS message rate metric evaluates the number of transmitted messages. This metric considers the amount of information that could be jeopardized in case the communication channel is compromised. Table 9.5 shows the parameters of the SMS message rate metric and their weights.

**TABLE 9.3**

Multimetrics Communication Channel Metric

| Parameter | GPRS with GEA/3 | SMS over GSM with A5/1 |
|---|---|---|
| **Criticality level of privacy** | 20 | 40 |

**TABLE 9.4**

Multimetrics GPRS Message Rate Metric

| Parameter | 0.5 | 1 | 2 | 5 | 9 | 20 | 60 | 120 | ∞ |
|---|---|---|---|---|---|---|---|---|---|
| **Criticality level of privacy** | 80 | 60 | 45 | 30 | 20 | 15 | 9 | 5 | 0 |

**TABLE 9.5**

Multimetrics SMS Message Rate Metric

| Parameter | 2 Messages | 1 Messages | 0 Messages |
|---|---|---|---|
| **Criticality level of privacy** | 9 | 5 | 0 |

**TABLE 9.6**

Multimetrics Encryption Metric

| Parameter | No Encryption | 64-Bit Key | 128-Bit Key | Not Applicable |
|---|---|---|---|---|
| Criticality level of privacy | 88 | 9 | 5 | 0 |

Finally, the encryption metric measures the needed strength to decrypt data, provided by the encryption component. It focuses on the key length used to encrypt the transmitted data. Table 9.6 shows the parameters of the encryption metric and their weights.

## Metric Integration

Every scenario in which the system runs has some SPD requirements, called $SPD_{Goal}$. Thus, the system has as many $SPD_{Goal}$ values as scenarios. In addition, for each scenario, the system can be configured differently, which results in a distinct SPD level for each configuration in every scenario. Therefore, the introduced methodology evaluates all configurations of every scenario, and it ends up comparing and selecting the configuration that best satisfies the established $SPD_{Goal}$.

The methodology is composed using the evaluation of components, subsystems, and the system. As shown in Figure 9.2, the methodology starts by evaluating each component, then the combination of components that make up a subsystem, and finally, the set of subsystems that forms the system.

Each component, each subsystem, and the SPD level of the system are computed for all possible configurations supporting a given scenario. The SPD of the system is evaluated individually. However, whenever the obtained SPD

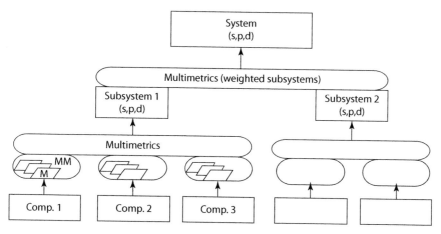

**FIGURE 9.2**
Component-level MM.

level has to be compared with the scenario SPD$_{Goal}$, all SPD elements are considered, and a visual representation is used for SPD$_{System}$ compliance. The main purpose of the visual SPD-level representation consists of simplifying the comparison between the SPD level of each configuration and the SPD$_{Goal}$ for a given scenario.

As stated above, three possible scenarios have been defined. Each scenario has a predefined SPD$_{Goal}$, and the evaluation of each configuration for each scenario results in a specific SPD level. Table 9.7 shows an example of the SPD levels for configurations A, B, and C for Scenario 1 (only the privacy levels are shown, to simplify the example).

In order to simplify the selection of the most suitable configuration, every element of the SPD level is substituted by a green, yellow, or red circle. The color is selected according to the numeric difference between the evaluated SPD level and SPD$_{Goal}$, following the next criteria:

- |SPD$_{Goal}$: SPD level| ≤9, green ●
- |SPD$_{Goal}$: SPD level| ≤9 and ≥20, yellow ○
- |SPD$_{Goal}$: SPD level| ≥20, red ●

MM is a simple process that evaluates the repercussions of each metric, component, or subsystem, based on its importance within the system. Its main advantage resides in its simplicity, which is based on the combination of two parameters: the criticality level for a given system configuration and its importance.

The criticality level varies between a metric, component, or subsystem, depending on the level in which MM is applied. Furthermore, the usage of the same operator along the whole SPD$_{System}$ evaluation simplifies the methodology by making it more understandable for people who are not experts in the field. The output of MM is a single number that shows the criticality level of the components and subsystems, and is easily translated into an SPD level.

The importance or significance of a specific metric, component, or subsystem, within the system SPD-level evaluation, is given by its weight. The weight is a value within a range of 0–90; the higher the number, the larger is the significance of the evaluated element. The definition of the weight of an

**TABLE 9.7**

Multimetrics SPD-Level Composition

| | | SPD$_{Goal}$ | | SPD Level |
|---|---|---|---|---|
| Scenario 1 | Configuration A | (s,80,d) | (s,90,d) | (s, ○ ,d) |
| | Configuration B | (s,80,d) | (s,80,d) | (s,● ,d) |
| | Configuration C | (s,80,d) | (s,80,d) | (s,● ,d) |

element depends on the role or function it performs, and is established by the system designer or an expert in the field.

The MM approach is based on two parameters: the actual criticality $x_i$ and the weight $w_i$. The criticality C is accomplished by the root mean square weighted data (RMSWD) formula:

$$C = \sqrt{\sum_i \left( \frac{x_i^2 W_i}{\sum_i^n W_i} \right)} \qquad (9.1)$$

There are three possible criticality-level outcomes: (1) component criticality, after evaluating the suitable metrics; (2) subsystem criticality, from the evaluation of components; and (3) system criticality, after performing the MM operation on subsystems. The actual criticality $x_i$ is the result of (1) the metric for a component evaluation; (2) the component evaluation, obtained by a previous RMSWD, for a subsystem evaluation; or (3) the subsystem evaluation, obtained by a previous RMSWD, for a system evaluation. All these values are for a given configuration in a specific scenario.

We apply the approach to the evaluation of the ES-BE communication subsystem, using privacy evaluation as an example. The ES and BE communication subsystem is composed of four components, listed below together with their weight ($w$) within the subsystem.

1. Port (C1), $w = 40$
2. Channel (C2), $w = 20$
3. Data transmitter (C3), $w = 35$
4. Encryption (C4), $w = 60$

The SPD$_{Goal}$ for privacy SPD$_P$ is presented in Table 9.8 and refers solely to the privacy goal of the given scenario. Furthermore, as the evaluation of the subsystem is just considering the privacy $p$, the security $s$ and dependability $d$ values are not shown.

The results obtained from the evaluation of the privacy for the four components by the five metric values and the subsystem are shown in the same table. Components C1, C2, and C4 each refer to just one metric, while component C3 has been evaluated by two metrics, M3 and M4. Since the ES-BE communication subsystem is composed of four components, the evaluation of its criticality includes all of them.

As the last step, the privacy SPD level, SPD$_P$, of the subsystem has been calculated, and in order to simplify the most suitable configuration selection, a color has been assigned based on the difference between SPD$_P$ and the subsystem goal SPD$_{Goal}$. The table shows that configurations A and C satisfy the requirements of Scenario 1, and that configurations D–F satisfy the requirements of Scenario 2.

**TABLE 9.8**

Multimetrics Composition Evaluation

| | Criticality | | | | | | SPD$_P$ | | |
|---|---|---|---|---|---|---|---|---|---|
| | C1 | C2 | C3 | C4 | Subsystem | | Scenario 1 | Scenario 2 | Scenario 3 |
| | | | | | | | (s,80,d) | (s,50,d) | (s,5,d) |
| SPD$_{Goal}$ | | | | | | | | | |
| Multimetrics elements | M1 | M2 | M3 M4 | M5 | C1 ... ... C4 | | | | |
| Configuration A | 30 | 20 | 0 | 5 | 17 | 83 | ◐ | ● | ● |
| Configuration B | 61 | 20 | 4 | 5 | 32 | 68 | | | ● |
| Configuration C | 41 | 20 | 9 | 5 | 23 | 77 | ◐ | | ● |
| Configuration D | 82 | 41 | 2 | 9 | 45 | 55 | | ◐ | ● |
| Configuration E | 52 | 41 | 18 | 9 | 45 | 55 | | ◐ | ● |
| Configuration F | 83 | 41 | 27 | 9 | 47 | 53 | | ◐ | ● |
| Configuration G | 82 | 42 | 4 | 88 | 70 | 30 | ● | | ● |
| Configuration H | 82 | 42 | 40 | 88 | 73 | 27 | ● | | ● |
| Configuration I | 83 | 42 | 72 | 88 | 79 | 21 | ● | | |

After examining the values, and before the most suitable configuration for each scenario can be selected, the security *s* and dependability *d* values should be evaluated using the same methodology.

The presented methodology considers all SPD aspects during the analysis of the most suitable configuration for each scenario. The picture obtained in the end shows under which SPD conditions the system will run for a given scenario and configuration. During the design phase of the ES, the possible scenarios and configurations can be foreseen and the same analysis can be performed, providing SPD by design. The result clarifies whether it is necessary to modify the design of some system aspects, to satisfy the established goals.

Furthermore, for an existing system, the same analysis provides a clear picture of the SPD$_{System}$ level in operation. This analysis identifies which configuration options or system parts are not behaving as expected, thus increasing the whole system's risk. The early correction of misbehaving configuration options could prevent further consequences.

The above methodology can be automated further by leveraging the self-learning capabilities of fuzzy systems, which, in addition, are able to accommodate the subjectivity and uncertainty of quantified SPD metrics in certain security scenarios. In particular, when devices and protocols participating in the scenario under analysis are characterized by high levels of heterogeneity and diversity, it becomes crucial to provide the overall security system with the functionalities required to define SPD directives in a fuzzy manner, along with optimization tools capable of optimizing the action rules established over such directives. In this hypothesized case, the overall system must exploit these new capabilities to blend together the experience and criterion of the security expert and his or her past supervision of the results derived from actions taken after the assessment of the security incidence, to ultimately enforce optimal rules according to a metric of performance, resilience, or compliance with imposed requirements (e.g., latency).

Elements from computational intelligence, soft computing, and machine learning are expected to play a crucial role in MM security assessment by their efficiency to cope with data-based learning setups characterized by high levels of uncertainty. Of particular interest is the hybridization of fuzzy inference and logic, pattern mining models, and heuristic optimization wrappers in order to integrate in a single automated MM approach all the processing and learning capabilities referenced and argued above: (1) to generalize, based on a historical record, the SPD values corresponding to monitored inputs of the system at hand; (2) to make compatible and match heterogeneous SPD metrics, possibly under subjectivity in their quantification; and (3) to optimize automated decision rule sets based on the past experience and assessment of the effectiveness of enforced actions by security experts. In all, the whole spectrum of computational intelligence tools will lay at the core of future-generation MM security systems, as the ever-growing complexity featured by heterogeneous systems of systems demands accordingly higher levels of automation, learning, and intelligence for their management and operation.

## Measurable Security in Smart Grids

This section provides an application of the MM approach (see Chapter 5) to a smart grid use case. Compared with the smart vehicle use case in the previous section, the focus here is on the impact of weighing on the SPD assessment of the system.

The use case was previously elaborated in Noll et al. (2015), focusing on the introduction of measurable security in smart grids. We have taken the smart grid infrastructure in Norway as an example, following the current rollout that all households in Norway will have a smart meter by January 1, 2019 (NVE, 2017). Although the current implementation is limited to hourly readings, each smart meter has a HAN port for communicating with the home infrastructure. This HAN port is expected to be available in late 2018 (IoTSec, 2017). Thus, the use case addressed here extends the smart meter reading with smart home functionality, such as alarms and home control.

The range of smart grid services includes

- Billing functionality, providing the consumption every hour (current Norwegian regulation); higher readings, such as 1/15 min; or extended information, such as max usage

- Remote home control for, for example, the heating system or a washing machine
- Intrusion detection and monitoring, for example, hacking attempts to the home
- Fault tolerance and failure recovery, providing a quick recovery from a failure
- Grid monitoring to achieve a grid stability of at least 99.96%
- Grid failure, for example, voltage peaks and grounding errors
- Alarm functionality, for example, fire, water leakage, and burglary

Figure 9.3 shows the high-level overview over the communication between the home appliances, the smart meter, and the smart meter infrastructure. Communication to the smart home can be achieved through either the smart grid or the Internet infrastructure.

**Use Case Goals**

In order to demonstrate the MM approach for the smart grid, the system is simplified. Given the overview in Figure 9.3, the smart grid system is composed of (1) smart readers with the concentrator infrastructure, (2) cloud services, and (3) remote access for monitoring and control. With respect to services, we focus on (1) billing, (2) fire alarm, and (3) home control.

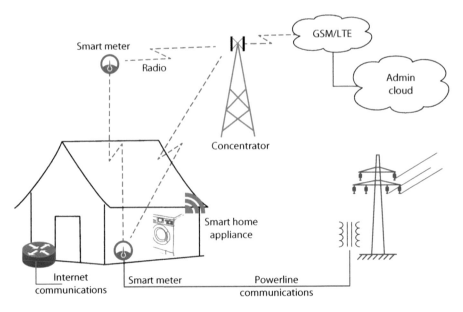

**FIGURE 9.3**
Smart grid high-level overview, indicating signaling via smart meters or power line.

In case of the *billing use case*, the meter reader provides hourly meter values and real-time energy-related alarms (sags, swells, and power faults). This information is first sent to the concentrator and kept there until it is sent to the control center. The concentrator provides the control center with the meter values, typically every six hours, while alarms are sent in real time. All communication from the concentrator to the control center is sent over the mobile network. The data are stored in a cloud infrastructure, where they are validated; missing values are estimated; and billing is prepared.

*Alarm services* are foreseen, including, for example, a fire alarm, which is communicated directly to the fire brigade and to the control center. In addition, the homeowner might be informed by an SMS depending on his or her profile settings. The current infrastructure first collects the alarm in the concentrator, and then directly forwards it to the control center. One part of the MM analysis will address the challenges of having just one communication channel.

The *home control use case* uses a bidirectional communication with both the control center and the cloud. In the case of the meter reading, it checks the status of different home devices and sends this information to the cloud. In the current centralized version, the cloud runs load control algorithms, and might send control signals to the home.

Applying the MM approach, the $SPD_{System}$ level is represented by a triplet representing SPD levels $(s, p, d)$. Depending on the use case, the system can be configured in multiple ways. Thus, for a given use case and system, there will be different configuration options. Besides multiple configurations, each system use case has a required SPD level, $SPD_{Goal}$. Each system or component configuration offers a different SPD level; hence, the proposed methodology evaluates all possible configurations, looking for the most suitable one.

Table 9.9 presents the goals for each of the use cases with respect to $(s, p, d)$. These goals, being subjective, have been established by Noll et al. (2015) through discussions with experts in the field. *Billing* and *home control* are focused first on security, leaving dependability as less important. This is mainly due to the fact that in both cases, it is necessary to avoid any kind of man-in-the-middle attack. Furthermore, the response time of the system does not have to be immediate, thus decreasing the dependability level. However, in the case of *alarm*, dependability is the principal target, then security, and finally privacy. A fire alarm needs fast delivery of the alarm

**TABLE 9.9**

$SPD_{Goal}$ for Each Specific Application

| Use Case | Security | Privacy | Dependability | $SPD_{Goal}$ |
|---|---|---|---|---|
| Billing | 90 | 80 | 40 | (90,80,40) |
| Home control | 90 | 80 | 60 | (90,80,60) |
| Alarm | 60 | 40 | 80 | (60,40,80) |

with high reliability, while protecting the privacy of the communication is of less importance.

## Multimetrics Approach for Smart Grids

Having defined the goals for the use cases, three more steps must be performed prior to performing the MM approach:

1. Identification of security functionality and metrics selection
2. Assignment of min and max values for the operation status within the metrics (see Figure 5.23 in Chapter 5)
3. Weighting factors for the importance of the component

### SPD Functionalities, Metrics Selection, and Operational Status

The automatic meter reader (AMR) uses the mesh radio link to communicate with the signal concentrator. The concentrator uses the mobile link to send and receive data from the control center. Thus, all three subsystems communicate with each other. Given the fact that different subsystems are interrelated, the whole system could be evaluated together. However, the analysis of the overall $SPD_{System}$ would became complex; thus, each subsystem is evaluated individually. The division of the system into subsystems, and each subsystem into several components, allows the easy identification and evaluation of the metrics.

The AMR is an ES installed in every house, and is tailored to measure, sense, and in the future, control the power consumption, fire sensors, and some other home parameters. While current AMRs are monitoring, they will be extended, allowing end users to control their home through the operator's infrastructure.

The mesh radio link is the communication channel used by the AMR and the concentrator to communicate with each other. Since the installation foresees one concentrator per group of houses, the communication between each AMR and the concentrator can be done directly or by multiple hops. In the case of direct communication, the transmit power for direct communication from each house to the concentrator will be higher than that for a mesh setup, increasing the chance of interference. In a nonsynchronized wireless network, communications from each AMR to the concentrator may also increase the probability of signal collisions. In a mesh configuration, data can be transmitted by using multiple hops. In this case, the probability of reaching the concentrator is bigger, since data can follow multiple routes.

The communication between the concentrator and the control center is performed by the mobile link subsystem. Being controlled by the service provider, the mobile communication system can choose between sending the data over SMS or GPRS.

The SPD level of the components that make up a subsystem can be measured by multiple metrics. The definition and selection of the necessary metrics require expertise in the field, and should be performed by a system engineer. One of the ideas behind this work resides in the creation and maintenance of a common metric database. The main benefit consists of reusing the metrics adopted to measure the same or equivalent component SPD level, or even use existing SPD values for subsystems or components with a given configuration. To the best of our knowledge, there is no metric database available, giving us the task of defining relevant metrics.

The definition of a metric starts by analyzing every component and identifying all parameters that could be used for its characterization. Those parameters must be evaluated from all SPD perspectives in order to end up with a criticality level for each of them. The next step is to evaluate the repercussions of the possible values of each parameter on the component SPD level. The output will be the weight, which could vary from one system to another, and thus needs to be defined or at least evaluated for every new system evaluation.

In case of the presented three subsystems, each of them is evaluated by two or three metrics, for a total of six metrics. The evaluation of the AMR subsystem is performed by remote access (Table 9.10), authentication (Table 9.11), and encryption (Table 9.12) metrics.

**TABLE 9.10**

Remote Access Metric

| Configuration | $C_s$ | $C_p$ | $C_d$ |
|---|---|---|---|
| Remote access ON | 60 | 60 | 40 |
| Remote access OFF | 9 | 20 | 50 |

**TABLE 9.11**

Authentication Metric

| Configuration | $C_s$ | $C_p$ | $C_d$ |
|---|---|---|---|
| Authentication ON | 9 | 30 | 60 |
| Authentication OFF | 80 | 70 | 40 |

**TABLE 9.12**

Encryption Metric

| Configuration | $C_s$ | $C_p$ | $C_d$ |
|---|---|---|---|
| Encryption ON | 9 | 9 | 60 |
| Encryption OFF | 80 | 80 | 40 |

The remote access metric evaluates the SPD level of the remote connectivity functionality of the system. As shown in Table 9.10, this metric establishes different criticality values for whenever the functionality is activated or not.

The authentication metric (Table 9.11) establishes the criticality level of having authentication activated in order to access the AMR. It considers both remote and local access to the AMR.

The third metric used to measure the SPD level of the AMR is encryption. This metric is used in all three subsystems to evaluate whether the transmitted data are encrypted. As shown in Table 9.12, it considers two different statuses, data encryption activated or not.

The evaluation of the mesh radio link subsystem is performed by mesh (Table 9.13), message rate (Table 9.14), and encryption (Table 9.12) metrics.

The traffic routing in a mesh link can be performed by sending the data directly or not. In the case of direct data delivery, a single hop is used, which is more secure and privacy-aware since data are not going through others, but it requires more transmission power and the dependability is not as high. Alternatively, multiple hop traffic routing is used whenever transmission time is not as urgent and there is a need to avoid collisions. Furthermore, the necessary transmission power is lower than that for a single hop and is more dependable since multiple paths can be used to deliver data.

The message rate metric measures the criticality level according to the frequency with which the messages are sent (Table 9.14). In this way, more messages per unit of time increases the security and privacy criticality and reduces the dependability criticality.

The encryption metric of Table 9.12 is used in the mesh radio link subsystem to evaluate whether the transmitted data are encrypted.

**TABLE 9.13**

Mesh Routing Metric

| Configuration | $Cs$ | $Cp$ | $Cd$ |
| --- | --- | --- | --- |
| Multipath routing | 60 | 60 | 30 |
| Single-path routing | 30 | 30 | 50 |

**TABLE 9.14**

Message Rate Metric

| Configuration | $Cs$ | $Cp$ | $Cd$ |
| --- | --- | --- | --- |
| 1 h | 20 | 20 | 70 |
| 20 min | 25 | 30 | 50 |
| 1 min | 40 | 50 | 30 |
| 5 s | 50 | 70 | 9 |

**TABLE 9.15**

Mobile Channel Metric

| Configuration | Cs | Cp | Cd |
|---|---|---|---|
| GPRS | 60 | 70 | 70 |
| SMS | 40 | 50 | 20 |

The evaluation of the mobile link subsystem is performed by mobile channel (Table 9.15) and encryption (Table 9.12) metrics.

Since the mobile link is under the service provider control, the system can choose which communication type it will send data over. Thus, the mobile channel metric establishes the criticality level of sending data over SMS or GPRS, as described in Table 9.15.

As previously explained, the encryption metric is also used to evaluate the criticality of the mobile link subsystem. The difference in its evaluation for each subsystem will be established by the weight.

This section introduces the three subsystems that compose the evaluated system, together with the metrics used for component criticality evaluation. Furthermore, the six metrics and their criticality values are presented. The result of evaluating the three subsystems is further explained in the following section.

### Smart Grid Weighting Factors

The MM approach is based on two parameters: the actual criticality $x_i$ and the weight $W_i$. The criticality $C$ is accomplished by the RMSWD formula shown in Equation 9.1.

There are three possible criticality-level outcomes: (1) component criticality, after evaluating the suitable metrics; (2) subsystem criticality, from the evaluation of components; or (3) system criticality, after performing the MM operation on subsystems. The actual criticality $x_i$ is the result of (1) the metric for a component evaluation; (2) the component evaluation, obtained by a previous RMSWD, for a subsystem evaluation; or (3) the subsystem evaluation, obtained by a previous RMSWD, for a system evaluation. All these values are for a given configuration in a specific use case.

The weight $w_i$ is provided by the expert in the field, and provides the significance level of each (1) metric within a component, (2) component within a subsystem, or (3) subsystem within the system evaluation. The weight value is in the range of 0–90. Thus, it follows the same approach as the criticality level, making the entire process under the same logic. However, a sensitivity analysis has shown that a linear significance level of the weight is not appropriate for ending up with representative SPD levels. Hence, the weight used in the RMSWD calculation of Equation 9.1 is $W_i$, calculated from $w_i$ through Equation 9.2 as

**TABLE 9.16**

Weights for Subsystems and Components

| Subsystem | Subsystem Weight | Component | Component Weight |
|---|---|---|---|
| AMS | 80 | Remote access | 70 |
| | | Authentication | 80 |
| | | Encryption | 80 |
| Radio link | 50 | Mesh | 60 |
| | | Message rate | 80 |
| | | Encryption | 40 |
| Mobile link | 20 | Mobile link | 70 |
| | | Encryption | 40 |

$$W_i = \left( \frac{w_i}{100} \right)^2 \tag{9.2}$$

The resulting value will be in the range of $9^{-4}$ to 1, maximizing the impact of high weight values toward the lowest ones. Table 9.16 provides the weights for subsystems and components.

The starting point for the system analysis was a set of 11 configurations (A–K) for the whole system, with, for example, configuration F having remote access, no authentication, mesh, a message rate of 1/min, SMS, and no encryption. Configuration J is similar to configuration F, but with authentication, a single link, GPRS communication, and encryption. The analysis performed by Noll et al. (2015) showed that none of the initial 11 configurations could satisfy the requirements for the billing and home control use case.

Thus, the analysis focuses on the maximum achievable security (*max*) in the system, given the security functionalities and the metrics we used to describe SPD.

As shown in Table 9.17, the *max* security configuration reveals an (s, p, d) level of the system of (84,77,42), in good correspondence with the SPD$_{Goal}$ of the billing use case, as well as with security and privacy of the home control use case. Dependability for the home control use case could not reach

**TABLE 9.17**

Use Case Goals, Configurations, Resulting SPD Level, and Goal Matching

| Use Case | SPD$_{Goal}$ | Configuration | SPD Level | SPD vs. SPD$_{Goal}$ |
|---|---|---|---|---|
| Billing | (90,80,40) | *max* | (84,77,42) | (●, ●, ●) |
| Home control | (90,80,60) | *max* | (84,77,42) | (●, ●,  ) |
| Alarm | (60,40,80) | J | (31,33,63) | (●,  ,  ) |

the desired goal. Following the same example, the results obtained for the alarm use case show dependability and privacy in yellow and security in red. However, for this use case, even if a configuration totally focused on dependability were created, the maximum value would be $D = 64$. Hence, in order to increase the dependability, the system needs to be redesigned by adding some other dependability-focused components, without decreasing the security and privacy values.

This section explained the MM approach and showed its applicability using three smart grid use cases as examples. As it has been shown that, in order to end up with a specific configuration that best satisfies the $SPD_{Goal}$ of each use case, it is necessary to set which SPD element is the major one and consider the rest complementary. The final result, $SPD_{System}$, is a triplet with the measured SPD values obtained from the application of a given system configuration.

### Evaluation and Overview

The MM approach provides a methodology for comparing SPD application goals with the achievable SPD level of the system. The smart grid system, presented in the previous sections, opens for energy management and a range of novel services. These services, as presented in Figure 9.4, will have their specific SPD requirements.

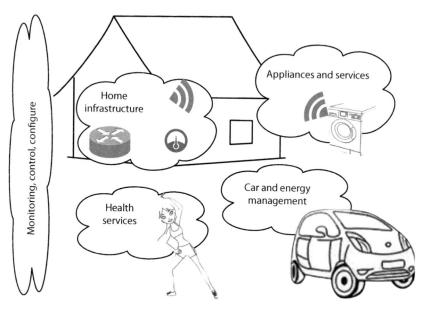

**FIGURE 9.4**
Services in the future home, requiring specific SPD measures.

As an example, the electrical car being integrated into the renewable energy concept will have a variety of services connected to the car. Security is of highest importance for a software update of the car, while privacy is of importance for knowing the charging and usage cycles of the car. Following the same example, the results obtained for the alarm use case show dependability and privacy in yellow and security in red. However, for this use case, even if a configuration totally focused on dependability were created, the maximum value would be $D = 64$. Hence, in order to increase the dependability, the system needs to be redesigned by adding some other dependability-focused components, without decreasing the security and privacy values.

Through the MM approach, we have a methodology at hand that allows a system engineer to include SPD considerations at the design phase, and check if the expected SPD functionalities can guarantee the application goals.

Given the novel applications with specific SPD requirements, the MM approach allows us to check which configurations are suitable to run the system with the expected SPD level, or which other measures are needed to improve the $SPD_{System}$ level. The results clarify if it is necessary to modify the design of some system aspects, in order to satisfy the specific application goals.

The advantage of the MM approach is that it provides the details for each component and subsystem in our smart grid example: a maximum security for the AMR subsystem of 75, a communication of 76, a mobile link of 64, and a whole system of 75. Thus, a system engineer will immediately recognize that the weakest security is in the mobile link, which, by the way, is addressed through an end-to-end encryption by AMR suppliers like Aidon and Kamstrup.

In the case of an existing system, the same analysis will provide a clear picture about the $SPD_{System}$ level in operation. This analysis will identify which configuration options or system parts are not up to expected level of SPD, and thus help to identify the critical subsystems. The early correction of misbehaving configuration options could prevent further consequences.

The applicability of the presented methodology is determined by the subjective weighting and criticality assignment. There is a need for standardization through an industrial interest board in order to establish the metrics, their criticality levels, and their weight in a system.

Our analysis shows that the MM methodology can be used to compare the SPD aspects for a given system under different configurations both during the design process or for an already existing system. The methodology adoption from the system developers side can bring several advantages, such as already evaluated metrics, components, and subsystems for different use cases. This would dramatically simplify the evaluation process and drive its adoption by the whole industry.

# References

Chui, M., J. Manyika, and M. Miremadi, Where machines could replace humans—and where they can't (yet), *McKinsey Quarterly*, July 2016, http://www.mckinsey.com/business-functions/digital-mckinsey/our-insights/where-machines-could-replace-humans-and-where-they-cant-yet (accessed April 16, 2017).

Garitano, I., S. Fayyad, and J. Noll, Multi-metrics approach for security, privacy and dependability in embedded systems, *Wireless Personal Communications*, 81, 1359–1376 (2015).

IoTSec, Smart meter connectivity, Norwegian IoT Security Project IoTSec.no, 2017, http://its-wiki.no/wiki/IoTSec: Smart_Meter_Connectivity (accessed April 14, 2017).

IoTSec, The denial of service attack from IoT devices, Norwegian IoT Security Project IoTSec.no, 2016, http://its-wiki.no/wiki/IoTSec: The_Denial_of_Service_Attack_from_IoT_devices (accessed May 5, 2017).

Noll, J., I. Garitano, S. Fayyad, E. Åsberg, and H. Abie, Measurable security, privacy and dependability in smart grids, *Journal of Cyber Security*, 3(4), 371–398(2015).

NVE (Norwegian Water Resources and Energy Directorate), Smart metering (AMS), (March 28, 2017), https://www.nve.no/energy-market-and-regulation/retail-market/smart-metering-ams/ (accessed April 14, 2014).

# 10

## SHIELD Technology Demonstrators

Marco Cesena, Carlo Regazzoni, Lucio Marcenaro, Andrea Toma,
Nawaz Tassadaq, Kresimir Dabcevic, George Hatzivasilis, Konstantinos
Fysarakis, Charalampos Manifavas, Ioannis Papaefstathiou, Paolo
Azzoni, and Kyriakos Stefanidis,

### CONTENTS

# Introduction

SHIELD technology demonstrators have been conceived to illustrate the functionalities of specific security, privacy, and dependability (SPD) technologies that, being domain independent, might be adopted in several application scenarios. The selected technologies have been introduced in Chapter 3, while this chapter provides more detailed information on their usage from a wider perspective. SHIELD technology demonstrators consist of laboratory prototypes and experiments that allow to test and evaluate all the functionalities offered by a SPD technology, rather than focusing on the specific functionality subset used in the domains presented in the previous chapters.

# SPD-Driven Smart Transmission

Smart SPD driven transmission refers to a set of services deployed at network level designed to ensure smart and secure data transmission in critical channel conditions. This technology is based on the reconfigurability properties of the software defined radio (SDR) and on learning and self-adaptive capabilities characteristic of cognitive radio (see Chapter 3). It has been conceived for SHIELD SDR-capable power nodes and has been developed in a specific SHIELD technology demonstrator called "SPD-driven smart transmission layer." The layer includes remote control of the transmission-related parameters, automatic self-reconfigurability, interference mitigation techniques, energy detection spectrum sensing, and spectrum intelligence based on the feature detection spectrum sensing.

## SPD-Driven Smart Transmission Layer Prototype

The smart transmission layer SDR/CR technology demonstrator consists of a number of secure wideband multi-role–single-channel handheld radios (SWAVE HHs), each interconnected with the OMBRA v2 multiprocessor embedded platform. A coaxial test bed architecture was assembled, allowing for implementing, developing, and testing of all relevant features and algorithms in controllable and repeatable environment.

SWAVE HH (from now on referred to as HH) is a fully operational SDR radio terminal capable of hosting a multitude of wideband and narrowband waveforms. It is equipped with all the interfaces needed to set up the prototype (10/100 Ethernet; USB 2.0; RS-485 serial, dc power interface) and provides support for both legacy and new waveform types. Two functional waveforms have been installed on the radio: SelfNET soldier broadband waveform (SBW) and VHF/UHF Line Of Sight (VULOS). Remote control of the HH is achieved via Ethernet, by means of SNMPv3 protocol. The serial connection needs to be converted from RS-485 to RS-232, and is used to transfer the spectrum snapshots from HH to the SHIELD power node (Figure 10.1).

With this test bed several cognitive functionalities have been developed and tested:

- *Self-awareness*: The network learns the current topology, the number and identity of the participants, their position, and reacts to variations of their interconnection.

- *Spectrum awareness*: The node collects information of the spectrum occupancy, either by spectrum sensing or indirectly using geolocation/database method, and shares the data with other nodes in order to create a map of existing sources.

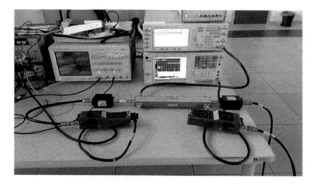

**FIGURE 10.1**
Smart transmission layer test bed architecture.

- *Spectrum intelligence*: The node analyses the spectrum information focusing on specific sources, trying to identify them by extracting significant signal parameters.
- *Jamming detection and counteraction*: The nodes recognize the presence of hostile signals and inform the rest of the network's nodes; the nodes cooperate in order to come with the optimal strategy for avoiding disruption of the network.
- *Self-protection*: The nodes cooperate to detect and prevent the association of rogue devices and to reject nodes that exhibit illegal or suspicious behavior.

Because of the high output power of the radios, two attenuators had to be included at each radio's front-end, while Hewlett-Packard 778D 100 MHz to 2 GHz dual directional coupler acts as a virtual channel with 20 dB nominal coupling, allowing for sampling and monitoring of the signal of interest. Agilent E4438C vector signal generator has been connected to inject a noise/interference signal to the network, while Agilent E4440A spectrum analyzer has been connected to facilitate the possibility of monitoring the radiofrequency (RF) activity.

### Energy Detector Spectrum Sensing

Obtaining information of the current spectrum occupancy is paramount for the CRs to be able to opportunistically access the spectrum, but may also aid them in recognizing anomalous or malicious activity by comparing the current state to those stored in their databases. There are three established methods for CRs to acquire knowledge of the spectrum occupancy: spectrum sensing [1,2], geolocation/database, and beacon transmission. HH currently only has the capability of performing energy detection spectrum sensing.

Every 3 s, 8192 samples from the analog-to-digital converter (ADC) are transmitted over the RS-485 port—this is a functionality hard-coded in the HH's field-programmable gate array (FPGA). Each sample is transmitted in two bytes: the first byte containing the six most significant bits (MSBs), with 2-bits sign extension on the left. The second byte contains the eight least significant bits (LSBs). In total, 16,384 characters are transmitted, making up for the interpretation of a 16-bit word. Currently, there is not a synchronization pattern; however, the idle interval between the two transmissions may be used, for example, to perform analysis of the received data. Transmission of a full window takes approximately 2 min.

The signal at the HH's FPGA input is a sample of raw spectrum. Raw samples are stored in a RAM buffer internal to the FPGA, and output through HH's fast serial port to the power node, where they can be processed.

Due to the high speed of the ADC (250 MSamples/s), serial port speed (115,200 bit/s) is not sufficient for the true real-time transfer, in addition, the processing capabilities of the power node would be completely devoted to the processing of received signal, leaving no room for higher-level applications. An adopted solution is to perform a quasi-real-time acquisition, that is, to collect a large "snapshot" of incoming spectrum, tens of kilo-samples, and to transfer the snapshot to the power node. When the snapshot has been transferred, a new collection may start. This is sufficient for proper analysis of the majority of RF scenarios: in practice, only fast pulsed signals might be completely missed.

A simple threshold-based energy detection allows us to identify the presence of signals in the frequency-band-of-interest.

In particular, the radio's firmware has been modified in order to provide to the serial port RS-485 a snapshot of 8192 samples that are directly coming from the RF front-end.

The snapshot provides the frequency content of a 120 MHz bandwidth around the carrier frequency used by the radio in the given moment.

More precisely, it is possible to observe:

$$f_{carrier} - 35 < f < f_{carrier} + 85$$

The frequency resolution is calculated on the basis of the maximum frequency of the ADC converter:

$$f_{res} = \frac{f_{max}}{N} = \frac{240 \times 10^6}{8192} = 29.3 \text{ kHz}$$

This frequency resolution, though, is not enough to recreate the spectrum shape in a satisfying manner (high probability of misdetection/misclassification). Moreover, the spectrum gathered by one snapshot is heavily affected by spurious signals created by the inner elements of the radio, making the acquisitions more challenging to interpret correctly. By increasing the number of snapshots up to 10, however, the same configuration gives the possibility to re-build the spectrum in a proper manner, which can be used to understand the occupation inside of the observable bandwidth, and the disrupting effects of the spurious signals significantly decrease (Figures 10.2 and 10.3).

With the current firmware configuration, the radio provides snapshots to the RS-485 port every 3 s, and due to the serial data rate each burst lasts for 2.3 s. Then, in order to collect 10 bursts, approximately 23 s are needed. Once gathered, these bursts are forwarded to the *spectrum intelligence block*.

*Feature Detector Spectrum Sensing*

Building upon the energy detector, certain features of the signals in the frequency domain may be extracted by performing a statistical signal characterization. First, an envelope of the wideband signal is created, then amplitude thresholding is performed and, by employing the maximum likelihood

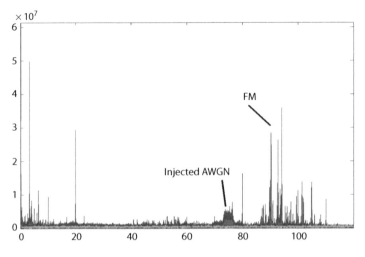

**FIGURE 10.2**
Spectrum reconstruction for varying numbers of consecutive snapshots—1 (green), 10 (red), 24 (blue).

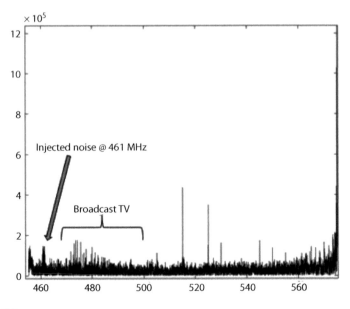

**FIGURE 10.3**
Spectrum reconstruction in the UHF band (note the broadcast TV around 480 MHz and the injected noise at 461 MHz).

decision rule, a number of frequency regions are identified corresponding to narrowband waveform candidates from the original wideband signal. Each of the identified candidate narrowband waveforms is then analyzed to extract its

- Maximum amplitude
- Center frequency
- Bandwidth

In detail, the spectrum sensing process is based on the following steps:

- Amplitude thresholding is performed and, by employing the maximum likelihood decision rule, a number of frequency regions (DFT bins) are identified corresponding to narrowband waveform candidates from the original wideband signal.
- The identified DFT bins that were previously grouped together (candidates for a particular waveform) are interpolated, in order to create an envelope for each of the identified narrowband signals:

```
while i<(length(X3_abs)-10)
    if X3_abs(i)>threshold
        X3_high(i)=X3_abs(i);
        no_of_high_bins=no_of_high_bins+1;
        group_found = 0; j=0; spacing=0;
        group_temp=[]; group_temp_amplitudes=[];
        while (spacing<10 && (i+j+spacing)<(length(X3_abs)))
            if X3_abs(i+j+spacing)>threshold
                vector_of_firstbins(i)=X3_abs(i);
                if j~=0
                    vector_of_duplicatebins(end+1) =
i+j+spacing;
                end
                group_found=1;
                X3_high(i+j+spacing)=X3_abs(i+j+spacing);
                group_temp(end+1)=i+j+spacing;

  group_temp_amplitudes(end+1)=X3_abs(i+j+spacing);
                j=j+1+spacing;
                spacing=0;
                else spacing=spacing+1;
                end
            end
            if group_found==1
                i=i+j;
            else i=i+1;
            end
        else
```

```
                            if (isempty(group_temp)==0)
                                group{end+1}=group_temp;
                                group_amplitudes{end+1}=group_temp_
amplitudes;
                            end
                            group_temp=[];
                            group_temp_amplitudes=[];
                    end
                    i=i+1;
            end
```

Program listing 10.1: Performing the thresholding and identifying groups of DFT bins corresponding to candidates for the narrowband waveforms.

- Each of the identified candidate narrowband waveforms is then analyzed to extract its maximum amplitude, center frequency, and bandwidth:

```
group_maximum_amplitude=[];group_bandwidth=[];
for l=1:length(group_amplitudes)
    group_maximum_amplitude(l)=0;
    if length(group_amplitudes{l})>min_no_of_samples
        for m=1:length(group{l})
            if group_amplitudes{l}
(m)>group_maximum_amplitude(l)
    group_maximum_amplitude(l)= group_amplitudes{l}(m);
            end
        end
    end
    group_bandwidth(l)=(group{l}(end)-group{l}(1)+1);
end
```

Program listing 10.2: Extracting the maximum amplitude and the bandwidth of the narrowband waveform.

Finally, this information is then fed to the *spectrum intelligence algorithm*.

### *Waveform Analysis*

Having a wideband spectrum analyzer allows for the monitoring of waveforms and analyzing their parameters. Two functional waveforms are currently installed on HHs: SBW and VULOS.

*SBW* is a wideband multihop Mobile Ad hoc NETwork (MANET) waveform, supporting operation in the 225–512 MHz part of the UHF band. The waveform provides self-(re)configurability and self-awareness of the network structure and topology, for up to 50 nodes and up to 5 hops. Furthermore, the possibility of simultaneous streaming of voice and data services is provided, with prioritization for voice streaming (in case of exceeded bandwidth). Allocated channel bandwidth is adjustable—up to

5 MHz—with channel spacing of up to 2 MHz. SBW uses a fixed digital modulation technique.

Self-awareness is exercised by monitoring the network topology for changes every "n" seconds (monitor interval is adjustable). Two quality of service (QoS) monitoring mechanisms are provided: bit error rate (BER) test and the statistics data for the transmitting/receiving side. These mechanisms are providing means for analyzing and comparing the quality of communication in regular and impaired channel conditions.

Figure 10.4 shows the envelope shape and properties of the SBW waveform, for the maximum signal bandwidth (5 MHz) and 1/10th of the maximum transmit power (–3 dBW), in frequency domain.

*VULOS* is a narrowband single-hop waveform designed for short-distance voice or data communication. It supports operation in both VHF (30–88 MHz) and UHF (225–512 MHz) frequency bands. The waveform allows for choosing between two analog modulation techniques: amplitude modulation (AM) and frequency modulation (FM), which may be configured on the fly, alongside with the modulation index. Channel bandwidth is adjustable up to 25 kHz, with channel spacing also adjustable up to 25 kHz. Furthermore, the VULOS waveform is able to utilize both digital and analog voice coder-decoders (CODECs) installed on the radio.

Figure 10.5 shows the envelope shape and properties of an FM-modulated VULOS waveform with the 25 kHz bandwidth, transmitted at 1 dBW in VHF band (30 MHz).

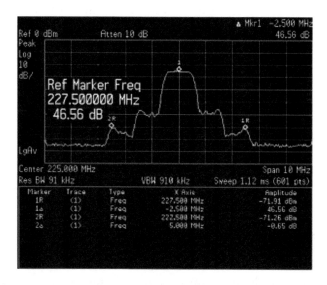

**FIGURE 10.4**
SBW waveform in the frequency domain—max hold.

**FIGURE 10.5**
FM-modulated VULOS waveform in the frequency domain—max hold.

### Database of the "Friendly" and "Malicious" Waveforms

Waveform analysis has an important SPD application: by creating a database of waveform types that are occurring in the system, it is possible to identify potentially malicious or misbehaving users. For test and evaluation purposes, the observed "friendly" waveforms are the SBW and VULOS, and their relevant features (maximum amplitude, bandwidth, and—in cases of assumed non-stationary users—center frequency) are stored in the database.

### Interference Influence Analysis

Various denial of service (DoS) attacks, especially jamming attacks [3], have, for a long time, been posing, and continue to pose, significant security threats to radio networks. RF jamming attacks refer to the illicit transmissions of RF signals with the intention of disrupting the normal communication on the targeted channels. RF jamming is a known problem in modern wireless networks, and not an easy one to counter using traditional "hardware-based" equipment. Additionally, SDRs and CRs bring the prospect for further improvement of the jamming capabilities of malicious users [4], and they also offer the possibility of developing advanced protection and counter-mechanisms.

One of the main focuses of the SPD-driven smart transmission layer is providing safe and reliable communication in jamming-polluted environments. For that, a detailed study of various jamming attack strategies and development of appropriate security solutions needs to be done.

The vector signal generator is presently used as means for creating disturbances in the communication channel, emulating a simple RF jammer. A set of measurements demonstrating how different types of generated interfering signals influence the performance of the communication on the channel was carried out.

In the first set of measurements, the aim is to show the correlation between BER and the radio's built-in link quality metric. Link quality is HH's built-in QoS feature, and is represented by an integer in the range of (0–200). The measurements are done with HHs having their signal bandwidths set to the maximum value (5 MHz), and repeated for two transmitting powers: −12 and 4 dBW. A generated interfering signal is a pulse signal that is generated at the same frequency as the frequency of the channel used for communication between radios (225 MHz). The amplitude of the generated interfering signal varies. The results are presented in Figure 10.6.

BER percentage is shown in the first half of the Y-axis (0–100), whereas the link quality level stretches throughout the whole Y-axis (0–200). The BER curves are mutually similarly shaped, with the expected offset due to differing transmission powers of the radio. The same goes for the link quality curve shapes. As can be seen, the occurrence of errors at the receiving side (area where BER > 0) corresponds to link quality levels in the range of 90–120. As expected, 100% BER corresponds to the link quality of 0, meaning that communication has become impossible.

In the second set of measurements, different types of interfering signals are generated by the signal generator, namely: a pulse signal, as in the first

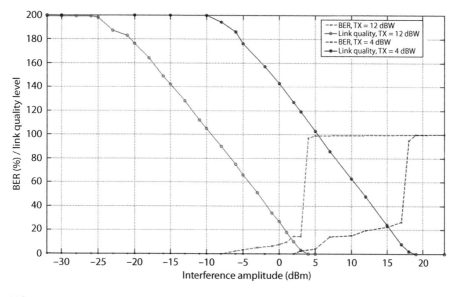

**FIGURE 10.6**

BER and link quality level vs. interference amplitude of interfering pulse signal.

measurement set; real-time I/Q baseband additive white Gaussian noise (AWGN) with the effective bandwidth of 5 MHz; real-time I/Q baseband AWGN with the effective bandwidth of 1 MHz, and a GSM signal. Once again, the central frequency of all of the interfering sources is the same as the frequency of the channel that the radios use for communication (225 MHz). The results are shown in Figure 10.7.

As expected, the pulse signal has the best interfering capabilities because it has the most concentrated power, and importantly that it has been generated at the exact frequency as the main carrier frequency of the transmitted signal. Even with small frequency offsets, the interfering impact of the pulse signal would drop significantly. For the same reason, the addition of AWGN results in higher link degradation in cases of smaller allocated bandwidth, due to the higher power density. The vector signal generator is only able to produce an AWGN signal of amplitude up to 20 dBm, hence the measurements for the higher values were not done.

### Optimal Anti-Jamming Measures

To alleviate the problems analyzed in the previous section, theoretical analysis of the electronic warfare was done, and basic anti-jamming principles based on spread spectrum techniques were studied for a future improvement of the prototype, that, for the time being is focused on the analysis of the alternative techniques, namely:

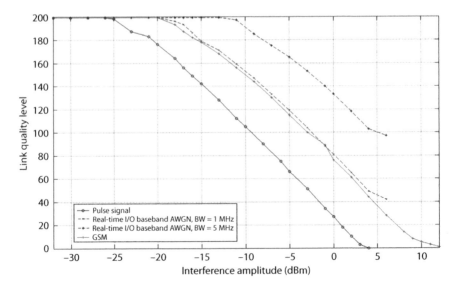

**FIGURE 10.7**
Link quality level vs. interference amplitude for different interfering signals.

- Retroactive frequency hopping (channel surfing)
- Altering transmission power
- Modulation (waveform) altering

Channel surfing refers to the process of changing the operating frequency of the radio in the presence of a strong interfering signal.

Increasing transmission power and/or gain may further improve the probability of detection and reconstruction of a signal.

Since it is well known that the effects of interference on a communication system are heavily influenced by the deployed modulation types at both the transceiving sides and the jamming sides, modulation altering may play important roles in alleviating RF jamming, provided that the radios are equipped with feature detection possibility, allowing them to recognize the features of jamming waveforms. The possibility of switching waveform type is open; however, at present it is uncertain whether the current spectrum sensing mechanism will, due to the limited number of samples (relatively low frequency resolution), be able to recognize jamming waveforms with satisfactory success.

Each of the proposed anti-jamming measures has its own advantages and drawbacks, and is thus suitable for deployment depending on the specific system parameters. These are summarized in Table 10.1.

### Remote Control of the Radios

Using Simple Network Management Protocol v3 (SNMP v3), several parameters of the HH radio may be externally controlled. For achieving this,

**TABLE 10.1**

Summary of the Anti-Jamming Measures

| Anti-Jamming Measure | Cost | Constraints | Optimality |
|---|---|---|---|
| Power alteration | Decreased battery life (for battery-operated radios) | Maximum transmit power | Against jammers with limited interference power |
| Frequency hopping | Hopping cost, typically conditioned by settling time of the radios' local oscillators, or by other hardware parameters such as processing time of the hopping command | Number of allocated channels | Against most narrowband jammers |
| Waveform alteration | Significant processing time for processing waveform alteration command (RF front end turned off in the meantime) | Number of waveforms to switch between | Against particular types of jammers that are utilizing waveforms of particular type (modulation) |

SNMP manager has to be installed and running on the power node. The host (power node) and the agents (HHs in the network) are connected through an Ethernet hub, and need to be on the same domain.

By utilizing three basic SNMP commands: GET, SET, and TRAP, it is possible to read the current value of the parameter, set a new value, or issue a message/warning if the current value satisfies a condition, respectively.

The controllable parameters and their corresponding features are stored in a management information base (MIB), which is loaded into the host's SNMP manager. The MIB table contains all the definitions that define properties of the controllable parameters, and describes each object identifier (OID), which is originally a sequence of integers, with a string.

A list of parameters that may be controlled externally, with the corresponding input data types, and the SNMP commands that may be invoked is given in Table 10.2.

**TABLE 10.2**

HH's Parameters That May Be Remotely Controlled via SNMP

| Parameter | Type | SNMP Commands |
|---|---|---|
| File transfer activation | String | SET/GET |
| File transfer type | String | SET/GET |
| FTP user name | String | SET/GET |
| FTP password | String | SET/GET |
| FTP address | String | SET/GET |
| Login username | String | SET/GET |
| Login password | String | SET/GET |
| Transmit power | Integer | SET/GET |
| Transmitter On/OFF | Integer | SET/GET |
| Currently installed waveform | String seq | GET |
| Waveform MIB root | String | GET |
| Waveform status | Integer | SET/GET |
| Audio message ID | String | SET/GET |
| Create new waveform | String | SET/GET |
| Activate preset | String | SET/GET |
| Activate mission file | String | SET/GET |
| Audio output gain | Float | SET/GET |
| Battery charge percentage | Integer | GET |
| File download status | Integer | GET |
| Trap receiver IP address | String | SET/GET |
| Number of units | Integer | GET |
| Radio channel | Integer | SET/GET |
| Zeroize all crypto keys | Integer | SET/GET |
| Crypto key loaded | Integer | GET |
| System end boot (failed, succeeded, in progress) | Integer | GET |

**TABLE 10.3**

HH's Parameters That May Be TRAPped

| Parameter | Description |
|---|---|
| NET radio OK | The notification is triggered when the visibility of the radio network is acquired |
| NET radio fail | The notification is triggered when the visibility of the radio network is lost |
| Critical alarm | The notification is triggered when the HH has sustained a critical operational error |
| End boot | The notification is triggered when successful boot-up of the HH has been verified |
| End file download | The trap notifies the end of the procedure of file download, indicating whether it was successful |
| Low power | The notification is triggered when the battery charge falls below a predefined limit |
| Create waveform OK | The notification is triggered when the waveform creation has failed |
| Create waveform FAIL | |

Accordingly, Table 10.3 provides a list of the parameters that may be TRAPped, with the short description of the conditions under which TRAPping messages are issued.

### *Automatic Reconfigurability of the Radio Parameters*

Using the standard SNMP client, HH's parameters may be reconfigured manually by the human operator. In order to enable the automatic reconfigurability of all of the relevant parameters, a C++ application based on the gSOAP toolkit was created.

gSOAP is based on a client–server interaction, supporting exchange of the XML and the simple object access protocol (SOAP) messages.

The SHIELD overlay (which serves as the top-most cognitive algorithm) is the client, and OMBRA v2/PC acts as the server. Periodically or on-request, XML/SOAP messages are exchanged between these entities, as denoted in the process diagram depicted in Figure 10.8.

### *Spectrum Intelligence*

By analyzing information provided by the energy detection spectrum sensing and feature detection block and comparing the relevant properties of the identified waveforms, the system may construct (and continuously update) knowledge regarding the spectrum opportunities, as well as the occurrences of the friendly/malicious transmission in the scanned band. This is referred to as the "spectrum intelligence" (SI) and is the principal step toward the autonomous cognitive functionality of the HH.

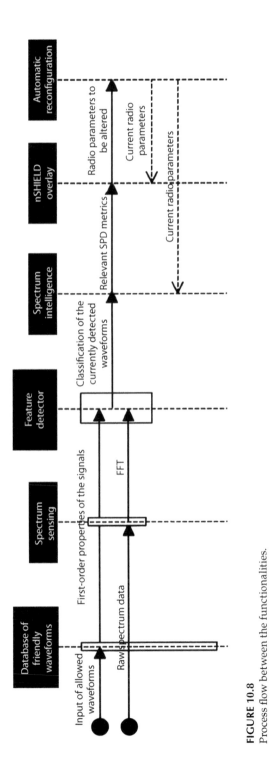

**FIGURE 10.8**
Process flow between the functionalities.

An SI algorithm keeps track of the occurrences of "known" and "unknown" waveform transmissions for each of the identified channels-of-interest, and subsequently triggers the corresponding action. For simplicity purposes, all "unknown" waveforms are considered as "potentially malicious," that is, corresponding to waveforms created by the jamming entities. SI invokes proactive channel surfing whenever the "unknown" transmission is taking place on a carrier frequency close to the carrier frequency currently used for transmission. SI chooses a new transmission frequency based on the identified spectrum holes, as well as the recent history of occurrences of identified "potentially malicious" waveforms in these spectrum holes. The new transmission frequency is instantly invoked by issuing the appropriate SNMP "change RF channel" command, both to the transmitter and the receiver side.

### Application to SHIELD Demonstration Scenarios

The SPD-driven smart transmission layer has been adopted in the dependable avionics scenario (see Chapter 6). Moreover, a specific technology demonstrator has been developed to demonstrate the following set of functionalities:

- Optimal anti-jamming measures
- Remote control of the radios
- Automatic reconfigurability of the radio parameters

The demonstrator has been set up as depicted in Figure 10.9.
Table 10.4 reports the list of commands used in the selected test mission.

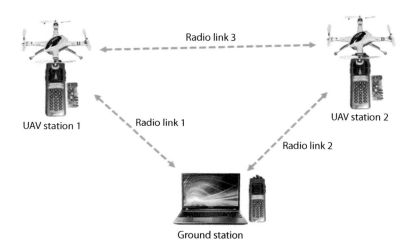

**FIGURE 10.9**
Overview of the smart transmission layer technology demonstrator.

**TABLE 10.4**

Reports the List of Commands Used In the Selected Test Mission

| Parameter | Type | SNMP Commands |
| --- | --- | --- |
| Transmit power | Integer | SET/GET |
| Transmitter On/OFF | Integer | SET/GET |
| Create new waveform | String | SET/GET |
| Battery charge percentage | Integer | GET |
| Radio channel | Integer | SET/GET |
| Crypto key loaded | Integer | GET |

The mission selected during test is based on the following steps:

1. The transmitter is turned on.
2. The appropriate waveform (either SBW or VULOS) is created. SBW is elected, which is a wideband digital waveform.
3. The status of the crypto-algorithm (on-off) is verified.
4. Intentional interference (jamming) is created on the channel currently used for transmission, causing the increase in the BER on the given channel and significant degradation of the communication link.
5. Once that jamming occurs, the overall SPD level is decreased and overlay decides to change either the operating radio channel in accordance with the predefined scheme (no spectrum sensing), or increase the transmission power (unless already set to maximum).
6. In addition, several parameters, such as battery charge of the radios, may be periodically read.

The connection of the prototype smart transmission layer to SHIELD architecture envisions both hardware and software interfaces.

Smart transmission layer communicates with the SHIELD overlay by means of the SHIELD gateway. Both control and information data are exchanged by the hardware interface, which is the Ethernet port.

For software interfacing, a client–server connection is established through the SHIELD data interchange (S-DI). The S-DI is used for communications between the SPD-driven smart transmission layer and the SHIELD gateway. The S-DI is a full-duplex protocol designed to handle asynchronous reads and writes from both parties: it does not specify any means of control flow, acknowledgements (ACKs) or negative-acknowledgements (NACKs), or command/response. These features must be implemented at the application layer. The S-DI protocol is based on simple object access protocol (SOAP).

SOAP is a protocol specification for exchanging structured information in the implementation of web services in computer networks. It relies on XML information set for its message format, and usually relies on other application layer protocols, most notably Hypertext Transfer Protocol (HTTP) or Simple Mail Transfer Protocol (SMTP), for message negotiation and transmission. Within this scenario, S-DI protocol uses SOAP-over-UDP standard covering the publication of SOAP messages over UDP transport protocol, providing for one-way and request/response message patterns.

Control commands and data exchanged between the smart transmission layer (consisting of HandHeld and OMBRA v2) and the SHIELD overlay are summarized in Figure 10.10.

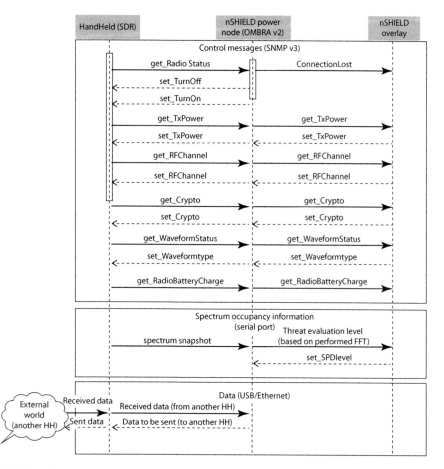

**FIGURE 10.10**
Inter-layer messages.

## SPD-Driven Smart Vehicle Management

The intelligence being built into various vehicle types will not only improve their safety and comfort but also enable new modes of transportation and new types of services, creating the corresponding markets. In all cases, it will be important to be able to monitor, preferably in real time, various parameters of the smart vehicles' condition. This section introduces a framework for the secure management of modern smart vehicles based on continuous monitoring of the SPD status of each vehicle and its subsystems, as well as real-time adaptations to preserve the desired SPD state. The proposed framework utilizes a formally validated approach to reason the composability of heterogeneous embedded systems (ESs), evaluate their current SPD levels based on predefined metrics, and manage them in real time. An implementation of EC is used in the Jess rule engine to model the ambient environment context and the rule-based management procedure. The reasoning process is modeled as an agent's behavior and applied on an epistemic multi-agent reasoner for ambient intelligence applications. By monitoring various operational parameters, which are assessed using SPD-related metrics, it enables real-time monitoring and interaction with a smart vehicle or a smart vehicle fleet, also integrating elements of a smart city infrastructure. A proof-of-concept implementation of the approach is developed and deployed on a complex test bed, demonstrating the successful integration of all the technological components into an actual working framework and its deployment on a real-world use case. Moreover, its performance overhead is assessed, validating the feasibility of the proposed approach.

### Smart Vehicles Domain

Smart vehicles are expected to constitute an important segment of the upcoming Internet of Things (IoT)-enabled world, where computing devices will not only permeate our lives. Modern vehicles already feature a number of embedded electronics that monitor and control the various automotive subsystems, aiming to enhance passenger comfort and safety, achieve energy-efficient operation, and maximize the vehicle's lifetime. Many types of accidents can be reduced through superior safety features, like early breaking and road lane departure warnings; this is also the focus of various governmental initiatives worldwide, which define stricter regulations for vehicle safety. Automotive legislation initiatives also necessitate the production of more eco-friendly vehicles, a target partly achieved by subsystems that monitor the vehicles' operation in real time and trigger adjustments to engine parameters. Based on the aforementioned stimuli, the electronics integrated into vehicles increase with every vehicle generation and are expected to rise steeply with the introduction of smart and, eventually, self-driving vehicles. This "intelligence" built into vehicles will also enable the introduction of a variety of enhanced services that everyone will enjoy, from end users (e.g., parents lending their vehicle to their child), to private entities (e.g., logistics or car-rental) and public entities (e.g., governments, smart cities, emergency services) operating vehicle fleets.

To facilitate the deployment and operation of these services in a secure and interoperable manner, we introduce a smart vehicle management framework based on the integration of novel primitives with standardized communication technologies and mechanisms. The proposed approach leverages the benefits of service oriented architectures (SOAs) to allow real-time monitoring of SPD, local and remote systems and their corresponding subsystems, using a formally validated reasoning process to monitor the changes in their SPD states. The proposed approach can provide real-time monitoring of a vehicle or a fleet of vehicles, collecting various operating parameters (e.g., regarding engine health or passenger safety), quantified using a set of SPD-related metrics, and allowing interaction with the vehicles (to adjust some parameters) in real time. The owner and other stakeholders can also set policies of fair use (maximum speed, allowed detours, etc.), tracking the driver's behavior based on these features. Thus, it allows individual users, companies relying on vehicle fleets for day-to-day operations, and even public entities (e.g., a smart city) to minimize the risks associated with passenger safety, protect vehicle investments, enhance productivity, and reduce transportation and staff costs.

To validate the feasibility of the proposed approach, a fully working implementation of the framework was developed and deployed on real vehicles. Moreover, a performance evaluation was carried out on a test bed consisting of devices expected to be found on smart vehicles and a typical backend infrastructure (i.e., a command and control [C&C] center), to investigate the performance overhead imposed by the components comprising the proposed framework. A full deployment of the proposed framework, with various smart city and smart vehicle elements, was used (see Chapter 9), while the various building blocks of the framework have been detailed in previously published works ([5–10]).

This section presents the motivation and related work, and illustrates the core technologies and the components of the proposed framework. Finally, the section analyzes a proof-of-concept scenario and the performance evaluation under a smart vehicle fleet management demonstrator that also integrates smart city elements.

## Motivation and Related Work

A typical car may currently utilize over 80 built-in microprocessors, providing advanced safety systems, emission monitoring, and in-car commodities [11], which aim to enhance the comfort and safety of their passengers, also protecting the vehicle's subsystems by providing early warning of failures and/or adjusting their operation accordingly. Typically, electronic control units (ECUs) manage and interconnect the distinct systems [12,13] and the infotainment infrastructure provides various enhanced facilities, like entertainment and navigation, to passengers [14].

The new generation of smart vehicles will be a mobile intelligent system within a larger smart city infrastructure, supporting communication with other vehicles, the city infrastructure, and backend systems. Prototype deployments of cars and road infrastructure exchanging information regarding lane state and traffic are already under construction by large manufacturers in the European Union (EU) and the United States. Public entities and private transportation organizations and businesses will also take advantage of the individual computational and communication capabilities of smart vehicles to achieve effective fleet management through real-time monitoring of the vehicle's state and the driver's driving behavior. These features will allow organizations to minimize the risks associated with vehicle investment, and promote strategies for increasing productivity and safety, while reducing transportation and staff costs.

Government regulations are decisive toward pertinent research efforts. The United Kingdom tries to minimize road deaths in business-owned vehicles; starting in 2008, road death is treated as an unlawful killing, enabling seizing of the company's records and bringing prosecutions against directors who fail to enforce safe driving policies for the drivers they employ. Therefore, fleet management is now imperative for the functional operation of an organization which owns a significant amount of vehicles.

The European Commission has also defined new regulations for vehicle safety. One such initiative is the EU-based eCall system, which is expected to become mandatory for every vehicle moving in the EU by 2018. This emergency service, described in regulation EN 15722:2011 [15], dictates that, when an accident occurs, the vehicle should be able to automatically relay essential information (the vehicle's location, its direction and speed before the crash, number of passengers, etc.) to appropriate public safety answering points (PSAP), providing information that an accident has occurred. By providing early notification and allowing efficient coordination of the emergency services, this will enable faster response to such incidents (expected to decrease response time by 50% in the countryside and 60% in built-up areas), drastically reducing the number of deaths and the severity of injuries for the thousands of people involved in road accidents every year. Moreover, the EU and the United States are collaborating to define a common subset of rules and standards related to smart vehicles and smart cities infrastructure.

However, security-related incidents have already been reported, where vulnerabilities in the infotainment infrastructure are exploited by attackers to remotely operate vehicle components [16]. More recently, attacks on production vehicles, exploiting vulnerabilities in Fiat/Chrysler's Uconnect system, enabled hackers to apply the brakes, kill the engine, and take control of steering over the Internet [17], with the company urging owners to update their cars' software to patch the identified vulnerabilities [18]. An attacker can control the breaking system, turn off the lights, or even lock the car while on the move. As more and more vehicles provide seamless connection to the Internet, security becomes an important aspect that must be considered at the design phase of any relevant proposal. To this end, systematic methodologies for developing secure and efficient vehicular ESs are needed.

Researchers within the Ford motor company have presented a methodology for modeling automotive systems in terms of security, privacy, usability, and reliability (SPUR) [19]. Every evaluated counterpart is analyzed based on the offered SPUR functionality and a qualitative value is assigned to each parameter (low, medium, or high). The method is applied on real system attributes, such as the valet key and the anti-lock braking system.

ScudWare is a semantic and adaptive middleware platform for smart vehicle spaces, presented in [20], with similar design goals as the approach presented herein. It constitutes a multi-agent system that synchronizes context-aware and adaptive components. The agent's reasoning process is based on first-order predicate logic and implemented in description logic (DL) with web ontology language (OWL) ontology rules. Nevertheless, the authors do not provide clear implementation details nor a performance evaluation, focusing on the theoretical background of ScudWare. The proposed framework provides a more robust reasoning process, and is based on modern standards, adopting a SOA approach. Moreover, it offers higher degree of context representation and reasoning which are further combined with quantifiable metrics for SPD.

In all cases, mechanisms should be included to allow the vehicle owner (e.g., a logistics company or a father lending his car to his son) to specify driving rules for the drivers, setting, for example, the maximum travel speed and/or the operating area (through geo-fencing). Furthermore, sensors could be used to monitor the vehicles health, informing the owner about engine malfunctions or other incidents in real time.

Finally, all mechanisms should ideally be developed with backward compatibility in mind, maintaining the ability to retrofit the necessary modules into existing vehicles. It is estimated that there are over 500 million vehicles already roaming EU and US roads alone; a huge market for anyone involved in retrofitting such modules to make existing vehicles "smarter" and let their drivers enjoy some of these new enhanced services.

The proposed approach aims to act as an enabler for addressing the preceding issues. The most important concern is the enhancement of passenger safety and the framework can help achieve this in various ways: it can assist in keeping vehicle fleets in good condition (by constant monitoring of vehicle health), it can monitor the drivers' compliance to good-driving practices, and it can enable faster response in cases of emergency. The adoption of this framework would, thus, allow a smart city or any other public or private organization with vehicle fleets to reduce the risks to their personnel and their vehicle investment, advancing productivity while reducing transportation and staff costs. Moreover, it can help achieve compliance with upcoming EU regulations and other government organizations both for vehicle safety and green infrastructure management.

## Core Technologies and Components

The framework consists of various core components, comprising a low-cost multi-agent system for smart vehicles, also allowing the integration of smart infrastructures (e.g., smart road and/or smart city subsystems) into the monitoring and decision-making process.

The proposed framework is a multi-agent system, implemented in the Java agent development framework (JADE) [21] which, via the open service gateway initiative (OSGi) [22] middleware, is able to monitor and manage devices profile for web services (DPWS) [23] devices. The AI reasoning process is an event-based model checker that is based on EC [24]. The context theory is aware of the SPD aspects of the underlying ES (component types and composition, implemented technologies and their different configurations, SPD metrics, etc.). Furthermore, strong access control mechanisms are also integrated into the management framework, allowing its operators to centrally manage access to the resources of smart vehicles, based on a predefined set of rules and policies. This secure policy-based access control, detailed in [8], exploits and extends the DPWS functionality already present on the framework's devices, to realize the necessary communication mechanisms and entities (e.g., policy enforcement points on vehicles). It relies on a standardized and proven technology for policy definitions, namely the eXtensible access control markup language (XACML) [25].

To effectively study and model ESs and their security characteristics, their architecture is segregated into four layers. These layers, from bottom-up, are: *Node* (represents all the embedded devices themselves), *Network* (consists of nodes connected in networks), *Middleware* (the management-software of the networks) and *Overlay* (formed by the framework's agents, who control distinct subsystems and exchange high-level security-, privacy-, and dependability-related information).

The proposed framework adopts the preceding layered approach. Agents, implemented using the JADE platform, are part of the overlay, while the necessary DPWS and OSGi mechanisms operate at the middleware.

The framework's entities are deployed using DPWS, allowing each device to expose the services and operations needed to transmit its current status (to allow real-time monitoring) but also to receive commands (for management purposes). DPWS can be used to expose not just the proposed framework's monitoring/management operations, but also those related to their functional elements (e.g., the various sensors and actuators), allowing system owners to leverage the SOA benefits across their whole infrastructure. By exploiting these benefits, the proposed framework facilitates the communication of critical information regarding the secure and privacy-aware and dependable operation of the devices.

This information is then aggregated to a control node, where it is composed using EC-based rules, in a formally validated manner. The system's agents base their reasoning process on Jess-EC [26], which can perform, among others, automated epistemic, temporal and casual reasoning for dynamic domains with real-time events, preferences and priorities; features that are necessary in the context of this work. The various SPD states and properties of each device are modeled as fluents, and the system changes as events of EC, all implemented using the previously mentioned extended Jess-EC. Rules that trigger the reasoning process of the agent are also supported, producing a metric-driven management of the infrastructure according to its SPD level. The updates on the current state as well as the changes that need to be enforced by the agent(s) are all realized via appropriate DPWS-based mechanisms present on all devices. These changes can be triggered by human interaction (e.g., from system operators) or by a predefined set of rules. Furthermore, in the case of large organizations or when monitoring facilities in segregated premises, a multi-agent system can be deployed, where each agent monitors the SPD levels of a subset of the infrastructure and communicates with the rest of the agents and/or a central agent to provide enterprise-wide monitoring and management.

The overall state, once calculated, is presented in a form which is usable and simple to understand by the system's operators. This enables real-time assessment of the security and dependability posture of the infrastructure (e.g., affected by the detection of an ongoing security attack or a component failure, respectively), allowing for a timely response to changes in the system state, either via human interaction or in an automated manner, to counter events that triggered said changes. Thus, the proposed framework simplifies threat assessment and risk management. Moreover, it enables the interfacing with more sophisticated, intelligence-driven, management mechanisms, potentially fully automating the management of the ecosystem (e.g., for automatic incident response).

The main technological building blocks of this architecture are detailed in the next section.

## Monitoring and Adaptation Agents

The agents and the multi-agent system are implemented in the JADE, which supports all the relevant FIPA standards for agent deployment, for example, the agent communication language (ACL) [27]. The reasoning process is based on the EC and is implemented in the rule engine Jess-EC. We model the ambient context in Jess-EC and develop reasoning services based on a formal theory that reasons about the composability and integration of the underling devices and technologies and verifies the current SPD level of the system, and a management theory of the ambient environment, also responsible for administrating the system in real time. The OASIS standard common alerting protocol (CAP) [28] is used to model the semantic information that is exchanged between the in-vehicle equipment with its agent, and the agent with other agents outside the vehicle. The external communication is achieved via mobile networks (2/3/4G). Agents transform the CAP alerts into CAP-EC events (and vice versa), which trigger the reasoning process.

At the smart vehicle end, an agent monitors the system and takes simple decisions regarding SPD and safety. The agent lies in the infotainment infrastructure communicating internally with the ECU and externally with other agents. The agent acts as an intermediate layer between the ECU and external systems, providing an abstract level of communication and protecting the internal system from attacks, as all respective changes are triggered via the reasoning process and not via direct interaction with the vehicle bus.

The C&C center is responsible for smart vehicle management. A master agent is deployed at the backend, collecting information from the smart vehicle and infrastructure agents. The master agent has global knowledge of the whole system and carries out the fleet management strategy. It can trivially be extended to use databases to store vehicle and driver history records, maintaining extensive log files and execution reports. Moreover, it can be used to display the positioning information of the smart vehicles into a geographic information system (GIS) in order to provide location-based services (e.g., geo-fencing). To implement this functionality at the backend, DPWS is integrated into the OSGi—a standardized middleware that constitutes a module system and service platform. The JADE agents are also integrated into the same platform and manage the vehicles' devices in real time through OSGi interfaces. ECU and supplementary devices model the provided functionality in DPWS and exchange information with the vehicle's agent through OSGi.

The software architecture of the agents forming the core of the proposed framework's intelligence is depicted in Figure 10.11.

## SPD Reasoning Process

In [5], we have presented a preliminary version of a composition verification and security validation reasoning system, where event-driven model-based methods have been proposed to describe the behavior of a dynamic system in a formal manner.

The work presented here extends this reasoning system by adopting the multi-metric formation and normalization processes [6,29] and the composition evaluation formulas of the medieval castle approach [30]. We propose the use of standardized mechanisms and protocols, both for context-aware modeling as well as real-time control and management of the devices. The resulting system implements a formal methodology in system composition verification and SPD validation, supporting a metric-driven reasoning

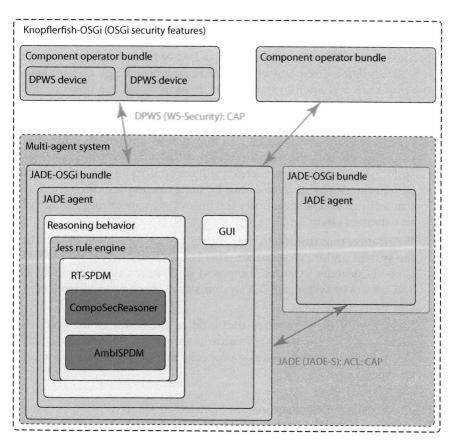

**FIGURE 10.11**
Software layers of the proposed framework.

process, via SOA-based interfaces that enable seamless access to the various devices and their functional elements.

The framework's agents consist of two components: namely the *CompoSecReasoner* and *AmbISPDM* [31]. The former uses the reasoning system presented in [5] to implement an extended composition validation and security verification, while the latter models the management theory of the ambient environment and manages the system in real time.

### CompoSecReasoner

In the proposed methodology, technologies and protocols are modeled as attributes. A security analysis is performed for every developed attribute and a relevant SPD is defined for each security level that is provided. Then, the evaluated metrics and properties of the system are determined, including how they may be affected by the underlying attributes at runtime. This information is encoded in the evaluation functions of the relevant metrics and properties.

CompoSecReasoner is a methodology that describes the SPD properties of a composed system, and how these features are affected by changes in the state-architecture. CompoSecReasoner can be used for composition verification, SPD validation, comparison of different system settings, and evaluation of the impact of a change in the system. It addresses three main issues of system development:

1. Extension of the composition and management processes with metrics. This enables the measurement of the SPD level for the final system. We consider that metrics will become an integral feature in system development as they can be used to compare different system settings and help quantify the impact of an incoming change in an existing system.

2. Verification that different components can be composed and form the system under examination. This is especially useful for heterogeneous systems (systems composed of different types of devices) and system of systems (SOS) (systems that are composed from several subsystems).

3. Validation of the properties that hold as the result of the composition of two systems. This feature is imperative for assuring that the composed system works properly and features the claimed specifications.

Technically, CompoSecReasoner is modeled as the reasoning process of JADE agents and ported in the OSGi middleware platform. The embedded devices implement OSGi bundles, which communicate information with the JADE bundles, and describe the provided services in DPWS.

The attack surface metrics [32] consider that an attacker uses a system's methods, channels, and data items to attack the system. These features, at the entry and exit points that may be exploited by attacks, determine the attack surface. A threat is successful if it enables direct or indirect interaction with a protected asset. Microsoft develops a relevant methodology, named the relative attack surface quotient (RASQ) [33], and mathematically quantifies the relative attackability of the Windows server operating system platforms. The metric's outcome is further validated against real exploits (based on relevant reports from the CVE and CERT databases).

Based on these studies and the attack surface formal basis, we propose the SPD surface to evaluate the properties of a system [6]. As the individual analysis of the three SPD properties produces erroneous conclusions regarding attackability, the SPD surface evaluates the aggregated effect, resulting in more accurate system analysis and better design.

In general, the surface metrics reveal the attackability of a system, but not the protection level. The SPD multi-metric methodology extends the system's surface with the *porosity* feature. The threat flows (TF) in the surface constitute the system pores where interaction between assets and threats is possible. Each TF is disassembled in security, privacy, and dependability pores representing its effect in the three SPD properties respectively. The system designer places controls (defense mechanisms) to protect the pore from attacks. Limitations form known conditions under which a control does not work properly. If all types of controls are placed for a pore and work correctly, full coverage of the interaction point is achieved.

The surface and the protection level analysis are integrated. The overall method estimates the real protection level by aggregating the risk analysis of the attackable points and the effectiveness of the deployed defense mechanisms.

The outcome is a triple vector of <Security, Privacy, Dependability> representing the total SPD of the system. The final SPD vector calculation is the perfect separation (100) minus the weighted summation of the limitations.

The main method to estimate the security level of a composed system is the medieval castle approach [30]. The system is modeled as a medieval castle with security doors, which are the target of an attacker. An attack is successful if it breaks through the doors and reaches a treasure room inside the castle—the resources that are protected by the security mechanisms. The difficulty in passing through the security doors and reaching a castle's inner treasure room indicates the castle/system security level. Each door is a security mechanism of the system and its resistance to attacks is measured by relevant metrics.

To quantify security, one has to quantify the security of the system subcomponents and then measure the overall security based on composition rules. Attack trees are constructed based on the main composition operations, representing the effort to attack the system from the outer layers of defense. More details regarding the medieval approach and the underling analysis can be found in the original paper [30].

The SPD methodology is utilized as the core metric to evaluate the protection level of the individual system components. Then, the basic security notion of the medieval castle approach is extended with the SPD feature. The SPD multimetrics are embodied to the medieval castle methodology. They act as a systematic way to calculate the defense of the castle's "SPD doors." The outcome is the total SPD of the currently composed system.

The composed metric advances the medieval castle composition approach and captures the requirements of two main real-time SPD modeling principles. The medieval castle models only static instances of a system. The proposed methodology evaluates dynamic systems with de/composition events altering the system's structure and SPD properties in runtime. For example, two subsystems may be secure (in some sense), but their composition need not be secure (in the same sense). The composed SPD multi-metric applies pre- and post-composition validation. At first, it determines if the composition can be succeeded. Then, it derives the composition's outcome and its side effects. Similarly, for decomposition.

The protection mechanisms at the different layers of a real system may interact with each other. This notion is not properly handled by the castle evaluation. In the proposed methodology, missing controls of the SPD multi-metric can be covered by relevant controls that are placed in the adjacent outer layers, forming the added protection of the interacting sequential defense layers. For example, an insecure data exchange application can be safeguarded by a secure communication service that is imposed at the network layer providing confidentiality, integrity, and authentication. The final SPD represents the overall control coverage that is achieved by the current setting.

For the system composition, verification, and security validation, the system is considered as a set of components of four layers (node, network, middleware, and overlay) and the components of each layer are composed of subcomponents of the adjacent lower layer.

The metrics and the properties are layer specific and evaluate their SPD at runtime based on evaluation functions. A component's SPD is the summation of the component's metrics and the minimum SPD of the underlying subcomponents (i.e., the weaker inner link). The component's SPD is constrained by the SPD of the higher layer component that contains it (i.e., the weaker outer link). Components of the same layer are composed by composition operations to form higher layer components. In order to perform such an operation, a component must be able to execute the operation's attributes (functional requirements) and achieve specific SPD for relative metrics and properties (nonfunctional requirements that are determined by the operation). When composition is successful, the composition verification and security validation process are revisited. A similar strategy is followed for decomposition operations.

### AmbISPDM

The core of the reasoning process is an event-based model checker which extends EC and is implemented in the Jess rule engine. CompoSecReasoner reasons about the system composition and SPD validation while AmbISPDM models the security and safety-related management strategy and the system's administration through real-time technologies. The whole reasoning process is transformed into a JADE agent's reasoning behavior and implemented as a multi-agent epistemic reasoner.

The agents are then encapsulated into OSGi bundles and are deployed on Knopflerfish [34], an open source implementation of OSGi. Each agent controls a distinct system via the OSGi middleware, while the various agents communicate with each other using the ACL language, with messages containing CAP data exchanged via JADE. The embedded devices specify their type and provided services using DPWS, which is also used to facilitate the exchange of CAP messages between managed devices and agents. Each underlying system component (node, network, and middleware) that communicates directly with the agent creates a component operator that controls the component's services. The agent and the component operators exchange well-formed information determined by the CAP scheme. The DPWS-related components are developed using the WS4D-JMEDS API [35] and are also integrated into OSGi bundles.

AmbISPDM is a formal framework for the ambient management of user safety and system SPD. It was applied in the nSHIELD project for the management of SPD-sensitive intelligent spaces.

- The proposed framework utilizes EC to monitor and manage ESs in real time. The provided abstraction level permits the modeling of a wide variety of domains.
- The reasoning process is modeled as an agent's behavior and applied on a multi-agent reasoner for ambient intelligence applications. A conflict resolution mechanism is also supported, that incorporates relational and epistemic reasoning, benefiting from both approaches.
- SPD metrics are entailed, enabling metric-driven reasoning and management.
- The overall service-oriented architecture in based on standards or widely used technologies and enables the easy deployment of heterogeneous embedded devices and their seamless function.

Agents monitor distinct ESs and manage them in real time. The reasoning theory of AmbISPDM manages the composed system based on ambient safety and system SPD rules. We propose the use of standardized mechanisms and protocols, both for context-aware modeling as well as the real-time control and management of the embedded devices. This implementation enables an efficient agent reasoning process with an enriched set of individual reasoning capabilities. Moreover, a conflict resolution mechanism is proposed that incorporates epistemic and relational multi-agent reasoning. Except from these logic processing features, the service-oriented architecture enables the seamless deployment and operation of heterogeneous ESs in ambient intelligence (AmI) applications.

At normal operation, AmbISPDM assists driving and in-cabin conditions. It supports the development of ambient intelligence applications and safeguards SPD. In cases of emergency, like a car crash, the system detects the event and performs the first actions toward managing the situation. The users and the responsible authorities are informed and a reaction plan is applied.

We use the Discrete Event Calculus Knowledge Theory (DECKT) [26]—an EC reasoner implemented in the rule engine Jess. DECKT extends the basic reasoning process with hidden casual dependencies (HCD). It enables reasoning with missing pieces of knowledge which may become known in a future time-point, like sensory measurements.

DECKT performs reasoning over virtual discrete time-points. We extend this process to express real-time events by integrating the Java Timer object.

For every time-point where an event with duration must occur, a timer is created that fires when the specific time elapses. Then, the reasoning process is automatically activated, deducing the current state of the system. A hash table of active Timer objects is maintained with the different time-points representing the hash keys. Each slot contains the Timer and the list of rules that cause the objects creation. A rule's effect can be cancelled in a future time-point before the Timer fires. Thus, the rule is erased from the Timer's list. When all rules are removed from a list before the Timer expiration, the object is deleted. We deploy several variations of these process, implementing different functionality for the time duration events, like create, pause, stop, kill, start, replay, and reset.

Each rule in Jess assigns a value, called salience, which determines the execution order of rules that fire simultaneously (the higher the salience, the most recent the rule is processed). The mechanism is exploited by AmbISPDM in order to model rules' priorities. Thus, high priority to specific rules is expressed by defining high salience values. When the first rule among a group of competitive ones fires, it assigns a relevant fact in the Jess engine internal memory to denote that the rest of the rules are blocked. For example, consider the case where the system must automatically inform the fire brigade in case of a safety-related incident, either via email or SMS. A relevant event fires that can trigger two rules for the two aforementioned types of transmission, respectively. If both types are currently active, AmbISPDM can select the communication form based on the user's preferences.

The agents form a peer-to-peer network and exchange information. The agents aggregate these pieces of knowledge and derive the global state of the system. When conflicts occur, a resolution mechanism will retain coherency.

### Examples of SPD Level Variation

With regard to the dynamic SPD levels, the smart vehicle can transmit subsets of the monitored information. Ideal SPD settings may not always be possible, as operation may be hampered by congestion, missing mobile network coverage or other variables. So, for example, if security mechanisms are omitted because of network issues or to preserve power, this can be represented as a lower security level in the SPD levels reported on the backend system.

Accordingly, in another case, the service user may decide not to report his exact location, but a larger area (i.e., one including more service users), in order to obstruct the tracking of his exact whereabouts and, thus, protecting his privacy. This will have a direct effect on the privacy level visible at the backend monitoring system.

Variations in the dependability level will be more common. The vehicle's health is monitored in real time (e.g., tire pressure, mileage, and warning messages from the ECU), and the dependability value should be adjusted accordingly. For example, when the car exceeds the planned travel mileage between services, the dependability value reported will be lower.

**Security**

*Access Control/Policy-Based Management*

Among the studied schemes for authorization, and policy-based management in general, proposed for systems with different requirements and properties, and a cross-platform solution that meets the requirements of all types of ESs and provides interoperability, crucial for next-generation pervasive computing devices, is XACML. XACML is an XML-based general-purpose access control policy language used for representing authorization and entitlement policies for managing access to resources. XACML is also an access control decision request/response language. As such, it can be used to convey policy requirements in a unified and unambiguous manner, hence interoperable and secure, if appropriately deployed.

The preceding fit well into the model of a network of heterogeneous ESs where access to resources is provided by nodes as a service, and into the management architecture developed by IETF policy framework. This typical policy-based management architecture combined with XACML, is mapped to a SOA network of nodes to provide protected access to their distributed resources. It consists of several components that run on different nodes of the architecture. These components are [36]:

- *Policy enforcement point (PEP)*: The system entity that performs access control, by making decision requests and enforcing authorization decisions.
- *Policy administration point (PAP)*: The system entity that creates a policy or policy set.
- *Policy decision point (PDP)*: The system entity that evaluates applicable policy and renders an authorization decision.
- *Policy information point (PIP)*: The system entity that acts as a source of attribute value.

Moreover, auxiliary entities may also co-exist alongside the previous, depending on the specific application and deployment at hand. Some of these entities, which will be showcased in later sections, may include the following:

- *Context handler (CH)*: Orchestrates the communications among the stakeholders, converts, if necessary, messages between their native forms and the XACML canonical form, and collects all necessary information for the PDP.
- *Obligation handlers (OH)*: Provide additional restrictions that should be taken into account when enforcing a decision, like the requirement to log any permitted access or to inform for unauthorized attempts.
- *Environment*: Provides additional information independent of a particular subject, resource, or action.

Considering that some smart platforms may not have the computing resources to accommodate expensive mechanisms, some of these roles (e.g., PDP) may only be undertaken by more powerful nodes expected to operate within the target node's (i.e., the PEP's) trusted environment. Thus, a node, depending on its capabilities and the available resources, might include one or more of these functional components. A basic deployment of the framework's access control entities is depicted in Figure 10.12.

The XACML handling and decision-making engine can be adopted from any open source implementation. Such open source resources include Sun's XACML implementation [37], PicketBox XACML (formerly JBossXACML) [38], the Holistic Enterprise-Ready Application Security Architecture Framework (Heras AF) XACML [39] and the Enterprise Java XACML project [40]. Closed source commercial alternatives exist as well, but these are not as modifiable and, thus, often have limited usability in custom implementations. Considering the previously mentioned options, Sun's XACML is the framework of choice for implementing the presented framework's access control engine, as it remains popular among developers and is actually the basis of various current open source and commercial offerings.

Thus, standardized and proven technologies are combined to provide policy-based management in the proposed framework. Leveraging this

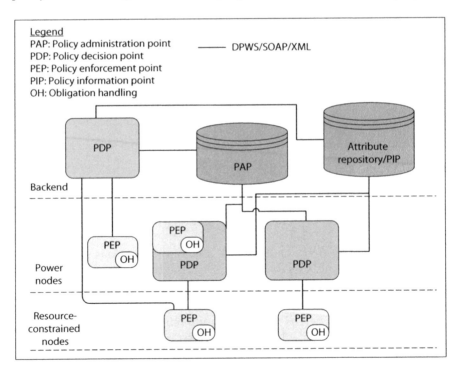

**FIGURE 10.12**
The policy-based management architecture.

approach, a multifunctional smart vehicle may feature various hosted services (e.g., a location service and a fuel consumption service), access to which will be controlled based on the active policy set. As an example, we can consider the case of a logistics company which allows drivers to track their designated vehicles' exact locations through their mobile phone. In this case the company can simply modify the active policies to give a specific individual access to a vehicle's location service; after returning the vehicle, the policy can be reset to withdraw the driver's access privileges.

### Message Protection

The protection of messages exchanged between agents is a critical concern, as a potential violation of the framework's security could give an erroneous view regarding the actual vehicle (or vehicle fleet) situation, cause unnecessary trouble for emergency response services or even endanger the safety of vehicle occupants. For this purpose, the framework protects the communication between the managed vehicles and their respective agents through the use of the WS-security standard, typically used alongside DPWS. This standard can provide end-to-end security, non-repudiation, alternative transport bindings, and reverse proxy/common security token at the application layer. The OSGi security features are additionally used to provide inner-platform security on both agents and component operator bundles (i.e., managed devices) by limiting bundle functionality (e.g., which bundles can be started/stopped, when, and by whom) to predefined capabilities. Furthermore, the JADE-S add-on is used to safeguard the ACL messages exchanged, by providing user authentication, agent actions authorization against agent permissions, as well as message signature and encryption.

### User Interface

While the technical aspects detailed in the previous sections are important, a decisive factor to the practical success of any framework of this nature will be its usability. The accessibility both in terms of the usability of its user interface, as well as the success of visualizing the system's current state in a user-friendly and intuitive manner are important, as well as ensuring that it imposes minimal requirements from the end user in terms of training, configuration, and maintenance [41].

The inner workings of the system are irrelevant to the operator, who may or may not be technically proficient. Moreover, its output should be accessible to higher management, who, in the context of businesses, are usually people with limited time and accustomed to handling interfaces with dashboards, gauges, and other high-level representations [42]. The preference for dashboard layouts is not restricted to higher management, but has been shown to be preferable to users in general], while recent research indicates there are significant benefits to this approach, helping address various business challenges (e.g., to provide a global view on resource occupation, aiming to enhance decision making for both human and non-human resource management at runtime]). Thus, this is the approach adopted for visualizing the agents' outputs and the corresponding user interface.

The SPD levels derived from the SPD composition are plain numbers, for example (56, 75, 89), since they are the results of the composition process detailed in the previous sections. Though these numbers are appropriate to be transmitted and computed by machines (i.e., "machine to machine" interfaces), when it comes to presenting them to human operators in user interfaces, there is the need for making their representation more intuitive. That would make the SPD level much easier to understand in order to allow operators to be aware of the situation and possibly operate manual countermeasures wherever appropriate. The appropriate presentation of information is highly dependent on the specific domain and application; however, a basic template could be applied to most domains. To this end, and in order to further facilitate the comprehension of the SPD state for a human operator, the use of a graphical scale and appropriate colors should be used (e.g., red for all values below 40, yellow for values from 40 to 70, and green for values from 70 to 100). An example of this approach, chosen for the proof-of-concept implementation of the backend monitoring software, as detailed in [7], appears in Figure 10.13.

Other than the main GUI detailed previously, which presents the aggregated SPD state of the monitored infrastructure, each agent also features its own backend GUI. This is aimed at advanced users/system operators, providing low-level information of the SPD status changes reported by each subsystem. The events appear as they happen on the screen, but these changes are also logged in the corresponding files for offline auditing. A screenshot of this user interface appears in Figure 10.14.

### The Technology Demonstrator

We have retrofitted the proof-of-concept implementation of the proposed framework onto an existing vehicle, using off-the-shelf components. This setup relies on the infotainment system—or the user's smartphone, if a smart infotainment system is not available, as was the case with the vehicle we used—for communication with the backend system. An Android-based application is deployed on the smart device, which performs the SPD and safety-related reasoning process, also communicating and receiving commands from the C&C. The application collects information regarding the car's sensors (e.g., fuel consumption, engine status) from the ECU via a Bluetooth-enabled OBD scan tool. These OBD tools are widely available nowadays for a very reasonable cost; the one used in the setup cost ~10 Euros. The application collects additional data from sensors already integrated into

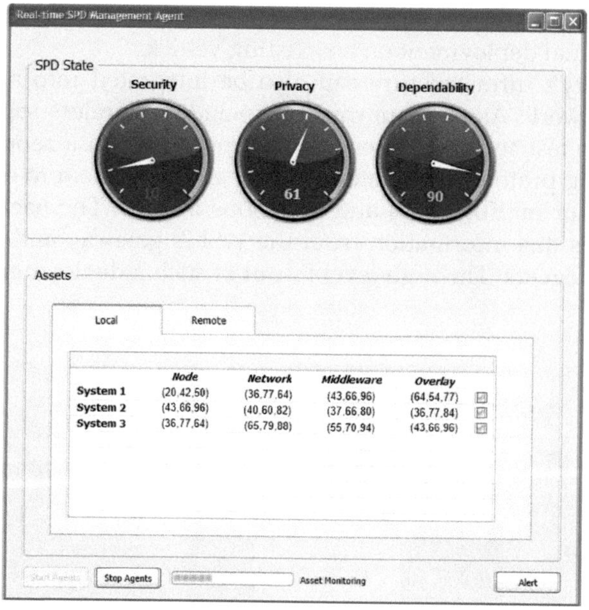

**FIGURE 10.13**
The proof-of-concept master agent GUI.

**FIGURE 10.14**
The backend GUI of the framework's agents.

infotainment or smartphone, like acceleration and GPS position. Figure 10.15 shows the actual deployment on an existing vehicle.

A smart city's infrastructure can also be integrated into and managed by the framework. As proof of this, we include a wireless sensor network (WSN) in the test setup. The sensors communicate via a reputation-based secure routing protocol (presented in [45]), allowing them to detect several types of sensor malfunctions and malicious attacks. The backend collects and processes this information from the WSN's gateway and can, in turn, inform other agents. The high-level layout of such a deployment appears in Figure 10.16.

**FIGURE 10.15**
Proof of concept retrofitted on existing vehicle.

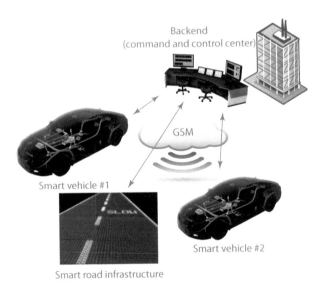

**FIGURE 10.16**
High-level layout of managed smart city and vehicle fleet infrastructure.

### Demonstration Scenario

We demonstrate the proposed framework using a smart city scenario, which includes the WSN-equipped smart city infrastructure described previously, the backend C&C and two access control protected smart vehicles with various sensing capabilities. The scenario architecture appears in Figure 10.17.

The use case scenario is designed to demonstrate the SPD variations of the involved systems in the cases of: attacks on the smart city infrastructure, engine malfunction on one of the vehicles and, finally, a car crash. The scenario steps appear in more detail in Table 10.5. The SPD values are visualized in the C&C master agent's GUI via traffic light icons, displaying green, yellow, and red to indicate normal, low, and critical SPD states, respectively (green for values from 70 to 100, yellow for values from 40 to 70, and red for all values below 40).

### Performance Evaluation

As a full-scale deployment on an actual smart city was not feasible, an experimental test bed was used to evaluate the proof-of-concept implementation. The Linux-based device was deployed on a Beaglebone embedded device (720 MHz ARM Cortex-A8 processor, 256 MB RAM, Linux OS), emulating the ECU. The Android-based version of the agent was deployed on a tablet (1.9 GHz quad-core processor, 2 GB RAM, 16 GB, Android OS), emulating the vehicle's infotainment. The backend system runs on a laptop PC, while a separate laptop was used as a gateway for the "smart city infrastructure" WSN which was based on IRIS motes (Atmel ATmega1281 16 MHz CPU, 8 kB RAM, 128 Kb program flash memory) running the reputation-based secure routing algorithm previously described. Finally, a desktop PC was used as a client to the services provided by the vehicle (e.g., to access its location) for benchmarking purposes. This setup, which successfully carried out all the demonstration steps detailed previously, appears in Figure 10.18.

The performance of the framework was evaluated focusing mostly on the resource-constrained devices, as documenting all performance aspects of all entities involved in a complex framework like this would not be practical. To investigate the framework's behavior, a benchmark client was developed that issued 300 consecutive requests to access the smart vehicle's location (from the infotainment device) and its engine temperature (from the ECU). Both these operations were protected by the access control mechanism detailed previously, thus before replying, the devices verified (by communicating with the backend) that the client was authorized to access this data, based on the active policy set. Furthermore, at random intervals, the benchmark client would communicate with the master agent to trigger changes in the SPD state of the prototypes (e.g., to increase the key length used for encryption), thus also evaluating the impact such changes can have on the responsiveness of the devices and the system as a whole.

The master agent is the most computationally demanding entity of the framework; its code size is 1.87 MB, it occupies 45 MB RAM and needs 1.6 s,

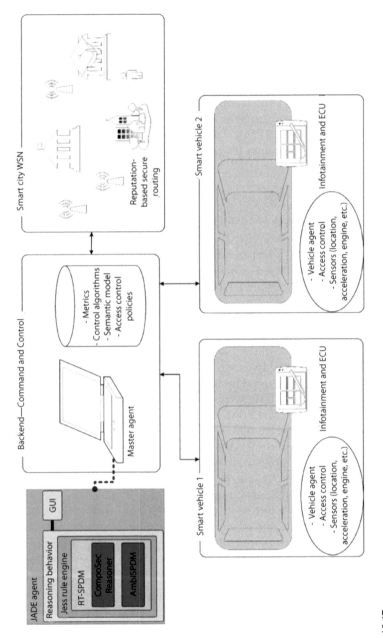

**FIGURE 10.17**
Scenario architecture.

**TABLE 10.5**

Scenario Steps

| Step | Events | Effect | Overall (S,P,D) Value | State Visualization |
|------|--------|--------|-----------------------|---------------------|
| 1 | Power-on of all systems and discovery/registration | Initial state | **80, 70, 65** | |
| 2 | A Black Hole attack[a] is detected at the WSN. MA is informed through WSN that an attack occurs and it sends a command to the vehicles to increase security | Security level decreases | **60, 70, 65** | |
| 3 | Security level is increased on both Car 1 and Car 2. MA is informed through the vehicle agents | Security level increases | **85, 70, 65** | |
| 4 | The WSN has counteracted the Black Hole attack, reporting the change to the MA. The MA asks Cars 1 and 2 to return to normal state (to conserve resources) | Security level returns to initial state | **80, 70, 65** | |
| 5 | ECU of vehicle 1 informs of increased temperature. MA is informed through the vehicle agent | Dependability decreases | **80, 70, 50** | |
| 6 | Car 1 crashes. Transmits eCall SMS. SA1 dependability decreases. SA1 privacy decreases (operator at C&C now has access to exact vehicle location). SA1 security decreases (encryption is disabled to facilitate emergency response and emergency services). The MA is informed of the changes and sends email to C&C operators/administrators and other stakeholders | S, P, and D levels decrease | **50, 50, 40** | |
| 7 | Car 1 is repaired. SA1 reports new state to MA | S, P, and D levels increase | **80, 70, 65** | |

[a] In its basic form, an attack whereby a malicious node that disrupts the routing process by constantly advertising that it is available to route data, but then refuses to forward the packet it receives].

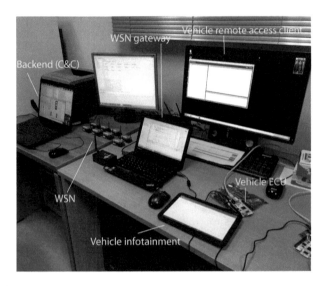

**FIGURE 10.18**
Test bed used for demonstration and performance evaluation. Proof-of-concept retrofitting (insert).

on average, to perform a reasoning process. Moving to the rest of the entities, we focused on the infotainment device (i.e., the Android application) and the corresponding Linux-based application developed for the embedded platform (emulating the ECU). Their CPU load during tests was not that significant, with an average of 4.8% recorded on the Beaglebone and an average of 4.1% on the Android platform. Memory consumption was also acceptable, averaging 33.46MB on the Beaglebone and 26 MB on the Android device. Perhaps the most interesting performance parameter is the delay experienced by the user attempting to access the smart vehicle's services and the latency of the system when changing between SPD states. These results appear in Figure 10.19, which shows the response time (recorded client side) for each of the requests issued concurrently to both the vehicle's infotainment (tablet) and its ECU (Beaglebone). As is evident from the graph, the Linux-based embedded device is more responsive, with an average response time of 93.16 ms compared to 198.7 ms for the infotainment. This is expected, as there are more processes running in the background on the Android tablet and, moreover, features a GUI which further impacts its responsiveness. Nevertheless, both devices demonstrated acceptable response times.

Also, evident in the graph are the spikes recorded when the devices had to change SPD states (and, thus, had to process the incoming master agent request, change their operating parameters accordingly, and inform the master agent once the changes were in place). This difference is more evident in Figure 10.20, where the average response time for each device is compared in the two states (i.e., normal operation and between SPD state changes). These

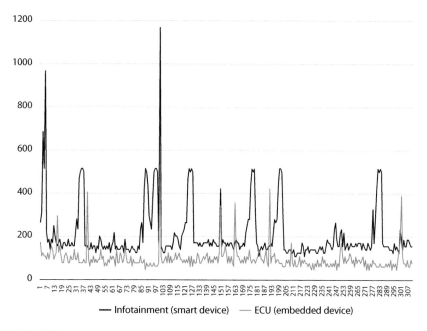

**FIGURE 10.19**
Response time (in ms) per request, for both test platforms.

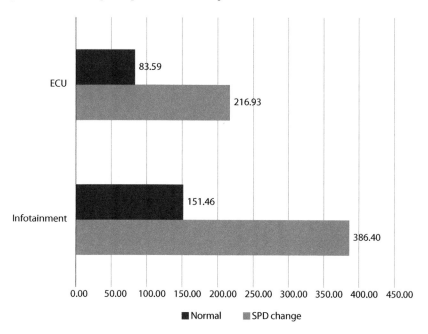

**FIGURE 10.20**
Average response time (in ms) for 300 requests. Comparison between different states (normal operation and during SPD state change) for each device type.

changes in SPD states do introduce significant delays, but they should not be a concern during normal operation, as such changes are expected to happen rarely (e.g., when the framework needs to react to some attack or when a vehicle crashes).

### Final Remarks

The framework for smart vehicle management that can be used to improve the quality of use and service of the vehicles or vehicle fleets, as well as to facilitate incident response in cases of emergency.

The proposed framework could be utilized by the public or private sector; for example, government services could monitor their fleet improving everyday workflow and vehicle assignments, helping fully exploit the vehicle investments. Private sector entities operating in transportation could also benefit by adopting the proposed framework. Bus and taxi companies could reduce the transportation time and design and, adopt better strategies to satisfy their passengers' expectations, for example, by better scheduling trips, thus providing quick and timely service.

Therefore, there is significant incentive in further improving the framework. To achieve this, the framework's agents will have to be retrofitted on actual vehicles, including motorbikes, in order to confirm its effectiveness in actual, large scale and heterogeneous deployments. This will be necessary in order to collect valuable feedback that will help to improve key areas of framework's operation, such as its scalability, and, most importantly, its ease of deployment, use and, maintenance.

---

## DDoS Attack Mitigation

### Introduction

Based on the marking schemes presented in Chapter 3, a filtering and traceback mechanism that can effectively stop ongoing DDoS attacks has been implemented. We design and implement the architecture of a highly configurable mechanism able to react in different attack scenarios and ensure the highest amount of legitimate user service under an ongoing DDoS attack. With precise tuning of this mechanism, an organization can provide robust while flexible protection against such attacks.

### Filtering Mechanism Overview

The two packet marking schemes presented in Chapter 3 give the victim the tools to perform two crucial operations when under a DDoS attack. First, to keep their service online and responsive by mitigating the effects of the attack. Second, to identify the true sources of the attack and possibly ensure via cooperation with the, usually unknowing, attacking network operators, the removal of the attack threat.

For the service availability operation, the victim must be able to identify and filter the attack traffic. The simplest case is by dropping incoming packets from networks that have been identified as part of the DDoS attack. This often results in denying service to legitimate hosts also for two reasons. The first reason is the inherent false positive probability of the packet marking scheme. Networks that are not part of the DDoS attack are falsely identified as such and denied communication with the victim's services. The second reason is the fact that the packet marking schemes are not capable of distinguishing nodes within the same network and under the same edge router. This means that, even with no false positives, when the victim filters out an offending network some legitimate users will probably be denied access too.

The filtering operation is a real-time operation since the existence of the packet marking allows the filtering mechanism to drop all packets that bear the same marking once the decision is made that packets originating from a certain network (i.e., having the same distinct marking) are part of an ongoing DDoS attack. The fact that the filtering mechanism relies only on the contents of each incoming packet and not at all on the source IP address, enables not only real-time filtering but also robustness against IP spoofing attempts.

For the traceback operation, the victim must be able to discover the true IP address of the offending network. Packet markings do not hold the source IP address but the marking itself together with the publicly known IP addresses of the upstream routers can be used to calculate the source IP address of the originating network. The traceback procedure is recursive and time consuming and should be used as a forensics mechanism after, or during, the DDoS attack to ensure its permanent discontinuation via cooperation of network operators from the originating networks.

### Architectural Design

Figure 10.21 illustrates the architecture of the filtering and traceback mechanism. It consists of three monitors that send the various measurements directly to a control center for decision making.

The control center together with a repository act as the decision makers and the results are propagated to a reconfigurable firewall that does the actual traffic filtering. The main modules of the mechanism are the following:

- DDoS load monitor
- DDoS detection monitor
- System load monitor
- Reconfigurable firewall
- Control center
- Repository

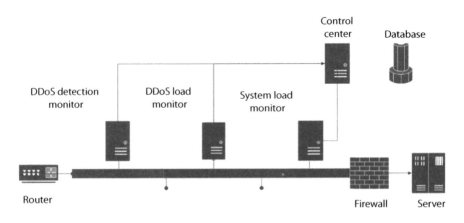

**FIGURE 10.21**
The filtering and traceback mechanism architecture.

The general function of the mechanism is that the three monitors sent periodic reports to the control center, the database holds information for specific networks that are of interest to the organization that hosts the mechanism and the control center evaluates constantly which networks should be filtered out and for how long. The actual filtering is done via reconfiguration of the firewall but, in our case, based on packet markings instead of source IP address.

A description of each module of the mechanism is outlined in the following sections.

### DDoS Detection Monitor

The DDoS detection monitor is the module that observes the network traffic and decides whether there is an ongoing DDoS attack and, also, determines which networks are part of this attack. There are various DDoS detection mechanisms that are in use by the industry and can be of use to such a module. Those mechanisms make use of statistical correlation, machine learning, or various other methods. The main difference is that the necessary correlation is done based on the packet markings instead of the source IP address. The correlation procedure usually involves a large number of samples that need to be analyzed. This is not a real-time procedure but once the packets originating from a specific network are identified as part of a DDoS attack, the packet marking of that network is reported to the control center. The actual filtering is done at a later stage.

### DDoS Load Monitor

The DDoS load monitor and the system load monitor repost in regular intervals the overall load of the system and the load of the system that originates from the networks that have been identified as parts of the DDoS attack. Those two reports help the control center to decide which networks to filter out and for how long.

### System Load Monitor

The system load monitor is a normal monitor that reports on the various health statistics of the underlying network and/or service. The measurements that we use in this mechanisms context is the system load of the underlying service that is of interest to the filtering mechanism.

### Repository

A third piece of information necessary for the decision by the control center is the contents of the database regarding known networks that are of interest to the organization that hosts the filtering and traceback mechanism. The database holds a simple table that lists the network name along with the packet markings that correspond to packets originating from that network. It also holds user-specific columns that act as weight factors on the decision of whether to block those networks, or not, if they find themselves as parts of an ongoing DDoS attack.

### Control Center

The control center is the module that gathers all the information from the various monitors and after consulting the contents of the database, decides which networks should be blocked in each iteration. The decision is passed in the form of a packet markings list to the firewall and the firewall is reconfigured based on that list to drop all traffic that holds those specific packet markings.

### Firewall

The firewall is the module that does the actual filtering based on the decisions made by the control center. It is a reconfigurable firewall and its filtering rules can be changed on the fly by the control center using the firewall's configuration API. The main difference from standard firewalls is that the filtering is based on the IP header fields that are used for the packet markings instead of the source IP address. The firewall rules that do not affect the filtering of DDoS attacks, do still use the source IP address for normal operations. Also, the firewall overwrites the packet marking, before forwarding the packet, and resets the IP header overloaded fields with their default values. The last operation is dependent on the packet marking scheme that is being used.

## Functional Design

The main function of the filtering mechanism is to decide whether or not to block incoming packets from a certain network. This decision is done in the control center module by using the measurements of the three monitors and the information about each know network that is stored in the mechanism's repository.

The repository is a simple relational database with one table that stores two types of information about each known, to the organization that hosts the filtering mechanism, network. It stores basic information about the network relating to its IP range and the packet marking that the packets originating from this network are expected to have. It also stores qualitative metrics for the networks in the form of impact weights for each metric. For the purposes of the demonstrator we use network size, customer base, and network region.

The repository and the qualitative metrics are necessary because there will be cases that a network is part of an ongoing attack but it also hosts legitimate users that are of relative value to the organization. Thus, the decision if those networks should be blocked, and under which circumstances, are a matter of what priority each network has for the organization. Sometimes, it is preferable to hold back on blocking certain networks, from a business point of view, until the system load exceeds a certain threshold.

Figure 10.22 describes the schema of the database. The main field names and descriptions are as follows:

- *Name [string]*: A human readable designation of each network.
- *IP range [string]*: The IP ranges of the network.
- *Marking [binary]*: The value of the packet marking that packets originating from this network are expected to bear.
- *Expected load [integer]*: The average volume of incoming traffic from this network.

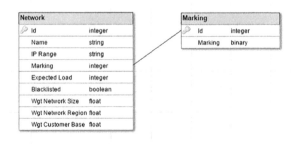

**FIGURE 10.22**
The database schema of the filtering mechanism repository.

- *Black listed: [Boolean]*: A flag that denotes that this network is either of no value to the organization or consistently hosts malicious hosts and therefore should always be blocked first in case of a DDoS attack.
- *Wgt network size [float]*: Weighted metric on the value of the network based on the network size. Larger networks get increased weight due to their potential larger customer base.
- *Wgt customer base [float]*: Weighted metric on the value of the network based on the actual customer base as is documented in the organizations records. Larger customer base means increased weight.
- *Wgt network region [float]*: Weighted metric on the value of the network based on the network's region. The value of the weight depends on the degree of customer penetration that the organization has on each region.

The decision is taken using the following procedure:

- If the offending networks are unknown, that is are not registered in the database, the control center will progressively block those until the system load reaches a user defined comfortable level. The control center does not block all the offending networks since there is always the probability of legitimate users belonging to those networks.
- If there are known networks that have been identified as part of the DDoS attack, the control center will block those networks based on the priority that is defined from the weight factors that are present in the database for each network. The goal here is to keep the high-priority networks unblocked, even if they are part of the attack, while maintaining a system load as close to a user defined level, but lower than that. This is done in regular iterations where based on the new reports from the three monitors, the list of blocked network changes accordingly. In that way, the control center ensures service availability while keeping the maximum amount of potentially legitimate users unblocked.

### Example Scenario

This section describes two typical scenarios of the progression of a DDoS attack that involves both known and unknown networks. The attack is recognized by the DDoS monitor and a spike in incoming traffic is measured by the system load monitor. Once the DDoS detection monitor finished processing the markings of each attack source, a number of distinct marking are reported to the control center. After a repository lookup, some of those markings belong to known networks and some of them are unknown. The control center starts filtering those networks in small iterations until the incoming traffic stabilizes below a maximum threshold (usually 100% system load) and above a minimum threshold. The existence of a minimum threshold ensures that the control panel will not block more networks that necessary for the normal function of the system thus avoiding blocking potential legitimate users.

In Figure 10.23, we see the start of a DDoS attack. The attack is comprised of three unknown networks and two known ones. The control center blocks all the unknown networks in an attempt to stabilize the system load. It then blocks the lowest priority known network and manages to keep the load between the min and max threshold. If the attack from one of the unknown networks stops then the system load drops. The control center unblocks the blocked known network and increases the load above the min threshold. After a period of time, new unknown networks join the DDoS attack. The control center repeats the decision-making procedure, blocks the necessary networks, and stabilizes the system load.

## Final Remarks

The demonstrator consists of mainly two components. The first component, the filtering mechanism, resides solely in the system level in the form of several software modules that interoperate with each other in a transparent way from the SHIELD point of view. The second component, the marking mechanism, requires a set of capabilities from the network infrastructure, the routers, for the marking procedure but, apart from that, does not interact with the middleware since the marking procedure is stateless and no information exchange is performed by the filtering and traceback mechanism. The marking procedure is exclusively simulated in the context of the demonstrator.

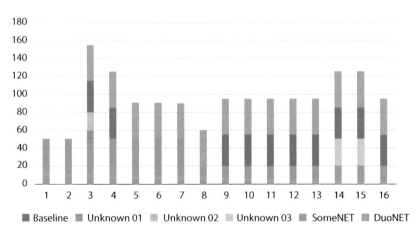

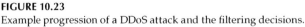

**FIGURE 10.23**
Example progression of a DDoS attack and the filtering decisions.

# References

1. M.O. Mughal, A. Razi, S.S. Alam, L., Marcenaro, and C.S. Regazzoni, Analysis of energy detector in cooperative relay networks for cognitive radios. *2013 Seventh International Conference on Next Generation Mobile Apps, Services and Technologies (NGMAST 2013)*, September 25–27, 2013, pp. 220–225.

2. P. Morerio, K. Dabcevic, L., Marcenaro, and C.S., Regazzoni, Distributed cognitive radio architecture with automatic frequency switching, *Complexity in Engineering (COMPENG)*, pp. 1–4, 2012.

3. K. Dabcevic, A. Betancourt, L. Marcenaro, and C. S. Regazzoni, A fictitious play-based game-theoretical approach to alleviating jamming attacks for cognitive radios. *IEEE International Conference on Acoustics, Speech and Signal Processing (ICASSP)*, Florence, Italy, 4–9 May, 2014.

4. P. Tague, Improving anti-jamming capability and increasing jamming impact with mobility control. *Mobile Adhoc and Sensor Systems (MASS), 2010 IEEE 7th International Conference on*, pp. 501–506, 2010.

5. G. Hatzivasilis, I. Papaefstathiou, C. Manifavas, and N. Papadakis, A reasoning system for composition verification and security validation. *Proceedings of the 6th IFIP International Conference on New Technologies, Mobility & Security (NTMS)*, Dubai, UAE, pp. 1–4, 2014.

6. G. Hatzivasilis, I. Papaefstathiou, and C. Manifavas, Software security, privacy and dependability: Metrics and measurement, *IEEE Software*, 33 (4), pp. 46–64, 2016.

7. K. Fysarakis, G. Hatzivasilis, I. Askoxylakis, and C. Manifavas, RT-SPDM: Real-time security, privacy and dependability management of heterogeneous systems. *Human Aspects of Information Security, Privacy, and Trust*, Springer, Berlin, pp. 619–630, 2015.

8. K. Fysarakis, I. Papaefstathiou, C. Manifavas, K. Rantos, and O. Sultatos, Policy-based access control for DPWS-enabled ubiquitous devices. *Proceedings of the 2014 IEEE Emerging Technology and Factory Automation (ETFA)*, Barcelona, Spain, pp. 1–8, 2014.

9. K. Fysarakis, G. Hatzivasilis, C. Manifavas, and I. Papaefstathiou, RtVMF: A secure real-time vehicle management framework, *IEEE Pervasive Computing*, 15 (1), pp. 22–30, 2016.

10. K. Fysarakis, O. Soultatos, C. Manifavas, I. Papaefstathiou, and I. Askoxylakis, XSACd—Cross-domain resource sharing and access control for smart environments, *Future Generation Computer Systems*, 59, 2016.

11. J.A. Cook, I.V. Kolmanovsky, D. McNamara, E.C. Nelson, and K.V. Prasad, Control, computing and communications: Technologies for the twenty-first century model T, *Proceedings of the IEEE*, 95 (2), pp. 334–355, 2007.

12. A. Doshi, B.T. Morris, and M.M. Trivedi, On-road prediction of driver's intent with multimodal sensory cues, *IEEE Pervasive Computing*, 10 (3), pp. 22–34, 2011.

13. J.F. Coughlin, B. Reimer, and B. Mehler, Monitoring, managing, and motivating driver safety and well-being, *IEEE Pervasive Computing*, 10 (3), pp. 14–21, 2011.

14. K. Lee, J. Flinn, T. J. Giuli, B. Noble, and C. Peplin, AMC: Verifying user interface properties for vehicular applications. *Proceedings of the 11th Annual International Conference on Mobile Systems, Applications, and Services—MobiSys'13*, Taipei, Taiwan, pp. 1–12, 2013.

15. Economic Commission for Europe, Telematic applications: eCall HGV/GV, additional data concept specification, September 2011. [Online]. Available: http://www.unece.org/fileadmin/DAM/trans/doc/2011/dgwp15ac1/INF.30e.pdf.
16. K. Koscher, A. Czeskis, F. Roesner, S. Patel, T. Kohno, S. Checkoway, D. McCoy, B. Kantor, D. Anderson, H. Snachám, and S. Savage, Experimental security analysis of a modern automobile. *Proceedings of the IEEE Symposium on Security and Privacy*, Berkeley/Oakland, CA, pp. 447–462, 2010.
17. A. Greenberg, Hackers remotely kill a Jeep on the highway—with me in it, *Wired*, July 2015. [Online]. Available: https://www.wired.com/2015/07/hackers-remotely-kill-jeep-highway/.
18. *The Guardian*, Jeep owners urged to update their cars after hackers take remote control, *The Guardian*, July 2015. [Online]. Available: https://www.theguardian.com/technology/2015/jul/21/jeep-owners-urged-update-car-software-hackers-remote-control.
19. T.J. Giuli, D. Watson, and K.V. Prasad, The last inch at 70 miles per hour, *IEEE Pervasive Computing*, 5 (4), pp. 20–27, 2006.
20. Z. Wu, Q. Wu, H. Cheng, G. Pan, M. Zhao, and J. Sun, ScudWare: A semantic and adaptive middleware platform for smart vehicle space, *IEEE Transactions on Intelligent Transportation Systems*, 8(1), pp. 121–132, 2007.
21. JADE Framework. JAVE Agent Development Framework is an open source platform for peer-to-peer agent-based applications. [Online]. Available: http://jade.tilab.com/.
22. Open Services Gateway Initiative (OSGi). The dynamic module system for Java. [Online]. Available: http://www.osgi.org/.
23. D. Driscoll, A. Mensch, T. Nixon, and A. Regnier, Devices profile for web services, version 1.1, OASIS, 2009. [Online]. Available: http://docs.oasis-open.org/ws-dd/dpws/wsdd-dpws-1.1-spec.pdf.
24. E.T. Muller, *Commonsense Reasoning*, M. Kaufmann, San Francisco, CA, 2010.
25. B. Parducci, H. Lockhart, and E. Rissanen, eXtensible access control markup language (XACML) version 3.0, OASIS Standard, 2013. [Online]. Available: http://docs.oasis-open.org/xacml/3.0/xacml-3.0-core-spec-cs-01-en.pdf.
26. T. Patkos and D. Plexousakis, DECKT: Epistemic reasoning for ambient intelligence, *ERCIM News Magazine Special Theme: Intelligent and Cognitive Systems*, issue 84, January 2011.
27. FIPA, ACL. [Online]. Available: http://en.wikipedia.org/wiki/Agent Communication Language.
28. OASIS, CAP. [Online]. Available: http://docs.oasis-open.org/emergency/cap/v1.2/CAP-v1.2-os.pdf.
29. I. Eguia and J. Del Ser, A meta-heuristically optimized fuzzy approach towards multi-metric security risk assessment in heterogeneous system of systems. *Proceedings of the 4th International Conference on Pervasive and Embedded Computing and Communication Systems (PECCS)*, pp. 231–236, 2014.
30. M. Walter and C. Trinitis, Quantifying the security of composed systems. *Parallel Processing and Applied Mathematics*, Wyrzykowski R., Dongarra J., Meyer N., Waśniewski J. (Eds), Lecture Notes in Computer Science (LNCS), vol. 3911. Springer, Berlin, Heidelberg, pp. 1026–1033, 2006.
31. G. Hatzivasilis, I. Papaefstathiou, D. Plexousakis, C. Manifavas, and N. Papadakis, AmbISPDM: Managing embedded systems in ambient environments and disaster mitigation planning, *Applied Intelligence*, pp. 1–21, 2017.

32. P.K. Manadhata and J.M. Wing, An attack surface metric, *IEEE Transactions on Software Engineering (TSE)*, 37(3), pp. 371–386, 2010.
33. M. Howard and Microsoft Corporation, Determining relative attack surface, *United States Patent, US 7299497 B2*, November 20, 2007.
34. Knopflerfish, Knopflerfish—Open Source OSGi service platform. [Online]. Available: http://www.knopflerfish.org/.
35. J. Müthing and B. Schierbaum. WS4D-JMEDS DPWS Stack. [Online]. Available: http://sourceforge.net/projects/ws4d-javame/.
36. A. Westerinen, J. Schnizlein, J. Strassner, M. Scherling, B. Quinn, S. Herzog, A. Huynh, M. Carlson, J. Perry, and S. Waldbusser, Terminology for policy-based management, RFC 3198, 2001. [Online]. Available: http://www.ietf.org/rfc/rfc3198.txt.
37. F. Najmi and S. Proctor. Sun Microsystems Laboratories, XACML. [Online]. Available: http://sunxacml.sourceforge.net.
38. A. Saldanha. PicketBox XACML. [Online]. Available: https://community.jboss.org/wiki/PicketBoxXACMLJBossXACML.
39. F. Huonder. Holistic Enterprise-Ready Application Security Architecture Framework (Heras AF) XACML. [Online]. Available: http://www.herasaf.org/heras-af-xacml.html.
40. Z. Wang. Enterprise Java XACML [Online]. Available: http://code.google.com/p/enterprise-java-xacml/.
41. J.D. Hollan and E.L. Hutchins, *Designing User Friendly Augmented Work Environments*, S. Lahlou (Ed.), Springer-Verlag, London, pp. 237–259, 2010.
42. W. Beuschel, Encyclopedia of decision making and decision support technologies, *IGI Global*, 2, pp. 1–1019, 2008.
43. A. Read, A. Tarrell, and A. Fruhling, Exploring user preference for the dashboard menu design. *42nd Annual Hawaii International Conference on System Sciences (HICSS)*, Big Island, HI, 2009.
44. I. Linden, Proposals for the integration of interactive dashboards in business process monitoring to support resources allocation decisions, *Journal of Applied Decision Sciences*, 23, pp. 318–332, 2014.
45. G. Hatzivasilis, I. Papaefstathiou, and C. Manifavas, ModConTR: A modular and configurable trust and reputation-based system for secure routing in ad-hoc networks. *IEEE 11th ACS International Conference on Computer Systems and Applications (AICCSA 2014)*, Doha, Qatar, November 10–13, pp. 56–63, 2014.

# 11

# *Applying SHIELD in New Domains*

**Paolo Azzoni, Francesco Rogo, Cecilia Coveri, Marco Steger, Werner Rom, Andrea Fiaschetti, Francesco Liberati, and Josef Noll**

## CONTENTS

## Introduction

The SHIELD methodology introduces, from an industrial perspective, a paradigm shift in handling the security of cyber-physical systems, trying to address the need for securing information instead of securing the infrastructure. This paradigm shift reflects the new horizons characterized by the dynamic interaction between businesses, such as the Internet of Things (IoT). From a research perspective, the SHIELD methodology has been adopted by the scientific community in several research initiatives, aiming at further extending the original approach and applying it in new vertical domains. This chapter provides an industrial view of security, privacy, and dependability (SPD) and a short description of research projects that adopted and extended the SHIELD methodology.

## Industrial Perspective

As anticipated in the introduction, we are at the beginning of a new age of business, where dynamic interaction is the driving force for our business. Today, the Internet-based service world is based on collaborations between entities in order to optimize the delivery of goods or services to the customer. The evolution toward the dynamic interaction between entities is ongoing (Figure 11.1).

In this context, the importance of the Internet as communication bearer is supported by the IoT as enabler for autonomous systems and machine learning as support for decision making. These three aspects are driving the

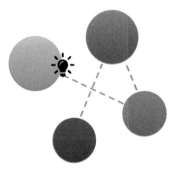
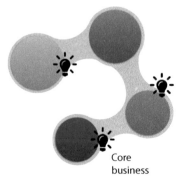

(a) Internet today—collaboration based     (b) Upcoming—dynamic interaction

**FIGURE 11.1**
Upcoming business world of dynamic interaction between entities.

process toward digital industries and digital societies. Knickrehm et al. (2016) presented the impact of digital services on behalf of Accenture Strategy and Oxford Economics. Their study points out that in the United States, already 33% of the contribution to the gross domestic product (GDP) comes from digital services, with the United Kingdom and Australia having around 30% of the GDP contribution from digital services. Even Germany, seen as a traditional industry, has more than 25% of its GDP contribution coming from digital services. Going into a sector specific analysis, 57% of the value creation in the *financial* sector comes from digital services. The corresponding numbers for digital value creation are 54% in the business segment and 47% in the communication sector. Given a GDP growth of around 1%–2% in the majority of the G20 countries, the growth numbers of the digital sector roughly double the GDP growth: the United States, the United Kingdom, Australia, and Spain had a digital growth of 3%–3.5%, while Germany, the Netherlands, and Brazil have a 2.5% digital growth. China has a digital growth of 6.5%, but starts from just about 10% value creation from digital services to the GDP.

Already in 2013, Det Norske Veritas (DNV) published the "Technology Outlook 2020" report, indicating the "change from passive data to automated processes," and the expectations that automation tools will have a significant share in industrial systems (DNV, 2015). Recent studies by Chui et al. (2016) for McKinsey compared the work situation in the United States, and estimated the probability of automation for these jobs. Three activities, covering more than half of the total working time, have a probability of 64% or higher for automation. As shown in detail in Figure 11.2, data and information collection accounts for 17% of the work time, with 64% probability for automation, and data processing work accounts for 16%.

This change toward automated systems is not limited to single segments of the industrial market; rather, it links the developments toward the grand challenges for Spaceship Earth (DNV-GL, 2016). The grand challenges are driving forces behind the sustainable development goals of Agenda 2030 (SDG 2030) of the United Nations (2015), and are adopted as guidelines for investments. Looking into the opportunities, we see that technology systems will become autonomous, controlling our environments. As an example, India has started the Smart City Initiative, with a total of 60 cities transforming into smart cities, affecting a total of more than 72 million people (Government of India, 2016). The transformative force in these smart cities demands an adequate water supply; an assured electricity supply; sanitation, including solid waste management; efficient urban mobility and public transport; affordable housing, especially for the poor; robust information technology (IT) connectivity and digitalization; good governance, especially e-governance and citizen participation; a sustainable environment; the safety and security of citizens, particularly women, children, and the elderly; and health and education. The majority of these challenges,

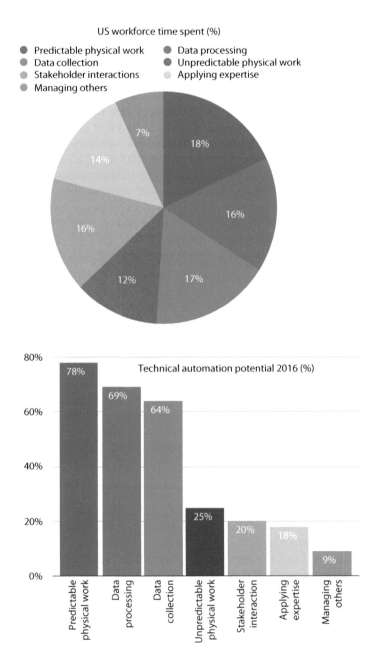

**FIGURE 11.2**
Technology potential of automation for typical duties performed in a working day. (Adopted from Chui, M., et al., Where machines could replace humans—and where they can't (yet), *McKinsey Quarterly*, July 2016.)

especially electricity, urban mobility, IT connectivity, digitalization, and e-governance, will require autonomous solutions with measurable SPD measures in mind.

In the new era of dynamic interaction between businesses, one of the real challenges is the disappearing borders between companies, and the exchange of sensor- and process-based information between entities. Given the second trend of dynamic modeling creating autonomous decisions, one of the big challenges is the lack of measurable security when exchanging information. "Is the information that your system receives from one of the suppliers (or competitors) reliable?" is one of the key questions that you need to answer if your process or business model depends on those data.

Security has impacted the way of making business. We expect that security will become a major factor for the ongoing business evolution cycle, following vertical domains, such as the avionics, health, automotive, industrial, and energy segments.

- In *avionics*, safety was the main driver for developing formalized methods and models for software engineering, needed, for example, for the "fly-by-wire operation" of modern aircrafts. Although this segment leads the developments, the market is somewhat limited. This means that most new helicopters and aircrafts (and in near future, unmanned aerial vehicles and drones) are controlling flight, route, engine, communications, and sensors for navigation (radar, electro-optical, radio, and satellite) by running millions of lines of software. A wide range of airborne equipment is available in today's aeronautical network, for different applications related to the safety of passengers up to an increasing variety of infotainment applications, maintenance tasks on the ground, and taxing and parking in the airport area.

- In the *automotive* segment, modern cars contain a million lines of software code. All that software is executed on 70–100 microprocessor-based electronic control units (ECUs) networked throughout the body of our car (https://spectrum.ieee.org/transportation/systems/this-car-runs-on-code/ [last accessed October 16, 2017]). Software in cars is only going to grow in both amount and complexity. This means that most new cars are executing tens of millions of lines of software code, controlling everything from brakes to the volume of the radio. A wide variety of communication systems is available in today's automotive network, for different applications ranging from body systems, engine control, driving assistance, and safety systems to a wide variety of infotainment applications. These embedded systems are interconnected and can also communicate with the devices of drivers and passengers, as well as other road infrastructure. Original equipment manufacturers (OEMs) are spending lots of money and effort in developing software that can provide the above applications and connectivity services. But the integration of the new embedded systems exposes cars to the Internet, which is susceptible to many other and new vulnerabilities and attacks.

- *e-Health* applications, with the main focus areas on postoperative monitoring, monitoring of chronic diseases, and positive surveillance of elderly people, have a clear demand for privacy and reliability. Current rollouts focus on specific solutions, supplied by a single supplier or a small group of partners in the supply chain. Little is done to ensure interoperability between actors, and privacy protection of personal data.

- Renewable *energy* is one of the drivers for the smart electric grid, with the automatic meter reading (AMR) being the first step in the rollout. Currently, no security is implemented, which we expect will become a subject of discussion once misuse is published.

- Sensors driving automated processes are the upcoming topic for development in the *control and automation industry*. Current applications are mainly based on corporate implementations, where only suppliers and users might share the sensor data. The oil and gas industry has a clear focus on standardized solutions.

The importance of the digital sector, as well as the expected growth of automation, and the interworking between industries, shows the necessity of measurable security. As previously argued, the attack-based solution (see also Chapter 2) is not scalable for the assessment of systems of systems, requiring the need for novel approaches. The SHIELD methodology has been conceived to address this lack of native SPD support in cyber-physical systems and to provide a concrete solution for measurable SPD.

## DEWI Project

The ARTEMIS project DEWI (Dependable Embedded Wireless Infrastructure, 2014–2017) was funded by the European Union (EU) with the vision to provide key solutions for wireless seamless connectivity and interoperability in smart cities and infrastructures, by considering everyday physical environments of citizens in buildings, vehicles, trains, and airplanes (http://www.dewiproject.eu/). The DEWI project consortium was formed by 58 renowned European partners from industry, as well as academia. Within DEWI, wireless solutions were developed in 21 industry-driven use cases coming from four industrial domains (aeronautics, automotive, rail, and building), adding clear interoperability and cross-domain benefits in the area of wireless sensor networks (WSNs) and wireless communication, in terms of reusability of technological building bricks and architecture, processes, and methods.

## DEWI Sensor and Communication Bubble

In DEWI, the concept of a locally adaptable wireless "sensor and communication bubble"—the DEWI bubble—was introduced. Such a DEWI bubble provides locally confined wireless internal and external access; secure and dependable wireless communication and safe operation; fast, easy, and stress-free access to smart environments; flexible self-organization, reconfiguration, resilience, and adaptability; and open solutions for cross-domain reusability and interoperability.

A typical DEWI bubble consists of three main elements: sensor and actuator nodes inside the bubble, bubble gateways providing a secure interface to other bubbles or the outside world, and humans or machines acting as internal (user is part of the bubble) or external (user accesses the bubble using the bubble gateway) DEWI bubble users. The developed DEWI bubble concept allows different communication technologies, such Wi-Fi, ZigBee, and Bluetooth, to interconnect the nodes within a DEWI bubble.

## DEWI Project and Security

The focus of the DEWI project was clearly on dependable wireless short-range communication and its application to different industrial domains. The DEWI bubble concept was utilized to design the data flow within the considered application scenarios: a DEWI bubble (e.g., a WSN within a vehicle) can only be accessed from outside via its DEWI bubble gateway. The latter can apply different authentication mechanisms, to ensure that only authorized entities can access the WSN, as well as encryption mechanisms, to protect sensitive data, which are exchanged between the WSN and other entities outside the DEWI bubble.

The DEWI bubble concept can be considered a basic security concept for dependable embedded infrastructure; however, neither the development of in-depth security mechanisms nor the design of advanced security concepts for embedded systems was a high-priority goal within DEWI. Instead, the DEWI project was aiming to cooperate with other EU projects with a focus on security in embedded systems. By providing security building bricks, as well as tools to perform a detailed system analysis with regard to security, the SHIELD projects were able to satisfy the needs of DEWI.

## DEWI Use Case: Wireless Update of Electronic Control Unit Software for Vehicles

A modern vehicle includes a growing number of ECUs, allowing us to incorporate new features and services. Enabling these features and services, however, requires elaborate and complex software installed on these

ECUs, potentially introducing a growing number of bugs in the automotive software. New concepts allowing efficient automotive software updates are required to fix such software bugs, as well as to enable the installation of new features and support the development and maintenance of modern vehicles.

Today, updating the software installed on ECUs integrated in a car is still a tedious and time-consuming issue and is usually done using a wired connection between the vehicle and a dedicated, as well as expensive, hardware device, a so-called diagnostic tester (DT). Within DEWI, wireless technology is utilized to update the software installed on vehicular ECUs in a safe, secure, efficient, and flexible way. The developed wireless software update system is thereby able to support the entire life cycle of a modern vehicle: from vehicle development, to vehicle assembly, to vehicle maintenance in a service center.

### Wireless Software Update Scenarios

Efficient and secure software updates can be beneficial over the whole life cycle of a modern vehicle, as such updates can significantly reduce the time needed to install the latest ECU software while the vehicle is developed, assembled, and maintained, and will also bring advantages to the customers (i.e., vehicle owners), as they do not need to take their car to the service station to receive the latest software versions (i.e., remote software update scenario).

The wireless software update system developed within DEWI provides solutions for the following scenarios:

1. Wireless software updates in the vehicle development phase
2. Wireless software updates in the assembly line
3. Wireless software updates in a service center (i.e., vehicle maintenance)
4. Wireless remote software updates in public or at home (i.e., connecting the vehicle to a local Wi-Fi network)

In the following, these scenarios are explained in more detail, the involved key users are stated, the environments where the software updates will take place are described, and the main benefits are highlighted.

During vehicle development, engineers will have to update the software of one or more ECUs of a test vehicle several times to test, compare, and evaluate newly developed features and variants. Therefore, the development engineers require a flexible and efficient system enabling wireless software updates, as well as vehicle diagnostics. Vehicle development activities will primarily take place in a restricted environment and will be performed by engineers (i.e., expert users). The wireless solution can be beneficial in this scenario because it allows fast and simple software updates. Updating or exchanging the software on the ECU can be done using a handheld device, like a tablet or a smartphone.

Vehicle assembly is typically performed in a highly automated and secure environment where most working steps are performed by machines and robots. Before a new vehicle can leave the assembly line, the latest software version should be installed on all of its integrated ECUs. This can require that the software of many vehicles be updated—ideally in parallel—to install the latest software on the ECUs of these vehicles. Due to the high number of vehicles, as well as the high degree of automation, scalability, reliability, and efficiency are key aspects of the wireless software update system.

In a service center, mechanics will diagnose, repair, and maintain several vehicles at the same time. To do so, a mechanic will connect a DT to a vehicle to run diagnostic functions, look for diagnostic trouble codes coming from the vehicle and its ECUs, and perform repairs if needed. Thereby the mechanic will also check if new software for an ECU of the car is available and install it. Parallel software updates are very beneficial, especially when large vehicle recalls (e.g., due to critical software bugs) are necessary: a mechanic can connect to several vehicles in parallel and install the latest software simultaneously. Additionally, the wireless solution can be very useful in this scenario because a mechanic will not have to use heavy diagnostic equipment (e.g., a PC and a battery contained in a solid metal case) anymore, but can use a lightweight handheld device to run diagnostics and software updates wirelessly.

Wireless remote updates can be performed either by utilizing a dedicated 3G/4G connection between the vehicle and the vehicle manufacturer (out of the scope of DEWI) or by connecting the vehicle to a local Wi-Fi (e.g., the home network of the vehicle owner). Today (2017) some vehicles already allow a connection to local Wi-Fi networks, and in future, this functionality will most likely be supported by all new vehicles. To install the latest software remotely, the vehicle owner will first connect the vehicle to a local wireless network and then be informed of the possible software updates. In the next step, the user can choose a suitable time slot for the software update to take place, as the vehicle cannot be used while a software update is performed due to safety reasons. Such remote updates are mainly relevant for future vehicles, as they have to be equipped with an integrated wireless interface with full access to the in-vehicle communication system. The main advantages of wireless software updates in this scenario are that (1) the vehicle owner does not have to take the vehicle to a service center anymore to receive new software binaries, and (2) the vehicle manufacturer is able to perform software updates for a whole fleet without large vehicle recalls.

In Table 11.1, the scenarios are summarized, the involved users and their education are stated, the most important properties are listed, and the required security levels are given.

**TABLE 11.1**

Scenarios for Wireless Software Updates

| Scenario | Users Involved | Important Properties | Security Level |
|---|---|---|---|
| Development | Development engineer, expert user | Flexible, efficient | Medium |
| Assembly line | Operator, expert user | Scalable, reliable, efficient | Medium |
| Service center | Mechanic, trained user | Efficient, backward compatibility | High |
| Remote updates | Vehicle driver or owner, untrained user | Reliable, easy to use | Very high |

### Safety and Security as Key Aspects of Wireless Software Updates

In a typical passenger car, dozens of ECUs are used, performing all different kinds of tasks within a vehicle, such as controlling the engine, supporting the driver when braking and steering, and operating the windows or door locks. Some of these tasks are critical to safety, and a system failure can lead to accidents threatening the well-being of the driver, passengers, pedestrians, and other road users in close proximity.

Such a system failure can occur if wrong, incomplete, or malicious software was previously installed on a safety-critical ECU or if the software update progress fails (e.g., due to a connection problem, but also due to an attack) while the new binary is being downloaded to the ECU. Suitable mechanisms have to be used to ensure that software updates are installed in a reliable way. Additionally, security measures have to be applied to prevent an attacker from installing malicious software on an ECU, tampering with the transferred data, or disrupting the software update process.

The wireless software update system developed in DEWI is based on a detailed and comprehensive security concept to avoid the possible attacks mentioned above. As shown in Table 11.1, security is a key aspect in all addressed scenarios. As software updates in the assembly line and in the vehicle development phase are normally done in a restricted and secured area, the required security level is not as high as in a service center, and especially for remote updates. However, strong security mechanisms are still required to avoid related threats, such as industrial espionage.

The security concept developed and used in this DEWI use case is based on a detailed system analysis utilizing the SHIELD multimetrics. In the "System Analysis Utilizing SHIELD Multimetrics" section, the analysis and corresponding results are described.

As already mentioned, a modern vehicle consists of different types of ECUs with, in the case of a failure, different impacts on vehicle safety. To address this in a structured way, an ECU classification was performed within DEWI:

- *Uncritical—Class 1*: ECUs typically related to the entertainment or infotainment system of a vehicle. Hardly any effect on vehicle safety and performance in the case of system failure. Lowest criticality level.

- *Body and comfort—Class 2*: All ECUs related to the body Controller Area Network (CANs) of a vehicle (e.g., window lift or heating, ventilation, and air conditioning). No (direct) influence on vehicle dynamics or driving. Medium criticality level.

- *Powertrain, chassis, driver assistance—Class 3*: Complex and safety-critical systems that control safety features of a vehicle and have a strong impact on vehicle dynamics and driving. (Very) high criticality level.

### Framework for Efficient and Secure Wireless Software Updates for Vehicles: System Overview

Wireless software updates first require a reliable and fast wireless network to transfer the data from the DT to a wireless gateway interface, which is needed to interconnect the wired in-vehicle communication system with the wireless domain. This gateway is then responsible for forwarding the data to the concerned ECU using the wired bus system (e.g., CAN bus) of the vehicle.

The sample architecture presented in Figure 11.3 shows the wireless interconnection of the DT and the wireless gateway interface, the so-called wireless vehicle interface (WVI), as well as the wired connection between the WVI and the ECUs using different bus systems in the vehicle. A central vehicle gateway (CGW) is used to interconnect these bus systems.

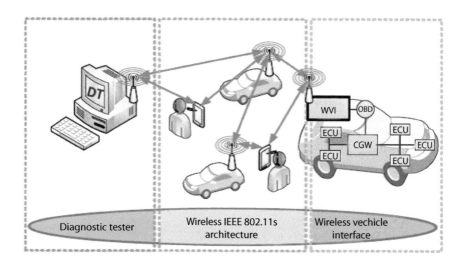

**FIGURE 11.3**
Architecture of secure and efficient wireless software updates.

The WVI can be either fully integrated in the in-vehicle communication system (e.g., a dedicated ECU or a smart gateway with a wireless interface) or realized as a plug-in solution, which can be connected to the vehicle using the onboard diagnostics (OBD) interface of the vehicle, as shown in Figure 11.3.

The DT holds the new binary, a description of the vehicle (e.g., an ODX description file), and the keys required to authorize a (wireless) software update. There is a secure backbone connection between the DT and an OEM, which can be used to get the latest software updates, vehicle descriptions, and sets of keys. This backbone link, however, is out of scope of the following considerations.

In DEWI, the DT is realized as a Java implementation, which can be used on PCs and laptops. Depending on the user and the scenario, the DT can either be directly used to install a new software binary on an ECU or act as a server, only providing the required data and keys. In the latter case, a handheld device, like a tablet or a smartphone, can be used to trigger a wireless software update.

The handheld device can be utilized by a user (e.g., a mechanic in a service center) to diagnose a vehicle; to get further information, such as repair instructions, about the vehicle; and to schedule or trigger wireless software updates. In the case of a software update, the handheld device will connect itself to the DT, as well as to the WVI, and will use these connections to initialize the data transfer. However, the binary data, as well as the required keys, will be exchanged directly between the DT and WVI. Thus, the new binary is neither sent to nor stored on the handheld device, which on the one hand reduces the power consumption, as well as the memory demand of the device, and on the other hand is a security feature, as the new software is never stored on a potentially compromised device. Additionally, an attacker cannot obtain the software binary by stealing the handheld device.

DEWI IEEE 802.11s, a communication protocol based on wireless mesh networks, is used to interconnect a DT and a WVI connected to a vehicle. Contrary to other standards out of the IEEE 802.11 protocol family, where an access point is used to interconnect the nodes of a network, the 11s protocol is based on a mesh network architecture where each node can either directly communicate with other nodes in its transmission range or use other nodes in between to transfer a data packet to its final destination via a multihop route. Due to the mesh characteristics of 11s, a data packet can use different paths when sent through the network. The resulting redundancy increases the reliability and availability of IEEE 802.11s networks, as poor links can be avoided by using another way to the final destination within the network.

### Threats and Security Aspects to Address with Regard to Wireless Software Updates

Wireless software updates, and the WVI in particular, potentially offer an attacker several attack vectors. An attack vector describes different possibilities to attack and finally break a system using weak spots in the security concept or in the utilized protocols or subsystems. In DEWI, different scenarios for wireless software updates were addressed, and corresponding threats had to be taken into account accordingly. The most important issues regarding security in the DEWI wireless software update use case are

- *Vehicle integrity*: Connecting a WVI to a vehicle means adding a wireless interface to the outside world, which potentially can be used to attack the integrity of the vehicle. The WVI and the entire architecture must ensure that only authenticated and authorized nodes can connect the vehicle, run diagnostics, and transfer data to the in-vehicle communication system.

- *Data integrity*: The ability to update the software of an ECU wirelessly can also be used by an attacker to flash malicious software on an ECU (i.e., by tampering with the transferred data). This malicious software can then be used to steal the vehicle, set the vehicle in a dangerous or undefined mode, track the driver (privacy issue), or in the worst case, remotely control the vehicle.

- *Data confidentiality*: The new software binary, which is transferred to the vehicle via the wireless link and installed on the ECU, is owned by the OEM or a supplier. An attacker must not be able to get access to the binary or obtain it by listening on the wireless link (i.e., eavesdropping). Additionally, user-specific data must be secured in a way that the privacy of the user is not endangered all the way through the wireless network.

- *Key exchange and trust*: Strong authentication between the nodes is very important to ensure a secured and trustworthy network. Therefore, different keys are required and must be exchanged as well as stored in a secure way so that an attacker is not able to guess, steal, or reuse such a key.

### System Analysis Utilizing SHIELD Multimetrics

The developed system enabling secure and efficient wireless software updates was described in detail in the last sections. In this section, the performed system security analysis utilizing the SHIELD multimetrics (see Chapters 5 and 9 for more insights) is described and the corresponding results are presented. Based on these results, the final chosen scenario-specific system configurations are explained.

It is important to mention that within DEWI, only system security was analyzed using the SHIELD multimetrics, but dependability and privacy were not. In the "Applying the SHIELD Multimetrics: Challenges and Lessons Learned" section, the reason is described.

*Goal Value Definition for the Considered Scenarios and Different ECU Types*

Wireless software updates will be performed in different phases of the life cycle of a modern vehicle. This will take place in different environments; will be operated, monitored, or triggered by different user types; and will be exploited by different sorts of (security) threats. In the "Wireless Software Update Scenarios" section, the identified scenarios were already described in detail. For the definition of the security goal values, these scenarios have to be examined once again with regard to the environment where the wireless software update will take place and the corresponding system security threats.

- Wireless software updates in the vehicle development phase

    When new systems and functions are developed, they must be tested on a test track in real vehicles before they can be used in series vehicles. A development engineer will therefore update the software of an ECU several times to validate bug fixes and compare the performance of different software versions.

    - *Environment*: Test tracks and company areas are restricted areas, the users are experts, and notebooks (plus dedicated hardware, if necessary) are used.

    - *Threats*: Private usage of company equipment to run unauthorized software updates (tuning and malicious software).

- Wireless software updates in the assembly line

    In a typical assembly line scenario, a lot of similar or equal vehicles in close proximity will be produced. All these vehicles will potentially receive the same software. By utilizing wireless software updates, these required software updates can be done in parallel.

    - *Environment*: Restricted and high-security area, high degree of automation and expert users, dedicated hardware

    - *Threats*: Low risk because it is really hard for attackers to enter the restricted area

- Wireless software updates in a service center

    In this scenario, one or several ECUs of a vehicle are updated and diagnostic functions are used to find and fix any kind of problem with to the vehicle.

    - *Environment*: Most parts of a typical service center are restricted areas; however, it is not unusual for customers to also access these parts from time to time. The users working in a service center are trained professionals, and dedicated hardware is used. In a big service center, different brands and all types of vehicles will be processed.

- *Threats*: Service center equipment can be stolen and used to run unauthorized software updates (tuning and malicious software). Additionally, an attacker can try to steal ECU software by listening to the wireless channel.

- Wireless remote software updates in public or at home

    In this scenario, the software of an ECU is updated by connecting the vehicle to the home network and using the Internet link to connect to the OEM backbone.

    - *Environment*: Done in public or at home using low-security networks (insecure environment), the users are not trained, and standard and potentially compromised hardware is used.

    - *Threats*: Attackers can exploit the low-security environment (e.g., using a malicious app on the smartphone of the user) to trespass the network. With that, the software binary can be easily stolen or, even worse, changed on the way to the ECU to install malicious software.

Based on the description of the scenarios above and in the "Wireless Software Update Scenarios" section, and the classification of different ECU types presented in the "Safety and Security as Key Aspects of Wireless Software Updates" section, the security goal values for each of the addressed scenarios were defined. The security goal values (i.e., values between 0 and 100) are presented in Table 11.2.

*Subsystems and Components of the Wireless Software Update System*

To analyze the system with regard to SHIELD SPD, the system must be divided into subsystems and components (described in more detail in Chapters 4 and 5), and for each of these components, different system parameters can be identified.

In DEWI, the subsystems were first identified by dividing the system architecture into three logical building blocks. Each block shown in Figure 11.4 represents a subsystem of the developed wireless software update system:

- *Subsystem 1*: Wireless IEEE 802.11s architecture
- *Subsystem 2*: DT including the OEM backbone
- *Subsystem 3*: WVI and the vehicle itself

**TABLE 11.2**

Goal Values for Each Scenario and ECU Type

| ECU Type | Service Center | Vehicle Development | Remote Update | Assembly Line |
|----------|----------------|---------------------|---------------|---------------|
| Class 1  | 50             | 60                  | 75            | 50            |
| Class 2  | 60             | 60                  | 80            | 55            |
| Class 3  | 70             | 60                  | 90            | 60            |

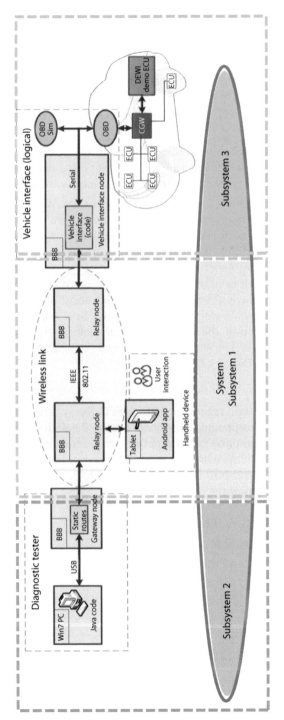

**FIGURE 11.4**
Identified subsystems of the wireless software update system.

**TABLE 11.3**

Subsystems and Corresponding Weights

| Subsystem | Weight within Subsystem (0–100) | Normalized Weight (0–1) |
|---|---|---|
| WVI subsystem | 70 | 1 |
| Wireless architecture subsystem | 60 | 0.73 |
| DT and OEM backbone subsystem | 50 | 0.51 |

**TABLE 11.4**

Components of the DT Subsystem

| Component | Weight within Subsystem (0–100) | Normalized Weight (0–1) |
|---|---|---|
| Data encryption | 90 | 1 |
| Network authentication | 90 | 1 |
| Master key exchange | 75 | 0.69 |
| Session key | 75 | 0.69 |
| Discovery strategy | 40 | 0.20 |
| Beacon period | 20 | 0.05 |
| Secure routing | 50 | 0.31 |
| Wireless channels | 30 | 0.11 |

According to the SHIELD multimetrics approach, a weighting on the subsystem level was performed. The weights give information about the impact of a subsystem on the overall system security and should be in a range of 0, meaning no impact at all, to 100, highest impact factor. In the next step, the weights have to be normalized to a range between 0 and 1, where the subsystem with the highest weight will be assigned to a normalized weight of 1. This weighting scheme is also applied on component level.

The weightings of the identified subsystems of the developed system enabling secure and efficient software updates can be found in Table 11.3. The WVI subsystem has the highest weight because this component is very critical with regard to the security and safety of the entire vehicle.

In the next step, the components of the identified subsystems can be extracted, and again, a weighting of these components is performed. In Tables 11.4 through 11.6, the components and the corresponding weights are presented. A more detailed description of the components of the WVI subsystem and some of its system parameters is given in the next section.

*Analyzing the Wireless Vehicle Interface Subsystem*

The WVI serves as a secure gateway interface between the wireless IEEE 802.11s network and the wired in-vehicle communication system. It is responsible for performing a strong authentication between the DT and the vehicle, will receive and temporarily store the new software binary, and have

**TABLE 11.5**

Components of the IEEE 802.11s Architecture

| Component | Weight within Subsystem (0–100) | Normalized Weight (0–1) |
|---|---|---|
| WVI type | 90 | 1 |
| Key storage | 70 | 0.60 |
| Binary storage | 20 | 0.05 |
| Reflashing WVI | 20 | 0.05 |
| Data verification | 40 | 0.20 |
| ECU supports encryption | 50 | 0.31 |
| Intrusion detection | 30 | 0.11 |
| Authentication | 60 | 0.44 |
| Session duration | 40 | 0.20 |
| Backup old binary | 10 | 0.01 |

**TABLE 11.6**

Components of the WVI Subsystem

| Component | Weight within Subsystem (0–100) | Normalized Weight (0–1) |
|---|---|---|
| Backbone connection | 75 | 1 |
| Logging | 25 | 0.11 |
| User authentication | 70 | 0.87 |
| Authorization (for updates) | 40 | 0.28 |
| Modes of operation | 50 | 0.44 |
| Session duration | 45 | 0.36 |

to control as well as monitor the process of transferring and finally installing the software binary on the ECU. Therefore, the WVI has to support diagnostic standards and protocols, such as Unified Diagnostic Service (UDS) or the Universal Measurement and Calibration Protocol (XCP). In DEWI, a (reduced) UDS stack was implemented on the WVI to be able to successfully perform a (wireless) software update according to a diagnostic standard.

In this section, some interesting components of the WVI subsystem and their system parameters are discussed. An overview of these components and system parameters is given in Tables 11.7 through 11.10. For each component, a brief description is given and its system parameters are listed. In each line, one system parameter is briefly described and the corresponding security as well as criticality level is presented.

The management and storage of secret keys is a very important issue with regard to system security. If an attacker is able to find a secret key, he or she will often be able to impair an entire system. To authenticate with the DT and encrypt or decrypt data, the WVI will have to permanently (e.g., a private key required to authenticate with the DT) as well as temporarily (e.g., a symmetric session key used to encrypt or decrypt data) store secret keys.

**TABLE 11.7**

Key Storage as System Parameter of the WVI Subsystem

| Component/ Parameter Name | Security Value | Criticality Value | Description |
|---|---|---|---|
| **Key storage** | | | **Describes different ways of storing keys on the WVI** |
| Software | 40 | 60 | Stored in the software running on the WVI |
| Normal memory (plain) | 30 | 70 | Stored in normal flash memory (plain) |
| Normal memory (encrypted) | 75 | 25 | Stored in normal flash memory (encrypted) |
| Secure memory | 80 | 20 | Stored in a secure memory |
| HSM | 90 | 10 | HSM is used to store the keys |

**TABLE 11.8**

Intrusion Detection as System Parameter of the WVI Subsystem

| Component/ Parameter Name | Security Value | Criticality Value | Description |
|---|---|---|---|
| **Intrusion detection** | | | **The WVI is able to detect attacks** |
| Shutdown, special restart | 90 | 10 | WVI will shut down and restart in a predefined (secure but limited/restricted) mode |
| Shutdown, reauthentication | 75 | 25 | WVI will shut down and reauthentication is needed afterwards |
| Block packets | 70 | 30 | Block packets coming from the attacker |
| No | 50 | 50 | Not possible at all |

In Table 11.7, the component *key storage* and its system parameters are presented. A developer has different ways to realize the key storage. A very easy but insecure way is to "store" a key in software by assigning the key bytes hard-coded to a variable or using a define directive. Instead, using a hardware security module (HSM) or a trusted platform module (TPM) is very secure way to store keys. An HSM can permanently store a private key (e.g., an ECC [Elliptic Curve Cryptography] or RSA [Rivest, Shamir, and Adleman] key), and an attacker will not have a chance to find the keys even if he or she has physical access to the WVI or HSM, respectively. Table 11.7 also represents this: storing the keys directly in software is very insecure, and therefore has a high criticality and, consequently, a low security value. In SHIELD, the criticality level (CL) can be transferred to a security level (SL) by calculating

$$SL = 100 - CL$$

**TABLE 11.9**

Authentication as System Parameter of the WVI Subsystem

| Component/ Parameter Name | Security Value | Criticality Value | Description |
|---|---|---|---|
| **Authentication** | | | **How authentication is handled by the WVI** |
| Physical access | 75 | 25 | Network-based authentication *and* the user of the DT must have physical access to the inside of the vehicle (e.g., press a button) |
| Location | 70 | 30 | Network-based authentication *and* the DT must be close to or inside the vehicle for authentication |
| Network | 50 | 50 | The authentication process of the DT and the WVI is only handled in the 11s subsystem |
| Multilayer authentication | 65 | 35 | Authentication mechanisms are used on the network as well as on the application layer |
| Multilayer plus physical access | 80 | 20 | Multilayer authentication *and* physical access to the vehicle is required for a successful authentication |

**TABLE 11.10**

WVI Type as System Parameter of the WVI Subsystem

| Component/ Parameter Name | Security Value | Criticality Value | Description |
|---|---|---|---|
| WVI type | | | **The WVI can either be a plug-in solution (e.g., via OBD) or a fully integrated solution (as ECU)** |
| Integrated | 85 | 15 | WVI as ECU: Part of the in-vehicle communication system. Will just work within one vehicle. |
| Plug-in raw | 45 | 55 | Plug-in device without location information. |
| Plug-in location | 75 | 25 | Plug-in device using location information as additional security layer (e.g., must be used within a service center). |
| Plug-in trusted node | 75 | 25 | Plug-in device as part of the trusted network. Will just work in connection with a trusted DT. |

Utilizing an HSM is a very secure way to store keys. Therefore, this system parameter, also shown in Table 11.7, has a low criticality level and, consequently, a high-security value.

Intrusion detection is another important feature of a WVI, as it actively monitors its own state, as well as the incoming data streams. If an anomaly, such as an attacker tampering with the data transferred via the IEEE 802.11s network, is detected, the WVI will take suitable countermeasures accordingly. In Table 11.8, possible countermeasures are presented and for each system parameter, the security as well as criticality values are shown.

One of the most important tasks of the WVI is to authenticate with the DT and, in some cases, also with other handheld devices of the involved users. The authentication must be strong to ensure that only trusted devices can connect to the WVI and have access to the vehicle and its in-vehicle communication system.

As already described in the "Framework for Efficient and Secure Wireless Software Updates for Vehicles: System Overview" section, the WVI can be realized as a plug-in device that can be connected to the vehicle using the OBD interface, or it can be fully integrated in the in-vehicle communication system. In the latter case, the WVI is realized as a dedicated ECU or a smart gateway and is hidden inside the vehicle. Because of that, an attacker will have a hard time manipulating or physically accessing the WVI, and thus the security value of the system parameter *integrated* of the WVI subsystem component *WVI type* is higher than a plug-in WVI type. This is also shown in Table 11.10, where all system parameters of the WVI subsystem component *WVI type* are listed.

For each component of the WVI subsystem, different system parameters have been identified within DEWI. A system parameter can be seen as a possibility to fulfill the needs of a component, and one possibility can be more or less secure compared with another one. A developer can now choose a system parameter for every component, thereby creating a possible configuration for the subsystem and—by doing the same for all other subsystems—for the entire system. A sample configuration for the WVI subsystem is shown in Table 11.11. For each component of the subsystem, one system parameter has been chosen.

The SHIELD multimetrics approach can now be used to compute the security score of the WVI subsystem. Refer to Chapter 5 for more information on

**TABLE 11.11**

Computed Security Values for a Sample Configuration

| Parameter | Weight | Normalized Weight | Configuration: Chosen Parameter | Criticality | Security Value | Weighted Parameter |
|---|---|---|---|---|---|---|
| WVI type | 90 | 1 | Plug-in raw | 55 | 45 | 1016.70 |
| Key storage | 70 | 0.60 | HSM | 10 | 90 | 20.33 |
| Binary storage | 20 | 0.05 | Never | 5 | 95 | 0.41 |
| Reflashing WVI | 20 | 0.05 | DT | 50 | 50 | 41.49 |
| Data verification | 40 | 0.20 | Hash | 25 | 75 | 41.49 |
| ECU supports encryption | 50 | 0.31 | Yes, all the way | 10 | 90 | 10.37 |
| Intrusion detection | 30 | 0.11 | Shutdown, special restart | 10 | 90 | 3.73 |
| Authentication | 60 | 0.44 | Location | 30 | 70 | 134.44 |
| Session duration | 40 | 0.20 | Timeout | 15 | 85 | 14.94 |
| Backup old binary | 10 | 0.01 | No | 50 | 50 | 10.37 |

**TABLE 11.12**

Overall Security Score of the WVI Subsystem

| Sum of weights | *Sum(NW)* | 2.98 |
|---|---|---|
| Criticality of WVI subsystem C_WVI | *Root(Sum(WP))* | 35.98 |
| Security score of the WVI subsystem | *100—C_WVI* | ~64 |

*Note:* NW, normalized weight; WP, weighted parameter.

the formulas used and mathematical background of the SHIELD multimetrics. In Table 11.12, the security score for the sample configuration shown in Table 11.11 is calculated. The sample subsystem configuration of the WVI has reached a security score of approximately 64.

*Identifying Suitable System Configurations*

Varying all system parameters of a system, like the wireless software update system, will lead to hundreds of different system configurations; however, a system engineer would choose the parameters in such a way that they perfectly fit the requirements of a scenario. Thus, he or she will not end up with hundreds of possible system configurations, but only a few appropriate system configurations geared to the specific needs of scenario. In this concrete wireless software update use case, for example, the system parameter chosen for the component *WVI type* will be different, depending on the considered scenario: while the service center will require a plug-in device, as this ensures backward compatibility and OEM independence, the remote software update scenario will require a fully integrated WVI, as it is unlikely that a plug-in device will be developed and supported by an OEM.

In the next step, the SHIELD multimetrics can help a developer or system engineer find the best-fitting system configuration for each addressed scenario, as it provides formulas to compute a security score for each identified and appropriate system configuration. These security scores can then be easily compared with each other, as well as with the predefined security goal value, and the most suitable system configuration for each scenario can be chosen.

*Suitable System Configuration for the Service Center Scenario*

In DEWI, possible system configurations for all addressed scenarios were identified and the most suitable system configurations per scenario were selected by means of the SHIELD multimetrics.

In this section, the final system configuration for the service center scenario is presented. The service center system configuration shown in Table 11.13 contains the three subsystems of the wireless software update system, all its components, and the chosen system parameters per component. Additionally, the security value of the system parameters, as well as the weight of each component, is given.

**TABLE 11.13**

System Configuration for the Service Center Scenario

| Configuration | | | Security Value | Weight |
|---|---|---|---|---|
| **WVI** | WVI type | Plug-in trusted node | 75 | 90 |
| | Key storage | Secure memory | 80 | 70 |
| | Binary storage | Session | 75 | 20 |
| | Reflashing WVI | Not possible | 95 | 20 |
| | Data verification | Hash | 75 | 40 |
| | ECU supports encryption | No | 50 | 50 |
| | Intrusion detection | Shutdown, special restart | 90 | 30 |
| | Authentication | Physical access | 75 | 60 |
| | Session duration | Duration | 70 | 40 |
| | Backup old binary | No | 50 | 10 |
| | **Subsystem** | | **70** | **70** |
| **IEEE 802.11s architecture** | Data encryption | Strong | 75 | 90 |
| | Network authentication | Authsae | 75 | 90 |
| | Master key exchange | Once per NFC | 60 | 75 |
| | Session key | Per connection with timeout | 80 | 75 |
| | Discovery strategy | DT: Advertisement beacon | 80 | 40 |
| | Beacon period | 5 s | 70 | 20 |
| | Secure routing | Normal multihop | 50 | 50 |
| | Wireless channels | Channel switch | 75 | 30 |
| | **Subsystem** | | **70** | **60** |
| **DT and OEM backbone** | Backbone connection | Internet over trusted local area network | 70 | 75 |
| | Logging | User-specific | 80 | 25 |
| | User authentication | NFC-Tag+PIN | 80 | 70 |
| | Authorization (for updates) | Active backbone connection | 85 | 40 |
| | Modes of operation | User profile | 75 | 50 |
| | Session duration | 1 h with inactive timeout | 80 | 45 |
| | **Subsystem** | | **76** | **50** |
| **Overall system score** | | | **71** | |

The calculated security score and the weight of each of the identified subsystems, as well as the overall security score of the system, are also included in Table 11.13. As mentioned and presented above, the SHIELD multimetrics were used in the first step to calculate the security score on the subsystem level. In the second step, these intermediate results can be used to calculate the overall security score for the entire wireless software update system. Before these results are discussed in detail, some selected system parameters that are dedicated to the service center scenario are discussed.

In a service center, potentially, vehicles of different types, brands, and ages will be maintained. Because of that, the properties *OEM independence* and *backward compatibility* are very important in the service center scenario, and as mentioned in the "Analyzing the Wireless Vehicle Interface Subsystem" section, a plug-in WVI device will be used to connect to the vehicle. *Plug-in trusted node* means that all plug-in WVI nodes, used within a service center, will be paired with the DT and the handheld devices before they can be used. This pairing process, based on near field communication (NFC), will also be used to exchange the master key required by the IEEE 802.11s architecture subsystem: the component *master key exchange* and its corresponding system parameter *once per NFC*.

In a typical service center, several mechanics will maintain vehicles and perform repairs. These mechanics will potentially have different privileges when using the wireless software update system. A new mechanic can probably only use the system to run wireless diagnostics on a vehicle but will not be allowed to perform a wireless software update (e.g., as this requires specific training). Experienced and well-trained mechanics will be able to perform wireless software updates and diagnostics. The head of the service center will additionally be allowed to add new devices (e.g., pairing a new WVI with the DT of the service center) and maintain the user profiles.

According to the results presented in Table 11.13, the system configuration for the service center scenario leads to a resulting security score of 71. This implies that the designed and developed system will be secure enough to fulfill the security requirements of the scenario for all types of ECUs, as the maximum goal value defined in the "Goal Value Definition for the Address Scenarios and Different ECU Types" section is 70 (for a Class 3 ECU). The subsystem results additionally show that the overall security score can be further improved, mainly by increasing the security level of the WVI, as this subsystem has the highest weight and the lowest security score.

*System Configuration Reaching the Highest Security Score*

A nice feature of the SHIELD multimetrics approach is that it can also give information about the highest security score that an analyzed system can reach. Therefore, a system engineer will choose the system parameter with the highest security value for each component and each subsystem and then apply the SHIELD multimetrics. The resulting overall system security score will give the engineer an upper limit for the system with regard to security.

In DEWI, this upper reachable limit was also analyzed and the corresponding results can be found in Table 11.14. The results presented in this table show that a maximum security score of 80 can be reached. According to the defined goal values in Table 11.2, in the "Goal Value Definition for

**TABLE 11.14**

System Configuration Reaching the Maximum Security Score

| Subsystem | Weight | Security Value/Score |
|---|---|---|
| WVI | 70 | 84 |
| IEEE 802.11s architecture | 60 | 74 |
| DT and OEM backbone | 50 | 83 |
| **Overall and maximum security score** | | **80** |

the Address Scenarios and Different ECU Types" section, nearly all security goals can be reached; however, a wireless remote software update for Class 3 ECUs will not be possible using the designed system. This means that a vehicle owner will have to take his or her car to a service center to install a new software binary on a Class 3 ECU used for powertrain, chassis, and driver assistance tasks, due to safety reasons.

## Applying the SHIELD Multimetrics: Challenges and Lessons Learned

Within DEWI, the SHIELD multimetrics were applied to real industrial use cases outside the SHIELD projects for the very first time. Following the single steps of the SHIELD multimetrics approach for the given examples was quite straightforward, and it was quite easy to get some preliminary results within DEWI. However, as a more detailed analysis of the wireless software update use case was performed, first problems and issues arose, but also some unforeseen advantages were found. In the following, these issues are discussed in more detail to help people willing to use the SHIELD multimetrics to have a smooth start and to know how to potentially tackle or avoid arising issues.

### How to Work with the Mixed-Term Dependability

Increasing the dependability of systems and (wireless) communication protocols was one of the main aspects within DEWI, and therefore the SHIELD multimetrics, able to analyze SPD at once, seem to be the perfect match. However, due to the mixed-term nature of dependability—depending on the definition used, dependability is a term combining reliability, maintainability, availability, and some more—a comprehensible evaluation was not feasible within DEWI, as it is hard to give score values for parameters with regard to dependability, as some of its attributes can be contrary: increasing the reliability and safety of a system (e.g., by maintaining an elevator every day) can decrease the availability of the system (e.g., the elevator cannot be used during maintenance).

This issue in not yet solved in the SHIELD methodology.

### Privacy as Part of Security in DEWI

In the first months of DEWI, all industry-driven use cases were analyzed in detail to identify their system requirements. These requirements were then collected, categorized, and analyzed. The results of this analysis revealed that only three requirements out of 90 collected security and privacy requirements were dedicated to privacy issues, and that these three requirements were still quite related to security; in summary: if user-specific data are transferred wirelessly, an attacker can try to eavesdrop on it, thereby compromising the privacy of the user. If strong encryption is used, the attacker will not have a chance to access the data and the privacy issue is solved.

Therefore, privacy was treated as a part of security with regard to the performed security analysis within DEWI.

### Common Understanding on How to Choose Security Values and Security Goals

Large distributed, cyber-physical systems are often designed and implemented by several engineers working in different departments or even different companies. The same is true for systems developed in DEWI, where engineers and researchers of industrial partners, research companies, and universities distributed all over Europe were working together.

When analyzing these developed systems with regard to system security, the gathered results, as well as the process of the analysis itself, must be understandable, replicable, comprehensible, and comparable. In other words, any DEWI partner needs to understand what a chosen security goal, security score, or security value means and what impact it will have on the system.

After analyzing the first DEWI use case using the SHIELD multimetrics, these preliminary results were presented to all partners at a DEWI general assembly meeting. Most attendees felt quite positive about SHIELD's structured approach to analyze the system security; however, some were wondering about the numbers chosen for the security goals, as well as the security values, and a clear need to foster a common understanding by utilizing different scales, and for textual guidance was identified.

### DEWI Scale to Assign Security Goal Values

One important step of the SHIELD multimetrics approach is to predefine a security goal value for each scenario. These goal values can be used for the most suitable system configurations by comparing the results of the security analysis (i.e., the security score of the system) with the predefined security goal values. The goal values can also be used to express which security level is needed for a certain application.

To support users of the SHIELD multimetrics in choosing these goal values, a linear scale was developed and some characteristic values were defined

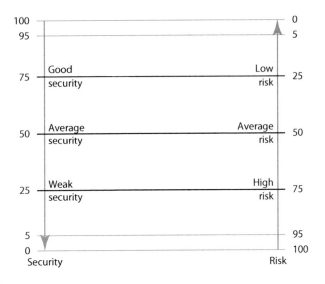

**FIGURE 11.5**
Defined scale helping a user to choose the security goal values for a certain scenario.

(Figure 11.5). For each value, the corresponding security level is described in Table 11.15. These descriptions focus on the attacker abilities and available resources:

- *Attacker level*: Who is able to attack the system? A hobby attacker using Google to find instructions for the attack, or a team of security experts trying to break into the system?
- *Used hardware*: Do the attackers use a normal PC or a supercomputer?
- *Time needed*: Will it take seconds, hours, or days to break the system?
- *Access to the car*: Is access to the in-vehicle communication system needed for the attack (the attacker must be inside the car)?

After presenting the scale shown in Figure 11.5 and the descriptions listed in Table 11.15, the acceptance within DEWI was increased significantly, as partners were able to understand why a specific goal value was chosen and what such a level means with regard to system security and security threats.

*DEWI Scale to Assess System Parameters*

To assess system parameters, a different scale is needed. This time, the criticality of a parameter is used for the assessment instead of the security level of a system parameter. As shown in Figure 11.6, a nonlinear scale for the assessment is utilized, as it better fits the human perception of criticality:

**TABLE 11.15**

Security-Level Descriptions for the Goal Value Scale

| Security | | Description | Criticality | |
|---|---|---|---|---|
| 100 | Fully secure | No chance to break the system even for a team of experts with infinite resources. | No risk | 0 |
| 95 | Very strong security | Very hard for a team of experts with hardware above SotA to break the system. | Very low risk | 5 |
| 75 | Good security | Experts need time *and* access to the vehicle to break the system. No chance for nonexperts. | Low risk | 25 |
| 50 | Average security | Experts need time *or* access to the vehicle to break the system. Nonexperts need time *and* access. | Average risk | 50 |
| 25 | Weak security | Nonexperts can break the system in affordable time *or* with access to the vehicle. | High risk | 75 |
| 5 | Very weak security | Breaking the system with state-of-the-art (SotA) hardware after short web research possible for anyone. | Very high risk | 95 |
| 0 | No security at all | Open system: No effort to break the system for anyone. "The key is in the ignition lock." | Open | 100 |

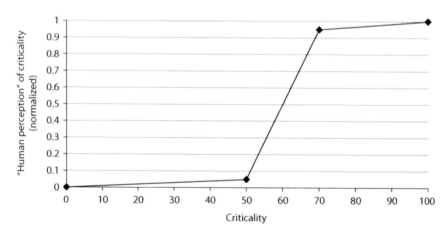

**FIGURE 11.6**
DEWI criticality scale for system parameter assessment.

- *0–50*: Normal, operational mode. In this mode, the system operates in the normal and expected mode. System faults (because of attacks) are unlikely.
- *50–70*: Critical mode. In this mode, the system is still up and running, but a small problem (e.g., a small delay or an unexpected message) can lead to a system fault.
- *70–100*: Failure mode. In this mode, the behavior of the system is unpredictable and faults (attacks) can happen any time.

Again, user acceptance within DEWI was increased after presenting and using this scale, as partners were now able to understand how system parameters were assigned, what a specific values means, and what impact a switch from one configuration to another can have on the (sub)system security score.

### Smart Framework for SHIELD Multimetrics

A system analysis based on the SHIELD multimetrics is a good approach, as its offers a clear, structured, and comprehensible way to perform a detailed security analysis. A system engineer will just have to follow the described steps: defining the goal values, dividing the system into subsystems and components, identifying the system parameters, and finally, applying the SHIELD formulas to compare different system configurations.

Currently, these steps are done within one or several Excel sheets, which is of course a good way to start; however, it also brings some problems. One issue that was found within DEWI was that it is hard to depict dependencies between single components. Table 11.16 contains three sample metrics or components that can be used to get the dependency issue across. Data transmitted via a communication channel can be sent unencrypted

**TABLE 11.16**

Sample Metrics to Highlight the Potential Issues with Dependencies between Components

| Metrics/Components of Communication Channel | System Parameters |
| --- | --- |
| Metric 1 **Data encryption** | Yes, Advanced Encryption Standard (AES) |
| | No data encryption used |
| Metric 2 **Encryption key length** | 128 bits |
| | 256 bits |
| Metric 3 **Validity of the encryption key** | Permanent |
| | Key per session (e.g., valid for 1 h) |

or encrypted (metric 1). If the data are sent in an encrypted way, the key length (metric 2) can be chosen—dependency 1—and will of course have an impact on the security score of a subsystem (i.e., communication channel). This issue can probably be fixed by merging the two metrics into one, leading to three system parameters: "No data encryption use," "Yes, AES with a 128-bit key," and "Yes, AES with a 256-bit key." However, metric 3 also has a dependency to metric 1, as the validity of a key is only relevant if encryption is used. As metric 3 will definitely influence the security score of the subsystem—even if encryption is not used—the system engineer "playing" with a system configuration in Excel will have to remove metric 3 if encryption is disabled.

However, a smart framework for applying the SHIELD multimetrics will probably (and hopefully) be able to depict such dependencies and therefore solve this issue.

### SHIELD Multimetrics and the New Security Standard SAE J3061

The Society of Automotive Engineers (SAE) recently released a new automotive standard called SAE J3061: "Cybersecurity Guidebook for Cyber-Physical Vehicle Systems."* This standard includes best practices intended to be flexible, pragmatic, and adaptable in their further application to the automotive industry, as well as to other cyber-physical vehicle systems.

The described methods and procedures are very similar to those in the well-established and well-known functional safety standard ISO 26262.† The characteristic V-model defined in ISO 26262 is also used in SAE J3061 as the overall cybersecurity (CS) process framework. The latter can be tailored and utilized within each organization's development processes to incorporate CS into cyber-physical vehicle systems, from concept phase, through production, operation, and service, to decommissioning.

The concept phase described in SAE J3061 is based on a flow of activities, and all required steps of this important phase are listed. Three very interesting steps are

19. Performing the threat analysis and risk assessment (TARA)
20. Defining the CS concept
21. Extracting the functional CS requirements

TARA is used to identify and assess the potential system threats and, furthermore, to determine the risk associated with each identified threat. The so-called *highest-risk potential threats* are then used to determine the CS goals

---

* SAE, SAE J3061: Surface vehicle recommended practice—Cybersecurity guidebook for cyber-physical vehicle systems, Technical Report, SAE International, Warrendale, PA 2016.
† ISO, ISO 26262: Road vehicles—Functional safety—Part 1: Vocabulary, Technical Report, ISO, Geneva, 2011.

of the system to develop, and then a CS concept can be elaborated. In the next step, the functional (high-level) requirements can be extracted based on the developed concept and the determined CS goals. SAE J3061 recommends an iterative process to further refine the results of the single steps: start with high-level requirements in the first iteration and refine these requirements in later iterations to finally obtain technical requirements.

The SAE standard provides an abstract, high-level description of the mentioned steps, as well as the expected results. However, no guidance on how to fulfill the single steps or how to obtain the required results is given (except a possible workflow in the appendix of the standard).

Within DEWI, the applicability of the SHIELD multimetrics to support the SAE J3061 standard was shown.* The corresponding publication illustrates how the SHIELD multimetrics can be used to obtain a secure system configuration, starting from the SAE J3061 CS goals. Furthermore, the results of a structured system security analysis can be used as a solid basis for creating the CS concept, as well as for extracting the related CS requirements.

In the following, the required steps defined in SAE J3061 are stated, the output of each of these steps is listed, and a brief description of how the single steps are supported using the SHIELD multimetrics is presented.

After initiating the CS life cycle (beyond the scope of the work performed in DEWI), the TARA can be performed according to the definition in the SAE J3061 standard to identify and assess potential threats. The output of this step is a list of the *highest-risk potential threats*. These threats can then be used to determine the CS goals of a cyber-physical system, and these goals can then be mapped to the DEWI security goal descriptions defined in the "DEWI Scale to Assign Security Goal Values" section. This mapping is illustrated in Table 11.17.

**TABLE 11.17**

Mapping of the SAE J3061 CS Level to the DEWI Scale for Defining Goal Values

| DEWI Security Level | | SAE J3061 Security Level |
|---|---|---|
| 90 | | Critical |
| 75 | Good | High |
| 50 | Average | Medium |
| 25 | Weak | Low |
| 10 | | QM |

---

* M. Steger, M. Karner, J. Hillebrand, W. Rom, and K. Romer, A security metric for structured security analysis of cyber-physical systems supporting SAE J3061, in *CPS Data—Second International Workshop on Modeling, Analysis and Control of Complex Cyber-Physical Systems*, Vienna, April 11, 2016, pp. 1–8.

In the next step, the SHIELD multimetrics approach can be used to perform the system analysis. The analysis will result in a secure system configuration for each scenario of a specific use case.

Further, these system configurations can be used to define a security concept, as well as to extract the security-related requirements of the system. A very simple way to extract these requirements is to determine the minimum system configuration (a system configuration that barely fulfills the predefined security goals) and transform the components and the chosen system parameters into requirements.

## ATENA

After the successful results of the pSHIELD and nSHIELD initiatives, a new project (that will prosecute these studies) has been funded by the European Commission under Grant Agreement No. 700581, named ATENA: Advanced Tools to Assess and Mitigate the Criticality of ICT Components and Their Dependencies over Critical Infrastructures. The main objective of the ATENA project is to improve efficiency and resilience capabilities of modernized critical infrastructures (CIs) and their related industrial automation control systems (IACSs), against a wide variety of cyber-physical threats, whether malicious attacks or unexpected faults, which may affect the whole IACS, corporate networks, or simple information and communication technology (ICT) devices.

ATENA will increase operator and customer awareness, thus facilitating a multiservices approach introducing new paradigms and architectures for their control systems. Nowadays, a single IACS serves a single CI, by means of a centralized approach, and relies on Internet Protocol (IP)–based communication links, connected to the Internet via a corporate network with lower cyber protection than IACSs. ATENA will help in understanding a new generation of IACSs, more distributed and interoperable (for smart metering purposes, distributed generation and storage, etc.), and will help them face cybersecurity issues raised by such innovations.

In particular, the results of the ATENA project will be gained throughout five incremental steps:

1. An extensive *comprehension* of CIs for the extraction of ATENA functional requirements and basic validation issues. The identification of adequate resilience and efficiency indicators is the basic step to proposing and developing a common modeling approach to assess, predict, and control their resilience and efficiency.

2. The *building of models* at an adequate level of granularity, to predict IACS and CI efficiency under cyberattack use cases, also relying on the paradigm of hybrid modeling. Such models will include asset

dependency models, allowing us to predict (and simulate) the risk propagation from one asset to depending assets.

3. The *design* of reaction strategies able to reach the required security and resilience, based on the concept of situation assessment reconfiguration.

4. The *development* of new ICT devices of IACSs, new cybersecurity solutions, secure information-sharing devices, risk predictors, and decision support systems (DSSs) that will improve CI resilience and efficiency.

5. The *introduction* of network function virtualization (NFV) component appliances, distributed across geographically dispersed infrastructure points of presence, such as national or regional data centers or fog computing nodes.

The project will leverage the SHIELD approach to promote composable security in CI, that is, to propose a formal way to build, deploy, and operate CI so as to ensure a desired security level against any potential threat. This is particularly challenging because the SHIELD methodology has proved to be effective in ICT and the embedded system–oriented domain: if its validation is also successful in the cyber-physical scenarios proposed by ATENA, the methodology could be considered mature enough to push for a real standard.

In order to integrate the SHIELD composability mechanism with more dynamic countermeasures specific for the CI domain, the ATENA concept has been split into two logical levels, with the design of two different "control loops":

1. A *static, off-line, long-term control loop* that is in charge of configuring the system according to the security assessment, and updating it on a periodic basis or when particular events require (new threats or discovery of previously undetected vulnerabilities)

2. A *dynamic, on-line, short-term control loop* that is in charge of promptly reacting to attacks and threats that may impact the operational life of the system

The *off-line loop* collects a set of *context information* (i.e., high-level information about the CI architecture, available technologies, countermeasures, and potential threats); on the basis of such information and of the *desired S/R* (security and resilience) level and *desired context* (parameters to be optimized other than the S/R level), a *secure configuration* is enforced on the CI by proper *configuration commands*. In addition to the achievement of the desired S/R level, the selected configuration may also endorse ancillary objectives (e.g., desired context optimization) and is derived in compliance with current standards (e.g., Common Criteria). The selected configurations are stored together with the associated context information in a dedicated *knowledge base*.

The *on-line loop* continuously monitors the CI system, based on the relevant *measurements,* as well as on the information stored in the *knowledge base,* and—using prediction and machine learning algorithms, rules, and policies—detects the presence of attacks, computes the mitigation actions, and enforces them back in the system or informs the operator about the actions to be enforced. Such on-line control is essentially based on two components:

1. A *detection/actuation mechanism* mainly based on NFV and *software-defined networking* (SDN) techniques, which provide support for multitenancy, allowing the distribution of responsibilities in a shared management model.

2. A *reaction or resilience mechanism* based on the new concept of *software-defined security* (SDS) that integrates IACS security design, distributed awareness, mitigation, and resiliency functionalities into a unique framework able to dynamically and proactively react to the evolving threats by enforcing the most appropriate security policies in each CI node.

As tangible results, the ATENA project will produce a *set of tools* that, implementing innovative *models, methodologies,* and *algorithms* for security assurance, and interacting with the available *smart components* of CIs, will increase the level of cyber-physical security and resilience of underpinning CIs and IACSs. These tools will be assessed and validated by means of field trials in industrially relevant domains and application scenarios.

## SCOTT

Having applied the SHIELD methodology in several smaller pilots, the 2017 initiative SCOTT—Secure Connected Trustable Things, created a pan-European effort with 57 partners from 12 countries (the EU and Brazil) to bring forth measurable security and privacy as the core of wireless development (https://scottproject.eu/). The ambition of the innovation action is to reach technology readiness levels 6–7 for trustworthy connectivity and interoperability. SCOTT focuses on wireless sensor and actuator networks and communication in the areas of mobility, building, and home and smart infrastructures, as well as health. This addresses numerous European societal challenges, such as smart, green, and integrated transport; secure and inclusive societies; and health and well-being. The 15 use cases address, among other things,

- Air quality monitoring for healthy indoor environments
- Managed wireless for smart infrastructure

- Secure connected facilities management
- Logistics management using collaborative robots and DevOps methodologies
- Secure cloud services for novel connected mobility applications
- Ubiquitous testing of automotive systems
- Trustable wireless in-vehicle communication network
- Secure car access solution
- Vehicle-as-a-sensor within smart infrastructure
- Secure wireless avionics intra communications for sensing and actuation
- Safe freight and traffic management in intermodal logistic hubs
- Autonomous wireless network for rail logistics and maintenance
- Smart train composition coupling
- Trustable warning system for critical areas (POI early warning system)
- Assisted living and community care

Focusing on "trustable things that securely communicate," measurable security is the core of the multidomain reference architecture, adopting ISO 29182: "ISO/IEC 29182 Sensor Networks: Sensor Network Reference Architecture (SNRA)." Trustworthy connectivity includes the concept of privacy labeling, providing the end customer with a privacy label (A+, A, B, …, F) for services and devices. The envisaged implementation of the privacy label will use the SHIELD multimetrics approach to assess the level of privacy.

The use case–driven approach addresses areas of high relevance to European society and industry; a specific focus will be put on cross-domain use cases and heterogeneous environments, emphasizing 5G and cloud computing aspects to build up digital ecosystems to achieve a broader market penetration. The envisioned technology roadmap will foster developments and market introduction of secure- and privacy-aware systems and services.

# References

Chui, M., J. Manyika, and M. Miremadi, Where machines could replace humans—and where they can't (yet), *McKinsey Quarterly*, July 2016, http://www.mckinsey.com/business-functions/digital-mckinsey/our-insights/where-machines-could-replace-humans-and-where-they-cant-yet (last accessed on April 16, 2017).

DNV (Det Norske Veritas), Technology outlook 2020, 2015, www.dnv.com/2020 (last accessed on June 14, 2015).

DNV-GL, Future of spaceship earth, September 2016, https://www.dnvgl.com/technology-innovation/spaceship-earth/ (last accessed on March 10, 2017).

Government of India, Smart cities mission, Solutions Exchange for Urban Transformation of India (SMARTNET), 2016, https://smartnet.niua.org/smart-cities-network (last accessed on May 10, 2017).

Knickrehm, M., B. Berthon, and P. Daugherty, Digital disruption: The growth multiplier, Accenture Strategy and Oxford Economics, 2016, https://www.accenture.com/_acnmedia/PDF-4/Accenture-Strategy-Digital-Disruption-Growth-Multiplier.pdf (last accessed on April 14, 2017).

United Nations, UNO sustainable development goals, 2015, https://sustainabledevelopment.un.org/sdgs (last accessed on April 10, 2017).

# Index